Locomotor disability in general practice

Oxford General Practice Series 5

Edited by

MALCOLM I. V. JAYSON

*Professor of Rheumatology and Director of
Rheumatism Laboratories, University of
Manchester*

and

RAYMOND MILLION

*General Practitioner, Salford
Hospital Practitioner, Rheumatic Diseases Centre,
University of Manchester,
Hope Hospital, Salford*

OXFORD   NEW YORK   TORONTO
OXFORD UNIVERSITY PRESS
1983

Oxford University Press, Walton Street, Oxford OX2 6DP

London Glasgow New York Toronto
Delhi Bombay Calcutta Madras Karachi
Kuala Lumpur Singapore Hong Kong Tokyo
Nairobi Dar es Salaam Cape Town
Melbourne Auckland

and associates in
Beirut Berlin Ibadan Mexico City Nicosia

OXFORD is a trade mark of Oxford University Press

British Library Cataloguing in Publication Data
Jayson, Malcolm I. V.
Locomotor disability in general practice. —
(Oxford general practical series; 5)
1. Neuromuscular diseases
I. Title   II. Million, Raymond
616.7   RC925
ISBN 0–19–261331–6

Library of Congress Cataloging in Publication Data
Main entry under title:
Locomotor disability in general practice.
(Oxford general practice series; no. 5)
Bibliography: p.
Includes index.
I. Locomotion, Disordered. I. Jayson, Malcolm I. V.
II. Million, Raymond.   III. Series. [DNLM: 1. Joint
diseases. 2. Neuromuscular diseases. 3. Movement
disorders. 4. Locomotion. W1 OX55 no. 5/WE 300 L819]
RD680.L6   1983   616.7   82–14422
ISBN 0–19–261331–6 (pbk.)

Typeset by Cotswold Typesetting Ltd, Cheltenham
Printed in Great Britain by
Thomson Litho Ltd, East Kilbride, Scotland

Preface

The wind of change is blowing through general practice. Never before have opportunities been so good for a career and for the practice of sound constructive clinical and social medicine. The co-ordination of the three aspects of British medicine (specialist therapeutic, general therapeutic and preventive, and specialist preventive) is at a new peak, though undeniably there remains substantial room for improvement and no room for sanctimonious complacency.

The undergraduate curriculum can no longer provide an adequate base of clinical knowledge in all the specialties, and even with the recent introduction of formal vocational training for general practice, many clinicians appear in the field insufficiently trained. Necessarily gaps in the knowledge of practitioners, young and old alike, exist across the spectrum of specialization and only good can accrue from efforts to reduce these deficiencies.

The locomotor disorders comprise a substantial proportion of everyday work in general practice. Encompassing rheumatic, peripheral vascular, and neurological disorders as they do, the volume of work generated exceeds 30 per cent of the whole, and therefore it is reasonable to expect a sound fundamental knowledge.

We have drawn together a panel of specialists who have set down their views on the basic principles of their particular interests, explaining what the general practitioner should know and should be able to handle and at what point he should seek the co-operation of his colleagues in helping their patients to as comfortable a conclusion of their difficulties as possible. References in the text have been kept to a minimum, but we hope that this will not prevent the reader from taking a deeper interest in the subjects presented.

We have deliberately omitted any discussion of the so-called 'fringe' medical techniques because we cannot satisfy ourselves that they have been rigorously and scientifically proven in practice to have additional value over conventional techniques.

We see our objective as increasing the knowledge of the practitioner and improving the lot of the patient. If either of these objectives is achieved we will be pleased.

Our gratitude is due to our colleagues in the Medical Illustration Department of Hope Hospital, who have been so patient and generous with their time and energy in providing us with excellent photographs and charts, and also to all our secretaries for typing the scripts.

We express our thanks to all our contributors who have shown considerable forbearance under our persuasive aegis and to the publishers for their interest in the subject.

Manchester   R. M.
June 1982   M. I. V. J.

Contents

Contributors

J. A. D. Anderson, TD, MA, MD, FFCM, MRCGP, DPH, DRCOG,
Professor of Community Medicine,
Guy's Hospital Medical School,
London. SE1 9RT

Elizabeth M. Badley, D.Phil.,
Deputy Director,
Arthritis and Rheumatism Council Epidemiology Research Unit,
University of Manchester,
Manchester. M13 9PT

W. Carson Dick, MD, FRCP,
Reader in Rheumatology,
Department of Rheumatology,
Royal Victoria Infirmary,
Newcastle upon Tyne. NE1 4LP

M. J. Dodd, B.Sc., MB, BS, MRCP,
General Practitioner,
Newcastle upon Tyne

C. S. B. Galasko, ChM, M.Sc., FRCS, FRCSE,
Professor of Orthopaedic Surgery,
University of Manchester,
Hope Hospital,
Eccles Old Road,
Salford. M6 8HD

J. A. Muir Gray, MD, MFCM,
Community Physician,
Radcliffe Infirmary,
Oxford. OX2 6HE

D. M. Grennan, MD, Ph.D., FRCP,
Consultant and Senior Lecturer in Rheumatology,
Rheumatic Diseases Centre,
University of Manchester,
Hope Hospital,
Salford. M6 8HD

Gerald A. Griffin, MB, BS, MRCS, LRCP, FRCGP,
General Practitioner, Orpington, Kent,
Clinical Assistant in Rheumatology,
Queen Mary's Hospital,
Sidcup, Kent.
Member of the Education Council,
Royal College of General Practitioners

Wayne Hall, B.Sc., Ph.D.,
School of Psychiatry,
Prince Henry Hospital,
Little Bay 2036,
Australia.

E. C. Huskisson, MD, FRCP,
Consultant Physician,
Department of Rheumatology,
St. Bartholomew's Hospital,
London. EC1A 7BE

Malcolm I. V. Jayson, MD, FRCP,
Professor of Rheumatology,
Rheumatic Diseases Centre,
University of Manchester,
Hope Hospital,
Salford. M6 8HD

C. Leon, MB, BS, FRCGP,
General Practitioner Scheme Organizer,
Northumbria Vocational Training Scheme,
Clinical Assistant,
Department of Rheumatology,
Royal Victoria Infirmary,
Newcastle upon Tyne. NE1 4LP

Roger M. Marcuson, M.Chir, FRCS,
Consultant Vascular Surgeon,
Hon. Lecturer in Surgery,
University of Manchester,
Hope Hospital,
Salford. M6 8HD

Raymond Million, MB, Ch.B, FRCGP, DCH,
General Practitioner, Salford,
Hospital Practitioner,
Rheumatic Diseases Centre,
University of Manchester,
Hope Hospital,
Salford. M6 8HD

Jonathan Noble, Ch.M, FRCSE,
Reader in Orthopaedic Surgery,
University of Manchester,
Hope Hospital,
Salford. M6 8HD

Louis A. Schmidt, FChS, SRCh,
Principal,
London Foot Hospital and School of Chiropody,
Fitzroy Square,
London. W1P 6AY

Philip N. Wood, MB, FRCP, FFCM,
Director,
Arthritis and Rheumatism Council Epidemiology Research Unit,
Honorary Reader in Community Medicine,
University of Manchester,
Manchester. M13 9PT

A. C. Young, MB, MRCP,
Consultant Neurologist,
Salford Royal Hospital,
Salford. M60 9EP

Section I

General topics

1 Epidemiology of locomotor disorders in general practice

Philip H. N. Wood and Elizabeth M. Badley

The critical property of the topic assigned to us is concern with locomotor disorders that are presented in primary contacts with health services. To set such contacts in context it is necessary to begin by identifying the nature of disablement and of the phenomena that contribute to its development. Brief note also has to be taken of experience in the community at large. We shall then proceed to indicate how many cases of which conditions an average general practitioner might expect to see in the course of a year. The general practitioner's response to these problems inevitably varies in relation to the nature and severity of the different conditions. Severity also influences what use is made of other resources, notably those for secondary referral, and so we shall include brief consideration of the availability and utilization of various services. Finally, having completed an outline 'diagnosis' of locomotor disability at the primary care level, we shall conclude with a discussion of 'treatment' by reviewing the implications of these findings for policy formulation as it affects general practice.

SCOPE OF THE PROBLEM

The dictionary defines locomotion as the act of moving from one place to another. However, more appropriate to the subject matter of this book is the definition of locomotor disabilities included in the recently published *International Classification of Impairments, Disabilities, and Handicaps* (ICIDH) (WHO 1980), namely, those which refer to the individual's ability to execute distinctive activities associated with moving, both himself and objects, from place to place.

THE NATURE OF LOCOMOTOR DISABILITY

For most people, semantic distinctions are akin to nitpicking. However, some precision in terminological usage is helpful in revealing the nature of problems in such a way as to display options for intervention. We therefore have to say something about certain concepts. When considering the consequences of disease it is possible to differentiate between three planes of experience. These are:

(i) *Handicaps*, which are the disadvantages for a given individual, resulting from impairment or disability, that limit or prevent the fulfilment of a role that is normal (depending on age, sex, and social and cultural factors) for that individual. Handicap therefore represents the social expression of disease consequences. The options to ameliorate handicap include modification of the underlying impairments and disabilities, adaptation of the environment in which the individual lives, and augmentation of the resources available to the individual so as to enable him better to handle those problems that confront him. In strategic terms, policies directed at the control of handicap concern social security and welfare, vocational rehabilitation, housing, education, and related non-health-service fields, as well as provision of health-service rehabilitation facilities. Locomotor disabilities might be expected to affect most of the dimensions of handicap, such as physical independence, occupation, social integration, and economic self-sufficiency, but it is disadvantages in regard to mobility that are especially to be anticipated (further details of these distinctions are provided in the ICIDH).

(ii) *Disabilities*, which include any restriction or lack (resulting from an impairment) of ability to perform an activity in the manner or within the range considered normal for a human being. Disability is therefore concerned with functional performance at the level of the person; its relation to both impairment and handicap is depicted more explicitly in Fig. 1.1. The objectives

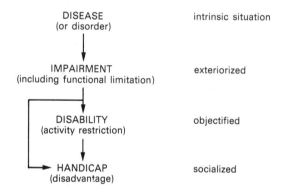

Fig. 1.1. The consequences of disease – a conceptual model.

of medical rehabilitation and remedial therapy in response to disability are to enhance function, such as by physical re-education; to supplement performance of an activity, as with an aid such as a walking stick; or to provide a substitute if it is impossible for any particular activity to be carried out (e.g. by supplying a hoist). Although thought has to be given to activities such as retrieval and reaching, kneeling and crouching, and stooping and other aspects of posture, the prime concern of locomotor disability is with three things, viz:

(a) ambulation – walking, traversing uneven terrain, climbing stairs and other obstacles, and running;

(b) confinement – transfer, including to and from a bed or a chair, and transport, such as getting on and off a bus and to inaccessible amenities;

(c) other aspects – including lifting and carrying.

(iii)   *Impairments*, which represent any loss or abnormality of psychological, physiological, or anatomical structure or function. Impairment therefore concentrates on function at the level of the organ, and attempts to correct matters usually fall within the ambit of treatment or management. The principal classes of impairment relevant to locomotor disability are as follows:

(a) visceral, where the problem is related to the availability of energy, either in a limb (e.g. as a result of peripheral vascular disease) or in the body as a whole (with cardiorespiratory disease);

(b) skeletal, which includes the two major types of local problem – those that are mechanical, due to restriction of motion (as with rheumatic disorders) or to absence of a part (amputations and congenital limb deficiencies); and those that involve control of a limb, notably paralysis and other interferences with motor function;

(c) generalized, sensory, and other, including impairments of continence, undue susceptibility to trauma, certain metabolic impairments, back pain, and conditions such as pregnancy and gangrene.

Having taken note of such a daunting range of relevant conditions, the reader may take some comfort from the fact that from now on we shall confine ourselves largely to rheumatic disorders, with some additional consideration of cardiorespiratory and neurological diseases.

THE GENERAL PRACTITIONER'S AWARENESS

Our perceptions of health phenomena tend to be rather selective, understandably being largely confined to the problems that present to our level of health services; the principal exceptions occur when our role changes and we ourselves join the ranks of sufferers. In many medical debates the general practitioner can justifiably claim to know more of a particular problem than reaches the awareness of medical colleagues working at other levels. However, even patients presenting to a general practitioner are selected, and they do not reflect the totality of health-related phenomena. It is therefore necessary to take some account of the reservoir from which a general practitioner's clinical experience is drawn.

Various symptoms are of widespread occurrence. Each individual who develops a symptom passes through an initial period of uncertainty. The symptom may appear to be trivial, and not too difficult to explain (the consequence of a dietary indiscretion or of unaccustomed activity, for example). If the experience appears to be self-limiting it poses a minimal threat, so that the situation may well be accepted and nothing may be done. However, the persistence of symptoms undermines initial responses and creates a much greater degree of uncertainty. Two further options are then likely to be considered; to attempt to ameliorate the complaints by self-treatment with restrictions, exercise, or medicaments, or to seek help from outside sources. Even of those who adopt the latter course, by no means all present in their general practitioner's consulting room. Instead, they may choose to attend a casualty department, an occupational health service, a remedial therapist, an osteopath or a chiropractor, or a practitioner of one of a diversity of other systems of belief. A few even try for the best of both worlds, seeking help from both their general practitioner and one of these alternative sources of advice.

We have some measure of what is happening in society at large in works such as *Medicine takers, prescribers, and hoarders* (Dunnell and Cartwright 1972). In this study adults averaged almost four symptoms in a 14-day period. Of those who complained, 29 per cent had aching limbs or joints, 21 per cent had backache, 19 per cent trouble with their feet, and 15 per cent breathlessness. Children in general seem to experience symptoms less frequently, the rates of complaint for locomotor problems being about ten times less than in adults – 3 per cent with pains in the limbs, 2 per cent with foot trouble, and 1 per cent pain in the back. The factors influencing whether someone with a symptom appears in a general practitioner's surgery are variable and complex. However, the hallmark of a 'patient' is that he or she acknowledges difficulty in sustaining an accustomed social role, be this due to anxiety or to more material interference with the ability to discharge those functions and obligations that are expected of the individual.

The discordance between symptom experience and consultation is often referred to as an aspect of the iceberg phenomenon. Another facet of the same phenomenon is presented by the disabled and elderly at home, and this, too, is another area where a general practitioner's experience is far from exhaustive – what we have referred to elsewhere (Wood and Badley 1978) as the forgotten disabled. Two important government surveys, *Handicapped and impaired in Great Britain* (Harris 1971) and *The elderly at home* (Hunt 1978), have shed light on this aspect. Of those who were impaired and not living in institutions, virtually a quarter (0.75 million people) had not seen their general practitioner in the preceding six months, the interval being longer than a year in almost half of these individuals. Among the elderly at home the figure was understandably somewhat higher, 60 per cent not having been visited by their family doctor in the previous six months. Thus although general practitioners are very much in the front line of response to health-related problems, their awareness of the range of difficulties is far from being complete.

THE DIAGNOSTIC SPECTRUM

The most useful information available about work loads in general practice is contained in the Second National Morbidity Study (OPCS 1974). From this source we have extracted some crude figures on the major body systems (Table 1.1), although these are obviously inflated by many complaints (e.g. upper respiratory tract infections) that are not of immediate concern. In order to emphasize this point, we have also shown estimates for four of the important causes of locomotor disability. Although these are not

Table 1.1 *Consultations relating to body systems responsible for major classes of locomotor disability. (From Second National Morbidity Study, OPCS 1974)*

Body system	Relevant disabling condition (with ICD rubrics, 8th revision)	Patients consulting for body system		Patients consulting for relevant disabling conditions	
		rate (per 1000 population)	proportion of total (%)	rate (per 1000 population)	proportion of total (%)
respiratory	*chronic bronchitis and emphysema (491–492)*	261	39	13	29
nervous system and sense organs	*stroke (cerebrovascular disease, 430–438)*	113	17	5	11
musculoskeletal system and connective tissue	*osteoarthrosis*	91	14	25	56
circulatory	*peripheral vascular disease (approximated by figures for arteriosclerosis, 440)*	66	10	2	4
All diseases and conditions	*selected conditions shown above*	672	100	45	100

comprehensive measures, they do indicate that disabling conditions form a much larger proportion of all complaints referred to the musculoskeletal system than is the case for any of the other three systems. This serves to alert one to the fact that rheumatic disorders are the dominant cause of physical impairment. Such data serve to provide a global picture, but they are not very helpful for indicating the relative frequency of specific conditions that may give rise to difficulty in differential diagnosis. For this purpose it is more helpful to focus on greater detail in regard to individual types of rheumatic disorder.

RHEUMATIC DISORDERS

Over the course of a year some 15 per cent of registered patients consult their family doctor with a rheumatic complaint. However, because some patients present with several different rheumatic problems during such a period, patients consulting with these disorders represent approximately 23 per cent of the total number of patients consulting. The crude case mix in five broad categories is shown in Fig. 1.2. These categories require a little further explanation, as follows:

The arthropathies, meaning disease of joints and designated 'arthritis' in Fig. 1.2. These include relatively common conditions such as rheumatoid arthritis, osteoarthrosis, gout, and, because it is classified this way in available statistics, ankylosing spondylitis, as well as a range of other, if individually rather less frequent, disorders affecting the joints of the limbs.

Back troubles ('backs' in Fig. 1.2), which include displacement of an inter-vertebral disc (colloquially, the 'slipped disc'), spondylosis, sciatica, lumbago, and other complaints of back pain.

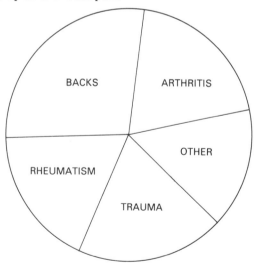

Fig. 1.2. Crude case mix in patients with rheumatic problems presenting to a general practitioner. (From Second National Morbidity Study, OPCS 1974.)

Non-articular rheumatism ('rheumatism' in Fig. 1.2), which refers to all the aches and pains in muscles and other soft tissues when the joints are not affected. Included, therefore, are conditions such as frozen shoulder and tennis elbow, as well as different forms of bursitis and tenosynovitis (including those connected with various occupations) and pains around joints and in soft tissue.

Soft-tissue injury ('trauma' in Fig. 1.2), the strains and sprains affecting muscles and other tissues around the joints.

Other rheumatic disorders ('other' in Fig. 1.2), which include rheumatic fever and rheumatic heart disease (unfortunately in these data the complications were not identified separately from acute rheumatic fever itself), the so-called connective tissue diseases (such as systemic lupus erythematosus, dermatomyositis, and systemic sclerosis), various other musculoskeletal disorders (such as Paget's disease of bone), and other rheumatic conditions (including symptoms, signs, and incompletely diagnosed complaints referred to the musculosketal system).

In considering differential diagnosis it is the likelihood of different individual conditions within these broad categories that is of more relevance. This is shown in Table 1.2, to the extent that this source allows such detail to be retrieved. Perhaps the most striking feature is the frequency of non-specific and ill-defined problems. Another important influence on differential diagnosis is age. Although it is unlikely that anyone escapes some form of rheumatic problem during the course of his or her life, the chances do vary with age. Overall, approximately 5 per cent of persons between the ages of 16 and 44 years have a rheumatic disorder, compared with 23 per cent of persons between 45 and 64 and with 41 per cent of those 65 years and older. Moreover, the various age groups tend to differ in regard to which type of complaint most frequently gives rise to trouble. These differences are shown in Fig. 1.3, and may be summarized as follows:

Soft-tissue injury is commonest under the age of 65, and in fact this is the only type of rheumatic complaint where experience in those under the age of 25 years is of the same order of frequency as in older adults.

Back troubles are notably a problem for adults between the ages of 25 and 64 years – their frequency in the elderly is somewhat less, and in the young much less.

Non-articular rheumatism is similar in frequency in the three older age groups but, again, much less common in the young. Certain specific problems are not revealed by these data. For instance, polymyalgia rheumatica is commoner than is often realized, and this is particularly an affliction of the evening of life. This condition can be distressing and incapacitating, and it may also lead to blindness – and yet effective treatment is readily available.

Other rheumatic disorders are much more heterogeneous and so the age patterns are much less meaningful, though for what they are worth they show a much greater similarity in frequency in all the age groups. However, specific note should be taken of conditions that become more frequent in the

Table 1.2. *Likelihood of different rheumatic disorders in general practice (percentage frequency of occurrence among patients consulting during a year; from Second National Morbidity Study, OPCS 1974)*

Arthropathies	20%	Non-articular rheumatism	18.5%
rheumatoid	3.4	bursitis (including frozen shoulder)	5.7
gout	1.0	knee problems	2.1
osteoarthrosis	12.0	foot problems	1.3
other	3.6	other	9.4
Back troubles	**27%**	**Other rheumatic disorders**	**15%**
spondylosis	4.7	rheumatic fever and RHD	0.8
discs	3.9	other musculoskeletal	7.6
lumbago and sciatica	6.8	ill-defined	6.5
other back pain	11.7		
		Trauma (soft-tissue injury)	**19.5%**

elderly – osteoporosis, osteomalacia, Paget's disease of bone, and fractures, particularly of the femur (Wood and Badley 1982), the latter problem not being reflected in these data.

The arthropathies are strikingly different, with a very steep increase in frequency leading to a high peak in the elderly. This pattern may appear to

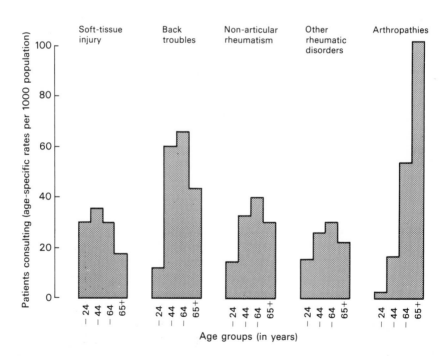

Fig. 1.3. Age-specific incidence rates for categories of rheumatic disorder. (From Second National Morbidity Study, OPCS 1974.)

Table 1.3. *Measures of work load from rheumatic disorders in an average general practice (expressed as annual rates for a practice with a list size of 2500; from Second National Morbidity Study, OPCS 1974)*

Rheumatic disorder	ICD rubric (8th revision)	Patients consulting (rounded)	Proportions (%) aged: 45–64 (yr)	65+ (yr)	Proportion female (%)	Episodes per patient (average)	Consultations (rounded)	Consultations per episode (average)
Arthropathies		**77**	**41.5**	**41.9**	**67.0**	**1.10**	**189**	**2.2**
rheumatoid arthritis and allied	712	13	45.9	41.1	66.1	1.18	47	3.2
osteoarthrosis (excl. spondylosis)								
spondylosis	713.0, 713.2	46	39.3	51.8	70.3	1.12	109	2.1
gout	274	4	53.0	27.6	19.9	1.13	10	2.1
other	rem. 710–718	14	41.5	20.4	68.6	1.07	23	1.6
Back troubles		**104**	**39.3**	**13.8**	**54.5**	**1.05**	**203**	**1.9**
spondylosis	713.1	18	48.9	21.5	60.4	1.07	34	2.1
displacement of intervertebral disc	725	15	40.4	5.9	44.7	1.10	43	2.7
sciatica	353	4	42.5	20.8	57.7	1.06	8	1.9
lumbago	717.0	22	36.4	12.7	47.6	1.06	38	1.7
other back pain	728	45	33.1	12.2	54.0	1.06	80	1.7
Non-articular rheumatism		**71**	**35.2**	**14.0**	**54.3**	**1.04**	**115**	**1.6**
bursitis, synovitis, and tenosynovitis	731	18	35.3	9.5	52.7	1.04	30	1.6
frozen shoulder	717.1	4	53.2	23.7	55.9	1.06	8	1.8
knee problems	724.1	8	26.6	5.7	35.1	1.10	16	1.9
foot problems	736–737	5	27.8	17.7	64.5	1.00	7	1.4
pain in joint	787.3	12	34.3	14.3	56.4	1.04	18	1.5
other rheumatism	717.9	24	36.7	17.4	58.4	1.05	36	1.5
Other rheumatic disorders		**57**	**30.9**	**12.4**	**56.4**	**1.05**	**92**	**1.6**
rheumatic fever and rheumatic heart disease	390–398	3	45.8	25.6	67.5	1.17	11	3.1
other musculoskeletal disorders	rem. 720–738	29	31.8	12.5	56.0	1.04	45	1.5
other and ill-defined	rem. 787	25	28.7	10.6	55.5	1.04	36	1.4
Soft-tissue injury (strains and sprains)	N840–N848	**75**	**23.5**	**7.3**	**43.6**	**1.08**	**117**	**1.4**
All rheumatic disorders		**380**	**34.2**	**17.8**	**54.6**	**1.04**	**717**	**1.8**
All diseases and conditions		1679	22.2	12.3	54.9	2.61	7524	1.7

contrast with the impression of rheumatoid arthritis being a disease of middle life (although that is in fact a feature of its onset, its course being fairly prolonged), and it is due to the numerical dominance of osteoarthrosis, which is commoner than all other types of arthritis taken together.

These patterns are based on patients consulting as proportions of the population in their particular age group. When looked at from the point of view of the general practitioner's consultation load it is inevitable that the emphasis is altered, although the broad perspective is not dissimilar. The preferential affliction of older people is indicated in Table 1.3. If the two adjacent columns identifying age (45–64 and 65 + years) are taken together, it can be seen that the proportion of all consultations occurring in those aged 45 years and more is only 35 per cent for all diseases and conditions, whereas it is 52 per cent for all rheumatic disorders. This proportion varies according to the type of rheumatic complaint, increasing from 31 per cent for soft-tissue injury to 43 per cent for other rheumatic disorders, 49 per cent for non-articular rheumatism, 51 per cent for back troubles, and a striking 83 per cent for the arthropathies – half of the latter figure being accounted for by those aged 65 and older.

So far we have considered experience in the two sexes taken together. When men and women are considered separately (of which an indication is given in Table 1.3), soft-tissue injury occurs more often in males, back troubles affect the two sexes similarly, and in other rheumatic disorders and non-articular rheumatism the problem is more frequent in females, a trend that is even more marked with the arthropathies.

As can be seen in Table 1.3, most patients consulting experience only a single episode during the course of a year. On the other hand, the recurrent and often more serious nature of the arthropathies is reflected respectively in a higher average number of episodes and a greater number of consultations per episode. Such differences are mirrored in how general practitioners tend to respond to different classes of problems, as can be seen in Table 1.4. For example, an above average proportion of consultations for inflammatory conditions, approximately a quarter, takes place in the patient's home rather than in the surgery. Similarly, the general practitioner more often refers patients with these disorders for additional help, such as for investigations, a specialist opinion in out-patient clinics, in-patient care, or assistance from local authority services. Overall, though, the proportion of all rheumatological patients referred is smaller than that for all diseases and conditions – only a fifth, compared with a third.

We have already remarked that by no means everyone with a complaint goes to their family doctor. Although we have no evidence specifically related to arthritis, there appears to be a lesser propensity for older people to make use of health services (Cartwright 1967), despite their greater morbidity and the higher consultation rates these would lead one to expect. These particular features of elderly arthritics we have reviewed elsewhere (Wood and Badley 1982).

Table 1.4. *Patterns of response to rheumatic disorders in general practice (from Second National Morbidity Study, OPCS, 1974)*

	Arthropathies			Back troubles			Non-articular rheumatism			Other rheumatic disorders		Soft-tissue injury	All rheumatic disorders
	Gout	RA	OA	Sciatica	DID	Lumbago and back pain	Frozen shoulder	Other rheumatism and limb symptoms	Bursitis and synovitis and other arthritis and rheumatism	Rheumatic fever and RHD	Other musculo-skeletal disorders	Sprains and strains	
ICD rubrics (8th revision)	274	712	713	353	725	717.0 and 728	717.1	787 and remainder 717 and 718	731 and remainder 710–718	390–398	remainder 720–738	N840–N848	
Episodes per patient (average)	1.13	1.18	1.10	1.06	1.10	1.06	1.06	1.05	1.05	1.17	1.05	1.08	1.08
Average number of consultations per patient*	*2.4*	*3.8*	*2.3*	*2.1*	*3.0*	*1.8*	*1.9*	*1.5*	*1.7*	*3.6*	*1.6*	*1.5*	*1.9*
Proportion of consultations taking place at patient's home (%)†	–	27	23	23	17	14	–	–	5	26	9	7	15
Proportion of patients referred for outside help (%)† includes:	–	34	23	17	29	15	–	–	14	29	20	13	21
direct admission to hospital	–	1.7	0.5	0.4	1.3	0.4	–	–	0.1	3.9	0.5	0.2	0.5
specialist opinion or service (incl. domiciliary consultation)	–	11.3	9.0	10.3	15.6	5.3	–	–	8.0	9.2	10.5	5.0	8.8
specified investigation	–	18.7	11.3	5.7	11.4	8.4	–	–	4.7	12.5	8.0	6.7	9.5
local authority services	–	0.5	0.6	0	0.2	0.1	–	–	0	1.1	0.2	0.3	0.3

– Insufficient detail reported in the source data.
*Figures in italics were not quoted in the source data and so have been computed approximately.
† Full details not available for all diagnostic classes.

EXPERIENCE IN CHILDREN

At the other end of the life span the situation is somewhat different. The relative infrequency of symptoms in children and the concern that their occurrence is likely to engender in parents could lead to a greater proportion of these complaints being brought to medical attention.

Be that as it may, the frequency with which children consult their family doctor with a rheumatic condition is shown in Table 1.5, where the data are expressed as rates per thousand children. During the year of this study 2.6 per cent of under fives and 5.8 per cent of those aged 5–14 years presented themselves to the doctor with some form of rheumatic complaint. To set this material in context it should be recollected that a general practitioner responsible for patients on a list of average size, 2500, is likely to be caring for about 750 children. The figures in Table 1.5 therefore need to be scaled down by a quarter in order to indicate the numbers of children likely to consult any individual doctor during the course of a year.

It is also interesting to take account of what proportion of the total work load such children account for. In those under five years of age the rate for children consulting with a rheumatic disorder amounted to 2.9 per cent of that for all diseases and conditions. Problems were commoner in older children, so that from the age of five upwards the burden, 9.1 per cent of the total, was nearer to that observed in adults. Another reflection of work load is the proportion of consultations that is made up by home visits – about a sixth overall. Rheumatic conditions exhibit a broadly similar pattern to that for all diseases and conditions, the rates tending to be slightly higher than that for all ages in the under fives but to be below the average in the 5–14 year olds. In general, though, and with the exception of certain serious diseases, the proportion of home visits for rheumatic complaints tends to be below average.

What makes up this burden of work, and what are the children suffering from? The overwhelming majority have pain in the limbs, and the greatest part of this, over 80 per cent, is accounted for by three types of disorder – strains and sprains, non-articular rheumatism (i.e. the joints are not affected), and the miscellany designated 'other diseases of bones and musculoskeletal system', the largest part of which is made up of various pain syndromes. Most of these complaints are not serious, and their duration is usually fairly brief. However, they are obviously associated with a lot of discomfort, and equally they often give rise to anxiety for parents.

The various forms of rheumatism are commoner in older children, the only condition encountered more frequently in the under fives being flat foot. It can be seen that injury plays a large part, so that from the age of five onwards strains and sprains contribute nearly half the overall rheumatic experience. Developmental abnormalities such as hallux valgus and flat foot are relatively infrequent, though a proportion of the back pain encountered may arise from analogous causes.

Table 1.5. *Children with rheumatic diseases consulting general practitioners (from Second National Morbidity Study, OPCS 1974)*

Type of disorder	Rheumatic disease	ICD rubric (8th revision)	Children consulting (sexes combined) rate per thousand children	
			0–4 yr	5–14 yr
Arthropathies*			**1.0**	**1.0**
	rheumatoid arthritis	712	0.5	0.3
	osteoarthrosis	713	0.5	0.7
Back troubles †			**0.4**	**3.4**
	lumbago	717.0	0.1	0.4
	displacement of intervertebral disc	725	–	0.4
	other back pain	728.7, 728.8, 728.9	0.3	2.6
Non-articular rheumatism ‡			**7.2**	**15.2**
	other non-articular rheumatism	717.1, 717.9	0.4	1.8
	other arthritis and rheumatism	710, 711, 714–716, 717.2, 718	0.3	1.3
	synovitis, tenosynovitis, and bursitis	731	0.5	2.9
	ill-defined rheumatic complaints	787	6.0	9.2
Other rheumatic disorders †			**8.4**	**11.9**
	rheumatic fever and rheumatic heart disease	390–398	0.1	0.2
	internal derangement of knee	724.1	0.2	1.7
	flat foot	736	2.9	1.8
	hallux valgus	737	0.1	0.4
	other diseases of bones and musculoskeletal system	720–723, 724.0, 724.9, 726, 727, 728.0–728.6, 729, 730, 732–735, 738	5.1	7.8
Non-traumatic rheumatic disorders			**17.0**	**31.5**
Soft-tissue injury§	strains and sprains	N840–N848	**9.3**	**26.7**
All rheumatic disorders			**26.3**	**58.2**

*Various other forms of arthritis (ICD 710, 711, 714, 715), including infections, could not be retrieved separately and so are shown under non-articular rheumatism.

† Other parts of the vertebrogenic pain syndrome (ICD 728.0–728.6) could not be retrieved separately and so are shown under other rheumatic disorders.

‡ Other arthritis and rheumatism includes polymyositis and dermatomyositis (ICD 716); see also note *.

§Including strains and sprains of sacroiliac joint and back.

Against this background of relatively minor disorders we can now set the more serious diseases in context. Three conditions stand out, though scarcely for their frequency – each contributes 1 per cent or less of all rheumatic children consulting. Acute rheumatic fever and the chronic heart disease to which it gives rise is much less of a problem than it used to be, and numerically it is the least important of these conditions. The higher figures for osteoarthrosis are rather surprising in this age group, and experience indicates that they have probably been given an inappropriate diagnostic label – children in this category are most likely to be suffering from mechanical problems, osteochondrosis (osteochondritis) and chondromalacia patellae being the most notable. Finally, there are children identified as suffering from rheumatoid arthritis or Still's disease (juvenile chronic arthritis), about which problem further information should be sought elsewhere. All told, however, only just over one child in every thousand is likely to be suffering from any one of these conditions.

CHALLENGES IN PRIMARY CARE

An obvious influence on how any doctor reacts in a particular situation is what is perceived about the severity of and the outlook for the complaint under consideration.

MEASURES OF SEVERITY

The simplest measure of severity, and one which at the same time provides a crude indication of prognosis, is the duration of any particular complaint. The best source of information on this is certified sickness incapacity for work. The distribution of different types of disorder contributing to rheumatic spells, shown in the first column in Fig. 1.4, obviously contrasts with the proportions depicted in Fig. 1.2. The explanation for the differences lies in two features of these data; first, only people of working age are covered by sickness incapacity provisions, and secondly, Fig. 1.4 is based on incapacity in men only, experience in women being represented less comprehensively in such data. Women experience proportionately more arthritis and less trauma than is shown in Fig. 1.4, and most of the conditions included under the heading 'other' tend to be much commoner in those past retirement age.

When the perspective is modified to take account not of the occurrence of spells, but of how long these spells last (i.e. the days lost from work due to these complaints), the pattern becomes that shown in the centre column of Fig. 1.4. Arthritis can be seen to be accounting for a much larger part of the burden when viewed from this point of view, which is what one would expect from the generally more severe nature of such problems. The same is true, only to a lesser extent, for the disorders lumped together as 'other'. Backs are responsible for slightly less of the whole, and the proportions due to non-articular rheumatism and soft-tissue injury are reduced more markedly.

We can take this a stage further by moving from general practice to hospital, considering new attendances at specialist rheumatology clinics. The data on which the right-hand column in Fig. 1.4 are based are not strictly comparable because both sexes and all ages are taken into account and, moreover, patients may well be referred to orthopaedic rather than to rheumatology clinics, particularly in view of the lesser availability of the latter in many parts of the country (Wood 1977). Nevertheless, this information does allow one to extend appreciation of the severity gradient. It is arthritis especially, but also some of the other conditions such as connective-tissue disorders, that are likely to give rise to most concern. However, an interesting diagnostic point should be noted as well, because the category of 'other' in out-patient referrals also includes non-rheumatological diseases masquerading as rheumatism.

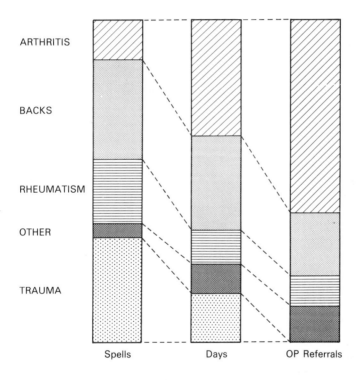

Fig. 1.4. Different perspectives on rheumatic disorders in Great Britain (proportions of total rheumatic burden attributable to different types of condition). Spells and days of sickness incapacity certified by general practitioners in males in 1977–1978; all rheumatic disorders were responsible for 86 spells and 3293 days per 1000 insured men. OP referrals to rheumatology clinics at university hospitals in Manchester in 1979 (sexes combined); there were 4.1 new attendances at rheumatology clinics per 1000 population of both sexes in England in that year.

Table 1.6. *Impact of rheumatic disorders on everyday life style (expressed as rates per thousand population at risk for the year 1971. Sources: Wood 1978; OPCS 1974; Harris 1971)*

Rheumatic disorders	ICD rubric (8th revision)	Impaired persons (adults ≥ 16 yr of age)		Disadvantaged in major dimensions of activity (proportions of impaired rheumatic sufferers, as %)							
				physical independence		mobility				occupation	
		all degrees	severe and very severe	confined to chair or bed	unable to bath themselves	wheelchair confined to	wheelchair use	housebound ‡	use sticks or crutches	have had to give up work	have had to change job
Arthropathies											
rheumatoid arthritis	712	3.4	1.3	7.8	25.0	3.3	10.8	17.4	32.2	31.3	2.4
osteoarthrosis*	713 353, 717.0, 725, and 728	18.7	3.0	1.8	17.9	0.4	3.4	13.9	36.4	18.9	1.7
Back troubles		2.0	0.2	0.4	6.4	0.3	0.9	1.2	12.7	17.3	1.7
Other rheumatic disorders	274, 390–398, remainder 710–738, 787, and N840–N848	8.1	0.9	2.1	12.1	0.3	2.2	6.4	25.6	16.8	2.1
All rheumatic disorders		27.8 †	4.9	2.6	17.1	0.8	4.0	12.5	32.7	20.0	1.8

*Arthritis unspecified has been included under osteoarthrosis because it resembles this latter category in most associated characteristics.
† The total of impaired persons with all rheumatic disorders is less than the sum of the individual rates because more than one disorder was present in a proportion of people.
‡ Includes those who are confined to chair or bed.

Turning now to disturbances in patterns of daily living, a different aspect of severity becomes apparent (Table 1.6). The picture is dominated by the arthropathies and, among these, it is osteoarthrosis that far and away leads the field – because it is much the commonest underlying pathological disorder present in the population at large. Most varieties of other rheumatic disorder, and particularly non-articular rheumatism and soft-tissue injury, are less evident in this context, mainly because of their generally more limited duration. However, even apparently less serious disorders may have a considerable impact in the elderly. For example, any rheumatic problem in the legs can interfere with the ability to get around. This reduced mobility in older people can initiate a progressive withdrawal from life that may be difficult to reverse, at times leading to dependence either on others or on institutions.

Table 1.7 provides further indications of disadvantages in those suffering from arthritis. It also shows two other things. First, the elderly generally experience hardship more frequently. Secondly, in an affected individual it is rheumatoid arthritis (RA) that is generally more disabling than osteoarthrosis (OA), largely because more joints tend to be involved and the changes in these are usually more destructive. However, this well-known fact must not be allowed to distract attention from the point we made in the preceeding paragraph – that of those disabled by arthritis, the underlying problem is OA in almost three-quarters and RA in only one-eighth.

The information in Table 1.7 presents a selective picture of disturbances in daily living, being limited to data that could be extracted from the work of the Government Social Survey (Harris 1971). A disturbing impact is nevertheless apparent. Many people suffering from rheumatic disorders, and especially those with arthritis, are subject to serious curtailment of their physical independence and mobility, they tend to be cut off from much that is going on around them, and they tend also to be at considerable financial disadvantage. The latter point emerges with particular poignancy in the elderly, and their relative poverty is likely to account for the fact that the proportions incurring extra expenses have not increased with age, despite the greater frequency of related disability – i.e. they did not have the financial resources to make life easier for themselves in this way.

IMPACT ON THE COMMUNITY

One other aspect needs to be acknowledged, that related to the severity of rheumatic disorders for the community, as part of the morbidity burden, rather than in terms of the impact of these conditions on affected individuals. The key figures are that more than eight million people consult their family doctor with a rheumatic complaint during the course of a year; that these conditions are responsible for almost one-fifth of all sickness absence from work, both for spells and for days lost (1.5 million spells and more than 61 million days lost in 1977–78, with an associated productivity loss in excess of £1 thousand million);

Table 1.7. *Disadvantages resulting from impairment by all rheumatic disorders and by the two major forms of arthritis (comparison of frequencies in the elderly, aged 65 years or more, with those in all adults, aged 16 years or more—after Harris 1971)*

Dimensions of handicap and underlying disabilities (occupation not taken into account)	Accomplishment or state (selected examples only)	Proportion of persons impaired by types of arthritis (%)			
		rheumatoid arthritis		osteoarthrosis	
		all adults	the elderly	all adults	the elderly
Physical independence					
Excretion disability	getting to and using w.c.				
	– difficult	33.2	34.5	23.3	23.9
	– can't do on own	7.7	8.8	2.8	3.2
Personal hygiene disability	having an all over wash				
	– difficult	23.4	25.2	19.1	29.0
	– can't do on own	27.1	38.7	1.8	18.6
Dressing disability*	putting on shoes and stockings				
	– difficult	46.8	60.7	37.6	43.2
	– can't do on own	7.1	6.6	7.4	8.3
Domestic disability	cooking† – doesn't do because of impairment	13.2	22.5	5.5	8.3
	laundry† – sends to laundry because of impairment	21.4	35.2	14.7	19.0
Mobility					
Confinement disability	confined to house				
	– bed or chair fast	7.8	10.7	1.9	2.4
	– otherwise confined to house	9.6	13.1	12.0	15.9
Ambulation disability	able to get out of house				
	– only if accompanied	15.3	21.1	7.9	9.9
	– on own but with difficulty	26.2	28.1	29.4	31.6
	– without undue difficulty by may take longer	41.0	28.1	48.8	40.4

Table 1.7. *continued*

Dimensions of handicap and underlying disabilities (occupation not taken into account)	Accomplishment or state (selected examples only)	Proportion of persons impaired by types of arthritis (%)			
		rheumatoid arthritis		osteoarthrosis	
		all adults	the elderly	all adults	the elderly
Social integration					
Solitary life	single or widowed	37.3	51.2	52.0	63.9
	live alone (cf. 5% of general population)	16.2	24.2	29.4	36.6
	of those living alone:				
	– do not have radio	0.0	0.0	7.6	7.9
	– do not have television (cf. 6% of those living with others)	41.1	49.3	21.4	21.7
Recreation disability	unable to go to clubs	16.4	55.4	11.3	51.2
	unable to go to events (church, party, etc.)	35.0	39.3	26.6	31.8
	given up things liked doing (e.g. hobbies)	71.4	69.0	59.6	59.8
	not had holiday in previous three years	38.5	45.6	31.8	35.4
Economic self-sufficiency					
Expense	extra expenses incurred for				
	– heating	32.4	31.2	22.2	23.1
	– domestic help	13.2	13.7	7.4	10.1
	– laundry	15.3	14.7	10.9	9.7
	– diet	7.8	8.1	7.8	8.2
	– travelling expenses	3.7	2.8	2.1	0.9
Income	median weekly income (cf. £18 in general population – 1969 figures)	£9.14	£6.87½	£8.00	£6.31

*Does not include those with very severe disability.
†Excludes those who would not normally carry out household tasks.

and that rheumatic disorders form the biggest single class of underlying problems in people who are handicapped and impaired, being responsible for one-third of physical disability in those aged 16 years or more and for virtually half in the elderly.

Two measures of rheumatic suffering indicate that the burden varies considerably in different parts of the country. The highest rates for spells of rheumatic incapacity are recorded in the Northern region of England, Wales, and Yorkshire/Humberside, and there are above average rates in Scotland and the East Midlands, whereas experience is well below average in East Anglia and the South East. Days lost vary in a broadly similar manner; Wales tops the league with almost double the national average and, in addition to the regions already mentioned, the North West also comes into prominence. A slightly different pattern emerges when these two pieces of information are inter-related in order to derive the mean length of spells. The below average performance of East Anglia in regard to both spells and days lost is nevertheless associated with the highest mean spell length in the country. Unduly long spells are also observed in Wales and the South West. Even the apparently favourable experience of the South East is associated with spell lengths that are only just over 10 per cent under the national rate, the shortest spells being reported from the East Midlands.

Although age obviously contributes to some of these differences, the pattern of different types of complaint also varies. Thus East Anglia and the South West have much less of a burden of non-articular rheumatism but are above average in regard to other rheumatic disorders. The excess of Yorkshire/Humberside is particularly with non-articular rheumatism, and this type of problem also stands out in the North West (together with arthropathies) and in Scotland (with soft-tissue injury as well). In the Northern region the only type of complaint that is not well above average is other rheumatic disorders, whereas in Wales, where all categories stand out, it is the arthropathies that are particularly noteworthy.

The other measure is provided by the 1.1 million people who are impaired by rheumatic disorders (Fig. 1.5), and this information can be considered at two levels of severity. There is no simple geographical pattern among those with less severe degrees of impairment, the rates varying from 21 per 1000 persons in Scotland and 23 in the East Midlands to 32 in East Anglia and 34 in the South West. East Anglia has the highest rate in women for any of the regions, whereas in Wales the rate is high in men but low in women. The variation is not dissimilar when attention is restricted to those with appreciable disability, proportions per 1000 persons ranging from 9 in Wales and 10 in the East Midlands and Scotland to 15 in the North and 17 in the South West.

Since those impaired by rheumatic disorders are mainly elderly, one would expect the regions with the higher proportions of elderly persons to have the highest rates of impairment. Certainly the South West has the highest proportions of both men and women aged 65 or more, and it has the highest

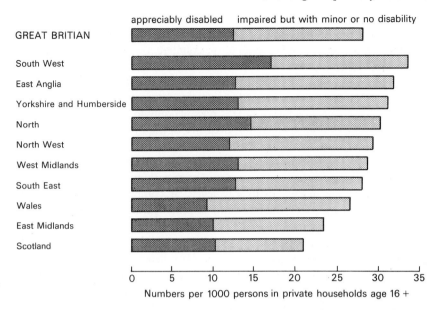

appreciably disabled   impaired but with minor or no disability

GREAT BRITIAN

South West

East Anglia

Yorkshire and Humberside

North

North West

West Midlands

South East

Wales

East Midlands

Scotland

0   5   10   15   20   25   30   35

Numbers per 1000 persons in private households age 16 +

Fig. 1.5. Regional variation in the frequency of persons impaired and disabled by rheumatic disorders. (After Harris 1971.)

proportion of men impaired by rheumatic disorders and the third highest proportion of women. Similarly, Scotland has the second lowest proportions of men and women beyond retirement age and the lowest proportions of men and women impaired. However, this relationship does not hold everywhere. The North, for example, has a below average proportion of women aged 65 or more but it has the second highest proportion of impaired women, and Wales has an above average proportion of elderly women but the third lowest proportion of impaired women. In general, then, the proportions of elderly persons in the regions explain some of the variation in the proportion of impaired, but by no means all.

AVAILABILITY OF EXPERTISE

The bulk of medical care for those suffering from arthritis and rheumatism is provided by general practitioners. We have already considered how general practitioners respond to these problems (Table 1.4), and now we must turn to the influences that have been important in determining these responses.

Three things have been important – how well the general practitioner has been prepared for this aspect of his work, whether appropriate help is available to him, and the particular types of problem associated with the different individual conditions. Before discussing these, though, it has to be recognized that what can be regarded as the contemporary approach to care is in fact the end-result of a complex process of change (Wood *et al.* 1980). What we are about to describe represents a synthesis from stages in this process, but the relevance of individual features is likely to vary between one doctor and another and between different parts of the country.

The most basic question is concerned with whether general practitioners have had training appropriate to the demands of their work. We have been associated with two studies enquiring into the rheumatological component of undergraduate medical education. It is possible only to make inferences about the circumstances under which the majority of today's general practitioners were trained. However, the conclusion is inescapable that teaching about one of the major systems of the body, the musculoskeletal system, has been notable more for its omission than anything else. The disturbing result has tended to be a self-perpetuating neglect, with low levels of competence in assessing the musculoskeletal system on the part of most general and specialist practitioners. Having received very little, if any, teaching about rheumatic disorders during their formal training, doctors have had to develop their own skills in helping patients suffering from these conditions. It is inevitable that learning from experience in an unguided manner is likely to be incomplete. This is reflected in consultants' opinions, which considered that, of rheumatic patients referred to hospital, some 26 per cent could have been managed by the general practitioner himself without help from specialist services (Wood 1977).

Younger practitioners should have benefited from a range of improvements made in recent years. However, there are continuing grounds for concern about the adequacy and appropriateness of rheumatological teaching in many medical schools. Most schools now require students to have some exposure to rheumatology but the nature of this experience, when it does take place, still leaves much to be desired. Among the shortcomings is the frequency with which an excess of factual material is offered while the development of clinical skills, such as techniques of examining joints and how to assess disablement status, is often neglected. Some optimism for the future is warranted because the General Medical Council has recently accepted the importance of rheumatological contributions to medical education by identifying the specialty specifically in its latest recommendations. The advent of vocational training schemes is also encouraging, although it is disappointing that the authorities responsible for planning these schemes have so far failed to recognize the importance of rheumatology as a specific component of this training. Opportunities for continuing education are becoming more widely available as well. However, evidence currently under scrutiny indicates that unfortunately the best use of these possibilities is not being made, further dialogue between rheumatologists and general practitioners being required to establish the complementary needs of each from such experiences.

If general practitioners have been handicapped by deficiencies in training appropriate to the demands of rheumatological work in primary care, the situation has been made a lot worse by the limited availability of supporting facilities and expert back-up. A general practitioner is likely to need facilities to augment the care he is able to offer. These include access to investigative facilities and the availability of supporting staff and services, such as nurses, health visitors, social workers, and remedial therapy. Although matters have been improving in recent years, by no means all these facilities are at the command of every family doctor; it is therefore understandable that the primary care team has not always been able to play as large and as helpful a part as should have been possible.

Access to expert help when the need arises should also be readily available. The most frequent requirement is likely to be for a consultant opinion by a specialist rheumatologist, and yet 26 of the existing 118 health areas in the United Kingdom are still without anyone to provide such a service in 1981; 're-reorganization' of the National Health Service is likely to make this situation worse as not only will 59 of the 240 new health districts be lacking in such specialist input, but less permissive attitudes to what is referred to as cross-boundary flow (i.e. the treatment of patients who reside in localities other than those falling within the boundary of the health district) are also to be expected. While it must be acknowledged that, in the absence of a rheumatologist, help will be sought from specialists in other disciplines, this situation is often unsatisfactory. For example, the expertise of consultants designated as general physicians tends to be not only variable but also unavoidably deficient in matters rheumatological.

Other forms of expert help include orthopaedic surgical treatment, which will be required for a small proportion of patients; delays in the availability of relief by these means, notably total hip-joint replacement, have been a cause for continuing concern. Two other types of specialist have a considerable contribution to make in regard to the age groups with which they are concerned, paediatricians and geriatricians, and it is cause for regret that liaison between rheumatology and these two specialties has not been as close as one might wish in the past.

So far we have discussed five categories of rheumatic disorder, whenever data have allowed us to distinguish such detail. However, when we turn to considering the particular types of problem associated with the different individual conditions (more than 200 altogether), it is simpler to deal with two broad classes. For this purpose we have drawn on the Brindle Lodge report, which provided an authoritative assessment of the present status of the principal rheumatic diseases in regard to needs for primary or consultative and hospital care under optimal conditions (Wood and Badley 1976). If at times our wording sounds prescriptive, it is because we have echoed the wording of this report.

(i) Non-inflammatory conditions, which include most soft-tissue injury, non-articular rheumatism, and back troubles, and many other rheumatic disorders as well. These conditions tend to be the most frequent but also the less serious, and the largest part of their management should be entirely within the capabilities of primary care.

The care of soft-tissue injury has been developed especially in sports medicine clinics and by medical advisers interested in competitive games and athletic events. Team obligations may well call for an expert commitment but most of the skills required are accessible in specialist writings and are within the capacity of family doctors, which is the only way the needs of the community at large are ever likely to be served. For localized non-traumatic soft-tissue complaints, physical therapy is generally unhelpful and therefore a waste of precious resources. Local injection is often the best method of treatment for most of these conditions. Injection facilities at hospital rapidly become overloaded and there is therefore a compelling need to devise ways of promoting special training facilities for general practitioners, so that the overwhelming bulk of this burden can be managed at the primary care level – with the important dividend of bringing speedier relief of sufferers.

More generalized forms of non-articular rheumatism, notably polymyalgia rheumatica, are rapidly developing a management status comparable to that for gout, being entirely manageable in general practice if there is access to laboratory facilities and only occasionally requiring recourse to a consultative opinion for diagnostic and management difficulties. Constitutional disorders, such as the hypermobility syndrome, mainly call for advice in the form of a single specialist consultation to establish a management programme. As regards back pain, and although the diagnostic and management problems differ in nature, the general approach to management should resemble that for osteoarthrosis (see below). There is urgent need to determine more precisely the duration of the natural history of self-limiting back problems in order to establish the optimum period for hospital referral, so that specialist facilities are not needlessly overburdened.

(ii) Inflammatory conditions, mainly composed of defined forms of arthritis but also including the collagen or connective tissue diseases, and also rheumatic fever. For convenience, we have included osteoarthrosis under this heading even though its inflammatory nature is currently subject to debate.

Optimum care of people suffering from osteoarthrosis consists of management at the primary care level. An initial rheumatological consultation may be helpful to establish a diagnostic assessment and to take account of complicating factors and other conditions that may be present. Thereafter recourse to specialist rheumatological facilities may be required if the need for rest arises, or if mechanical problems develop that call for orthopaedic intervention. The great majority of patients are unlikely to need surgery, whereas for the more severely affected the situation is in a state of flux owing to the rapidity of developments in the field of joint replacement.

Active early polyarthritis should be manageable in general practice, depending on the interest, ability, and therapeutic resources available to the practitioner. If various procedures and appliances are available only through a hospital, then there is probably little point in the latter not assuming overall management responsibility. It is estimated that 90 per cent of patients could be managed by an interested general practitioner with adequate back-up facilities, whereas uninterested practitioners probably refer 80 per cent to hospital. Note has to be taken that reasonable access to supporting resources includes the increasing role of clinical laboratories in monitoring the effects of drugs such as immunosuppressive agents. What can cause uncertainty for a general practitioner is the fluctuating course of a disease such as rheumatoid arthritis (RA). Whereas mild cases are only occasionally likely to require consultative advice, more severe cases, perhaps one-fifth of the total, may well need specialist help under three circumstances – (a) when the disease first becomes severe (point 1 in Fig. 1.6), when expert advice, both on medical aspects and for assessment in regard to other matters such as activities of daily living, may be called for; (b) at the peak of an exacerbation (point 2 in Fig. 1.6) when in-patient care may be required, especially for women because of the difficulty of their obtaining adequate rest at home; and (c) further expert advice when deterioration occurs (point 3 in Fig. 1.6), crises of care arising not only from changes in the disease but also from alterations in resources or the environment, such as the illness or death of a caring relative.

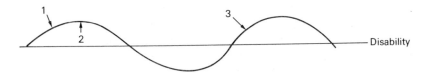

Fig. 1.6. The fluctuating course of polyarthritis, and occasions when specialist help may be particularly necessary.

The general principles of management outlined for RA would also apply to ankylosing spondylitis. However, since treatment differs from that for most other rheumatic diseases and the outcome depends on early diagnosis, which can be difficult, ready access to a rheumatological opinion is valuable for most cases. It is depressing that, even after extensive local propaganda, it is estimated that the average duration of disease before patients are seen in Manchester is 5–10 years, and yet the work of de Blécourt in the Netherlands has shown how important early and appropriate control is as regards ultimate outcome. We have already noted the current status of gout. As regards the potentially most lethal of rheumatic disorders, the collagen diseases, these are likely to need, because of their relative rarity and severity, to be seen at hospital on a regular basis, including intermittent periods of inpatient care.

TOWARDS A BETTER FUTURE

Advantages in scientific understanding of rheumatic disorders and the development of various derivative technologies have been such that the discrepancy between what can be done and what actually is done appears to be getting worse, at least in some parts of the country. Much can be done that formerly was not possible, in regard both to controlling the ravages of disease and to ameliorating its consequences. In these initiatives many people have a part to play, but for the person afflicted by a rheumatic disorder it is the general practitioner who functions, or should do, as the linchpin of such endeavours.

We have already indicated the heritage of too many general practitioners, an inadequate basic grounding in the skills necessary to diagnose and treat rheumatic disorders while they were in medical school and deprivation of the supporting facilities required to allow them to function competently and autonomously. Recent developments have been encouraging, with investment to remedy both of these deficiencies. However, much remains to be done, both educationally and in regard to service provision and organization. Elsewhere we have indicated that present delays in secondary referral for consultation and advice are probably the biggest single obstacle in the path towards devolving and improving care for rheumatic patients (Wood *et al.* 1980) – after all, from the points of view of both patient and general practitioner, when a second opinion is necessary it is generally required not some months hence but now.

There is one aspect of the challenges of primary care that requires more specific identification, and this takes us back to what we said at the beginning about the nature of locomotor disability. We illustrated the dimensions of experience by patients that should be taken into account, distinguishing between more biomedical aspects such as impairments and their functional consequences in terms of disabilities and handicaps. Unfortunately we have little grounds for confidence that the implications have yet been fully assimilated – few general practitioners appear to have much interest in rehabilitation, embryo doctors are taught very little about the assessment of disablement status, and our contacts with academic departments of general practice have not encouraged us to believe that disability problems are being integrated into teaching about primary care. This is not the place for us to try to remedy such shortcomings, even should we presume to do so. Most of the skills and resources are there to be tapped if only the imagination is exercised. The basic need is for a clearer and more sensitive appraisal of the nature of disablement experiences, so that realistic aims for the rehabilitation of rheumatic patients can be pursued (de Blécourt *et al.* 1981; Badley *et al.* 1981). The key policy goal in rehabilitation should be formulated in terms of *enabling* the individual to handle his problems, enable being the antithesis of disable. Such a strategy requires that skilled assessment of both clinical status and the individual's particular situation should be taken into account when exploring the basis for co-operation with the patient.

REFERENCES

Badley, E. M., Bury, M. R., and Wood, P. H. N. (1981). Arthritis and rheumatism. In *Going home* (ed. J. E. P. Simpson and R. Levitt) Chap. 6. Churchill Livingstone, Edinburgh.

de Blécourt, J. J., Wood, P. H. N., and Badley, E. M. (1981). Aims of rehabilitation for rheumatic patients. In *Rehabilitation in the rheumatic diseases* (ed. D. L. Woolf) *Clinics in rheumatic diseases,* Volume 7. Saunders, Philadelphia.

Cartwright, A. (1967). *Patients and their doctors. A study of general practice.* Routledge and Kegan Paul, London.

Dunnell, K. and Cartwright, A. (1972). *Medicine takers, prescribers and hoarders.* Routledge and Kegan Paul, London.

Harris, A. I. (1971). *Handicapped and impaired in Great Britain. Part 1.* Social Survey Division, Office of Population Censuses and Surveys. HMSO, London.

Hunt, A. (1978). *The elderly at home—a study of people aged 65 and over living in the community in England in 1976.* Social Survey Division, Office of Population Censuses and Surveys. HMSO, London.

Office of Population Censuses and Surveys (1974). *Morbidity statistics from general practice. Second National Study 1970–71.* The Royal College of General Practitioners, Office of Population Censuses and Surveys, Department of Health and Social Security. Studies on Medical and Population Subjects No. 26. HMSO, London.

Wood, P. H. N. (ed.) (1977). *The challenge of arthritis and rheumatism, a report on problems and progress in health care for rheumatic disorders.* British League against Rheumatism, London.

—— (1978). Epidemiology of rheumatic disorders. In *Copeman's textbook of the rheumatic diseases,* 5th ed. (ed. J. T. Scott), Chap. 3. Churchill Livingstone, Edinburgh.

—— and Badley, E. M. (1976). *Options in the delivery of medical care—a report of a working conference sponsored by the Department of Health and Social Security.* (The Brindle Lodge Report.) Arthritis and Rheumatism Council, London. (Reproduced as Appendix D in Wood (1977).)

—— —— (1978). An epidemiological appraisal of disablement. In *Recent advances in community medicine,* Vol. 1 (ed. A. E. Bennett). Churchill Livingstone, Edinburgh.

—— —— (1982). An epidemiological apparaisal of bone and joint disease in the elderly. In *Medicine in old age: bone and joint disease* (ed. V. Wright). Churchill Livingstone, Edinburgh.

——, Bury, M. R., and Badley, E. M. (1980). Other waters flow, an examination of the contemporary approach to care for rheumatic patients. In *Topical reviews in rheumatic disorders,* Vol. 1 (ed. A. G. S. Hill) Chap. 1. Wright, Bristol.

World Health Organization (1980). *International classification of impairments, disabilities, and handicaps, a manual of classification relating to the consequences of disease.* WHO, Geneva.

2 Psychological aspects of locomotor disorders: rheumatoid arthritis and non-specific back pain

Wayne Hall

There are many causes of locomotor disability but only two will be discussed in this chapter: rheumatoid arthritis and chronic back pain. The reason for restricting attention to these disorders is that they have been the most extensively investigated from a psychological point of view. Other causes of locomotor disorder such as, osteoarthrosis, spondylitis, and the seronegative arthropathies, have been neglected by psychologists. Although regrettable these lacunae in current knowledge are not too serious because rheumatoid arthritis and back pain are among the most common causes of locomotor disability (Wood 1978). What is said about the psychology of these two disorders will be true of other inflammatory and degenerative disorders to the degree that their symptomatology and functional restrictions resemble those caused by rheumatoid arthritis and chronic back pain.

RHEUMATOID ARTHRITIS

There are two ways in which psychological factors – broadly personality, character, temperament – might be related to rheumatoid arthritis. First, it has been suggested that psychological factors can operate causally be determining in part who it is that develops rheumatoid arthritis, the particular time that they develop it, and the clinical course that their disease takes. Psychologists and others interested in the psychological aspects of rheumatoid arthritis have been preoccupied with psychological contributions to the aetiology of the disease. Their motive has been the discovery of ways in which psychological methods might be used to prevent or ameliorate rheumatoid arthritis. The psychological consequences of rheumatoid arthritis have been largely ignored. The consequences of the disease will be emphasized in this chapter because knowledge of these is most likely to be of use to those who treat the patient with rheumatoid disease. In the interests of completeness research on the causal aspects of the psychology of rheumatoid arthritis will be briefly reviewed first.*

*A more comprehensive review which provides the justification for the assertions made in the following sections is available on request from the author.

PSYCHOLOGICAL INFLUENCES ON RHEUMATOID ARTHRITIS

Historically there have been three ways in which it has been suggested that psychological factors influence rheumatoid arthritis. First, advocates of the psychosomatic approach to disease (e.g. Alexander 1950) conjectured that certain personality characteristics predispose to the development of rheumatoid arthritis. Second, since at least the sixth century AD (see Rimon 1969) it has been believed that events of psychological significance, such as death or illness in the family, or marital problems, can bring on rheumatoid arthritis in susceptible persons. Third, more recently it has been suggested that psychological events may affect the clinical course of rheumatoid arthritis for better or worse (Rimon 1969).

The possibility of a predisposing personality has been extensively investigated by means of psychological assessments of patients with rheumatoid arthritis (Moos 1964; Wolff 1971; Hoffman 1974.) According to this research, patients with rheumatoid arthritis tend to be depressed, self-sacrificing, masochistic, conscientious, resentful, and unable to express their anger (Moos 1964; Wolff 1971). The evidence on which this portrait is based is more equivocal than some reviewers allow. First, the majority of the research has been done on patients with well developed and often severe disease. When attention is restricted to early cases (ones who have had the disease for less than a year) there are no differences between patients and normals (Crown *et al.* 1975); a finding which suggests that the previously reported characteristics are a consequence of the disease. Second, the personality characteristics that have been reported are not specific to rheumatoid arthritis. They are commonly found in patients with other forms of chronic illness, such as hypertension, ischaemic heart disease, and low back pain (McDaniel 1977; Shontz 1975). All these patient groups are prone to depression, irritability, resentment, and preoccupation with their symptoms; behaviour that is an understandable consequence of a chronic disease.

Although the role of psychological events in bringing on rheumatoid arthritis has not been as extensively investigated as patients' personalities the results are clearer. Compared to normals, persons who develop rheumatoid arthritis do not show any excess of major life events such as death, accident, or serious illness in the family prior to disease onset (Empire Rheumatism Council 1950). There is, however, suggestive evidence that chronic stress of the sort involved in living with hostile in-laws or an alcoholic husband may bring on the disease (Meyerowitz *et al.* 1968; Rimon 1969).

Evidence that psychological factors can affect the course of rheumatoid arthritis is anecdotal, retrospective, and scant. The adverse effect of psychological events has been suggested by a case-history (Cobb 1959), a patient series (Rimon 1969), and retrospective studies that compare the characteristics of patients with good and bad outcomes (e.g. Moos and Solomon 1964). Evidence for the good effect of psychological events has been supplied by a study of the effects of placebo medication on rheumatoid arthritis (Traut and Passarelli

1957). In this study 82 per cent of patients with disease that had proved resistant to conventional treatment showed a good response to a placebo tablet or injection; a good response was observed on both 'subjective' and objective indices of disease activity.

The brief reivew just provided is sufficient to show that the evidence in favour of a psychological aetiology for rheumatoid arthritis is less than compelling. There is no good evidence that personality predisposes to rheumatoid arthritis and only suggestive evidence that psychological events can bring on the disease or affect its clinical course. The evidence that is presently available does not warrant routine professional psychological involvement in the clinical management of rheumatoid arthritis, certainly not of the type or on the scale envisaged by advocates of the psychosomatic approach to the disease. It is doubtful in any case whether psychological services could be practicably offered to the large numbers of individuals who are afflicted. Only thoughtful research on well-defined populations will enable one to say what role psychological services have to play and in which patients they may be of some use. This does not mean that psychological measures have no role to play in the management of rheumatoid arthritis. As the following sections will show simple psychological measures that involve the use of encouragement, explanation, good humour, and reassurance may assist the arthritic to come to terms with his or her disability.

SOCIAL AND PSYCHOLOGICAL CONSEQUENCES OF RHEUMATOID ARTHRITIS

A patient's personality, family relationships, and friendships will be affected by rheumatoid arthritis, and in turn will affect his or her adjustment to the disease. This seems self-evident. Perhaps because of its self-evidence no one has seen an area deserving of investigation. Some of the questions of interest have been recognized. 'Exactly how do patients with rheumatoid disease experience their suffering? What is worse – the pain, limitation of activity, reduction in social life, curtailed sexual life? And how exactly do these and other factors produce personality change and in whom?' (Crown and Crown 1973, p. 194). But answers to these questions are not available. The answers that can be gleaned from the scanty literature are reviewed in the following sections. For the sake of simplifying presentation the account has been divided into a series of stages of illness in which different types of problem present. These stages will be called: disease onset and presentation, diagnosis, medical management, and adaptation to disease.

Onset and presentation

Because the onset of rheumatoid arthritis is often insidious (Buchanan 1978) patients may have difficulties in recognizing that they are ill and in deciding whether their illness is sufficiently serious to warrant medical treatment. Aches and pains are such common experiences that the patient may not realize that anything unusual has happened, and even if the symptoms are noticed he may be reluctant to bother his doctor for fear of being thought a complainer (Bury and Wood 1978).

Before a patient presents to a doctor there may be a period of uncertainty lasting for weeks during which he seeks advice about his symptoms. During this time the patient may refer the problem to their spouse and family members, to neighbours and other lay 'informants' (Friedson 1961). These sources may supply advice about possible diagnoses and remedies. Only after these pieces of advice have been followed and found wanting, or if the consensus is that the problem is serious, may the patient take the problem to their doctor.

The patient's experience with their general practitioner will depend in part upon the extent to which their disease is clinically manifest. Patients in whom many of the classical features are present (e.g. synovitis, arthralgia, and swelling) are unlikely to be misdiagnosed. It is doubtful, however, that all patients wait until such signs become manifest before presenting themselves to their doctor. Indeed, there is reason to believe that some patients, particularly women with early disease, are misdiagnosed because they lack the classical features of rheumatoid arthritis (Markson 1971). One of the reasons for this may be a general prejudice against taking women's complaints seriously which has been complained of by feminists. An alternate explanation may be found in the characteristics of patients who present early. There is suggestive evidence that patients who present without serological changes are more anxious than those with seropositive rheumatoid arthritis (Crown and Crown 1973; Gardiner 1980). Anxious patients may notice symptoms earlier, construe them as signs of serious illness, and hence take themselves off to their doctor before the disease becomes clinically manifest. The conjunction of anxiety and lack of signs in such cases may incline the general practitioner to diagnose 'functional' pain, hysteria, or 'psychogenic rheumatism' (Halliday 1937). The frequency with which such diagnoses are entertained is unknown but it is a subject that warrants further investigation, especially as a similar problem has been reported with multiple sclerosis where women who present with early disease may be diagnosed as hysterics. Given that the diagnosis of hysteria or functional pain has consequences (namely, suspicion about subsequent symptoms, failure to refer and less than thorough investigation) it is advisable to be cautious about using it (Merskey 1978). One must also remember that even if the diagnosis is warranted hysteria does not confer any immunity to rheumatoid arthritis.

Diagnosis

Once the diagnosis of rheumatoid arthritis has been made with some confidence this fact and its implications need to be made clear to the patient. The value of good information early in disease – as shortly after diagnosis as possible – is probably considerable although research evidence in its favour has not been provided. The prophylactic value of information in recovery from surgical assault has been amply documented (viz. Egbert *et al.* 1964; for review see Ley 1977). Advance preparation for surgery not only reduces patient distress but appreciably reduces postoperative narcotic use (by an average of 10 mg/day of morphine) and leads to earlier discharge (two days on average).

The information supplied in booklets by the Arthritis Research Council can be profitably used. When available these booklets can be given to patients but it would be unwise to rely exclusively upon them. The initial 'prescription' of information is probably best supplied by the doctor who makes the diagnosis. He or she should give a summary of current knowledge about the disease, its treatment, and prognosis. This ought to be tailored to the patient's level of comprehension and to their prior knowledge of the disease. The necessity to match explanation to patient education is well known (see Ley 1977). The advisability of taking patient beliefs into account is less often appreciated. It is desirable because patients are likely to have had different experience of the disease and to have had access to different sorts of information about it. For example, the patient who has known someone severely incapacitated by rheumatoid arthritis may need to be reassured that such an outcome will not necessary be theirs while the patient for whom 'rheumatism' is a synonym for the aches and pains of old age needs to be given some pessimistically corrective information about the likely course of their disease and the necessity to make some changes to their way of life.

The early provision of information about rheumatoid arthritis and its causes may ameliorate the problems of guilt, depression, and resentment that are said to be characteristic of patients with rheumatoid arthritis (see above, p. 31). Patients who are inclined to put moral constructions upon illness may see rheumatiod arthritis as a punishment visited upon them for past acts of omission or comission. Firm reassurance that the disease was caused by events beyond their control may disabuse them of this notion. Patients who are unconvinced by such reassurance may benefit from discussion with a knowledgeable clergy-man. Patients who are most assured of their own worth and rightness may become resentful: 'Why me?' 'Why do I suffer while others less deserving prosper?'. Referral to a clergyman may be the most appropriate way of dealing with the patient who retains some religious belief but finds current aetiological theories wanting in these respects. In the case of the unbeliever, doctors who are comfortable about discussing the issue can counsel the patient. Where this is not possible referral can be made to a social worker, occupational therapist, physiotherapist, nurse, or psychologist who may have acquired some experience in dealing with this issue.

It is advisable to involve the patient's spouse from the beginning in the process of learning about the disease because a complaint that recurs in studies of patients' experience of rheumatoid arthritis is a lack of sympathy and under-standing on the part of spouses (Bury and Wood 1978; Wiener 1975*b*; Wright and Owen 1976). Spouses and family may become sceptical about complaints because of the capricious and arbitrary occurrence of symptoms; pain and stiffness may come and go daily or unaccountably remit for varying periods.

The capriciousness of the disease may be regarded as a characteristic of the patient who is accordingly resented for exploiting his or her illness. To escape these suspicions the patient may become caught up in the self-defeating process of attempting to disguise their disability (Wiener 1975*b*; Wood and Bury 1978). This may be less of a problem for the patient whose spouse is forewarned about these characteristics of the disease.

Even in patients who subsequently make a good adjustment to rheumatoid arthritis it is probably reasonable to expect bouts of depression, irritability, resentment, and preoccupation with the disease. All these are easily understood. The question of when such symptoms warrant either pharmacological treatment or psychiatric referral is difficult to answer because it has not been seriously addressed. In the case of depression the decision is made more difficult by the fact that the systemic effects of rheumatoid arthritis are similar to those of depression. Given the good effects that have been observed with antidepressants in rheumatoid arthritis (Kuipers 1962; Scott 1969) it is probably wise to institute treatment if symptoms of depression persist for longer than two months.

Medical treatment

From the point of view of the medical practitioner compliance with treatment, or the lack of it, is a major problem in the management of rheumatoid arthritis. Non-steroidal anti-inflammatory drugs, corticosteroids, and antirheumatic drugs have been developed which can control or at least ameliorate the disease in many patients even if the effects of the drugs are not disease specific (Dick 1978). The problem is that not all of the patients who are prescribed these drugs comply with the physician's prescription; some only comply partially and others not at all. The incidence of non-compliance in general practice is unkown but in hospital samples it has been as high as 50 per cent (Geertsen *et al.* 1973).

There is a temptation to believe that the reason for the lack of compliance is irrationality on the part of the patient (Stimson 1974). This is a belief of great antiquity, being found in Hippocrates (see Friedson 1961, p. 184). Modern variants of Hippocrates' theory of treatment compliance have not been well supported by recent research. Patients who comply do not seem to be any more rational or less neurotic than patients who do not comply (Ley 1977). Poor communication of instructions is more often to blame for lack of compliance. Instructions are not given clearly and written material often presupposes a higher level of reading comprehension than is possessed by many patients (Ley 1977). Compliance can accordingly be improved by simplifying instructions and providing aids to the patient's memory such as a copy of the drug dose and the times to take it.

With effort compliance can be improved but it is probably unreasonable to expect to achieve total compliance with the drug treatments currently used in rheumatoid arthritis. Many of the commonly used agents have unpleasant side-effects, e.g. nausea, indigestion, constipation, dizziness, headache, weight gain

(Hart 1974). Some patients may prefer to suffer the symptoms of the disease rather than the side-effects of the drug prescribed. Provided that they have had the consequences of their decision explained it ought to be respected.

Compliance with exercise programs is probably of the same order as compliance with drug treatment. In a hospital sample only half of the patients were fully compliant (Carpenter and Davis 1976). Perhaps we should not be surprised that compliance is so modest when exercise programs often involve a great deal of painful, effortful activity and the benefits in the short term are difficult for the patient or therapist to discern. Compliance might be improved by making progress more apparent. This can be achieved by instructing patients in how to keep simple graphical records of their performance over time. For example, the distance walked in a week, or the number of exercises performed without pain, could be recorded over a month and graphed so that the cumulative effects of modest daily improvements are made more manifest (see Fordyce 1976, pp. 198–204).

The obverse of non-compliance with prescribed treatment is the use of non-prescribed treatments. This is evident in the widespread resort to folk remedies such as diets of various sorts; exotic concoctions of seaweeds, herbs, and shell-fish; and a plethora of rubs and ointments (Higham, personal communication). Any number of these remedies may be used concurrently with prescribed treatments. Because physicians are often disapproving of such remedies patients are often reluctant to admit to using them. Moral censoriousness is inappropriate and counterproductive. Who would not try a putative remedy that required little effort and held little risk of harm as for example, drinking cider vinegar or wearing a copper bracelet? The use of expensive remedies of doubtful worth is not to be encouraged but patients are most likely to be dissuaded from using them if they feel free to discuss them with their doctor.

We should not necessarily assume that patients who try home remedies are dupes. When the trouble is taken to ask patients about their efficacy there is a substantial degree of agreement about what does and does not work (Higham, personal communication). Even when the use of a home remedy is followed by a period of remission many patients show the appropriate scepticism about the home remedy's responsibility for the remission. They may hope that the remedy was responsible but are all too well aware of the strong probability that the use of the remedy and remission were coincidental (Wiener 1975*b*).

Adaptation to illness

This title is meant to cover that immense terra incognita that is the process involved in the patients 'learning to live' with their disease. Here there is even less guide than was the case with the earlier periods, none of which was remarkable for its comprehensive literature. All that I can do is sketch some of the problems that have emerged in the research that has so far been done.

The more immediate and personal problems of 'learning to live' with rheumatoid arthritis have been investigated by Wiener (1975*b*). She identified the central problem as one of uncertainty about the course of the disease. The main sources of uncertainty had to do with the likely severity of disability, the extent of involvement of the joints, the frequency of flare-ups, and the degree of pain and swelling that would have to be tolerated.

In dealing with the uncertainty Wiener reports that patients alternate between hope and dread. When ill they hope for remission; when in remission they dread the return of the disease. Their hope is manifest in their resort to home remedies and in the attention they give to details of diet that they believe may be influencing their disease for better or worse. Their dread of progression often takes the form of a fear of becoming a burden upon their families which finds expression in attempts to disguise the extent of disability by keeping up with their normal activities. The strategy of trying to keep up entails the disadvantage that 'when the arthritic cannot get by, it is harder to justify inaction to others . . .' (Wiener 1975*b*, p. 100). With experience patients learn to pace their activities, working in bursts punctuated by periods of rest. Eventually they are forced to lower their expectations about what lies within their capacity. Increased frequency and duration of flare-ups may force such a reappraisal. Of necessity this process involves coming to request and to accept help from others, something that may cause problems to persons who greatly value their independence.

Some of the patient's uncertainty can be reduced by discussion but the basic cause of the problem is difficult to address. Patient uncertainty about the likely course of their disease is an accurate reflection of our current capacity to predict the natural history of the disease. Statistically speaking the prognosis is 'good' in that about half of all patients can expect to escape severe incapacity (Buchanan 1978). The source of the problem is the difficulty of deciding in any individual case to which half the patient belongs. The long-term solution is better prognostic studies of rheumatoid arthritis and more attention to prognostic statification in drug trials (Feinstein 1967). If a fraction of the effort currently devoted to research on the aetiology of rheumatoid arthritis were diverted to studies of its natural history doctors might be better able to advise their patients about the likely course of their disease.

The process of adapting to rheumatoid arthritis is undoubtedly influenced a great deal by the circumstances of the sufferers: the housing in which they live, the transport to which they have access, the type of work they do, the support and help they can call upon from family and friends (Bury and Wood 1978). The influence that these things have on making the process of adaptation more or less difficult has only begun to be investigated (e.g. Bury and Wood 1978; Chamberlain *et al.* 1979; Harris 1971; Wright and Owen 1976).

The type of housing in which patients live can have a major effect upon their adaptation to the disease. Stairs which are difficult or impossible for arthritics to manage limit their access to many areas of their homes. In some cases the person is effectively restricted to the downstairs rooms of a two-storey house. Lack of access to transport that can be used by arthritics can imprison patients, especially the elderly and incapacitated. Chamberlain *et al.* report that only a third of their sample of elderly arthritics got out of the house once a day. Many were housebound, a state of affairs that was complained of by half of all patients they interviewed. The general finding of the study of Chamberlain *et al.* was that the most discontented patients were those who were the most disabled, the most housebound with the fewest contacts.

The patient's occupation may also greatly affect the extent of disability that rheumatoid arthritis causes. Bury and Wood (1978) have provided an example of the different ways in which rheumatoid arthritis affected a businessman and a manual worker. In 'the short term difficulties were more acute for the businessman who had to maintain a search for new orders for his firm and who, unlike the manual worker, could not rely on welfare benefits. On the other hand in the long run the manual worker was threatened with permanent financial hardship unless he returned to work, whereas the businessman had been able to use the early period of his illness to rearrange his affairs on a fairly permanent basis, allowing him to take time off work to rest' (p. 31).

Rheumatoid arthritis may present a very different set of problems to the housewife. A woman with children to rear and a house to look after is not able to take time off for illness on welfare benefits. Her work does not get done if she does not do it; something that husbands do not readily tolerate if Wright and Owens' finding of friction between spouses because of the husband's impatience is generally the case. Rheumatoid arthritis seems to place considerable stress on a marriage when the wife is affected and the husband's understanding of the disease is poor (Wright and Owen 1976); a finding which indicates the desirability of providing good information to the spouse as well as the patient.

The sorts of problems that have just been discussed do not all fall within the purview of problems that are medically remediable. They are 'social problems' that require for their solution various forms of 'piece-meal social engineering' (Popper 1945) such as rehousing, changes in public transport (Brookes 1974), provision of home help, and so on. Doctors may be wary of becoming involved in broadly political activity to achieve these changes, arguing that political activity is not part of their brief. But doctors have the same political rights as anyone else and the advantage of personal familiarity with the problems caused by rheumatoid arthritis. The addition of their voices to campaigns in favour of practicable changes may make a difference to their chances of being effected.

CHRONIC NON-SPECIFIC BACK PAIN

PSYCHOLOGICAL CAUSES OF CHRONIC BACK PAIN

Chronic back pain, back pain lasting more than six months, can be a symptom of degenerative, infectious, inflammatory, or neoplastic disease (Jayson 1978). In many cases, however, the cause of the chronic back pain is unknown. The term 'non-specific' back pain has been suggested for these cases (Jayson 1978). Some authors (e.g. Merskey and Spear 1967; Sternbach *et al.* 1973) have argued that some, perhaps many, of the patients with 'non-specific' back pain are suffering from 'psychogenic' pain – pain which is of psychological origin. The main warrant for their argument is that in addition to their pain lacking a known cause some of these patients present with symptoms of depression, anxiety, and bodily preoccupation (Merskey and Spear 1967; Wolkind and Forrest 1972).

Accounts of the mechanism of 'psychogenic' pain vary. The simplest explanation is a mechanistic one: 'stress' produces anxiety that leads to muscle tension which if prolonged causes pain (Merskey and Spear 1967; Holmes and Wolff 1952). This explanation possesses the considerable virtues of being plausible and acceptable to patients in need of some explanation of their pain. But apart from the cases reported by Wolff and Holmes there is surprisingly little evidence in its favour (Crown 1978) and that which indicates that muscle tension and pain are associated is just as consistent with the alternate explanation that muscle tension is a local response to pain.

Depression is another emotion which has been assigned a causal role in back pain (Forrest and Wolkind 1974; Merskey 1978; Sternbach 1974). On this account depression is more than a consequence of chronic pain; it is in fact the disorder of which back pain is a symptom. Depression which manifests itself as a somatic symptom has been termed 'masked depression' (Kielhotz 1973). The somatic symptoms that occur in masked depression include: 'sleep disturbances, headache, feelings of constriction in the throat and chest, cardiac symptoms, and particularly gastorintestinal disorders, cervical and shoulder syndromes and pain in the neck and vertebral column . . .' (Kielholz 1973, p. 12). The mechanism whereby depression gives rise to these symptoms is unclear though disturbances in levels of biogenic amines (Birkmayer *et al.* 1973) and the endorphins (Terenius 1979) have been suggested.

Two other types of explanation, the psychodynamic and the behavioural, lay more emphasis upon patient motivation. The psychodynamic explanation originates in the classical psychoanalytic account of 'conversion hysteria' provided by Breuer and Freud who attempted to provide a psychological explanation of somatic symptoms of unknown cause, the most common forms of which were pain, parasthesia, anesthesia, and paralysis. Patients who displayed these symptoms were said to be experiencing an unconscious conflict between competing motives, one of which was invariably sexual according to Freud. The symptom of which the patient complained was said to be a symbolic expression of the underling conflict; a conversion of the conflict into somatic form. The conversion symptom provided a solution of sorts to the conflict by preventing one of the motives from being expressed. The solution thus achieved comprised the 'primary gain' of the symptom. Once established the symptom could also bring 'secondary gain' – the benefits that accrue to the sufferer as a consequence of being ill, such as avoidance of responsibilities and control over others. These secondary gains sometimes interfered with the therapeutic task of discovering the underlying conflict and exorcising the symptom by making the patient conscious of it.

In the behavioural theory of psychogenic pain secondary gain has been given the central causal role. According to this theory, back pain may originate in any number of ways (e.g. trauma, degeneration, infection) but it persists beyond the normal period of repair because of various rewards which ensure its continuance (Fordyce 1976). The rewards range from the elicitation of the solicitous concern of spouse and family, the escape from unwelcome responsibilities, and the avoidance of work, to the more tangible gains such as compensation awards in the case of industrial injuries (Fordyce 1976). These rewards ensure the persistence of pain by means of 'operant conditioning', a process in which the reward automatically brings about an increase in the frequency of pain complaints and other forms of 'pain behaviour': the forms of behaviour through which we make our pain manifest to others. The therapeutic task is one of reducing the frequency of pain behaviour by altering the 'contingencies' (relationships that hold between pain behaviour and rewards) that maintain the pain behaviour. This is achieved by a four- to six-week inpatient admission during which: the patient is withdrawn from all analgesic medications, the patient's pain complaints are selectively ignored, the patient's spouse is re-educated about how to respond to the patient's complaints of pain, and the patient is put through a graduated exercise program (Fordyce 1976).

Because of the psychodynamic and behavioural accounts are ones that are broadly motivational they are similar to the commonsense idea that the patient who complains of pain without apparent cause is a malingerer who simulates pain for financial or other gain. There are differences, however. The behavioural theory differs from that of malingering in proposing a causal mechanism whereby rewards promote the 'conditioning' of pain. Because there is a causal mechanism patients are absolved of personal responsibility for their pain. The psychodynamic account differs from that of malingering in that the motive for the symptom is outside the patient's awareness. The subtlety of the differences between these explanations and malingering is not always appreciated by those who deal professionally with back pain patients. The consequence is that the disapproval which may be appropriate in dealing with the 'genuine' malingerer is carried over into all dealings with patients whose pain lacks a discernible cause and who appear to have a motive for being in pain.

The evidence commonly offered in favour of psychological causes of back pain consists of information on the characteristics of back pain patients seen in specialist orthopaedic, rheumatology, or pain clinics. Demographic and clinical data are often supplemented by scores on standardized psychological tests such as the Minnesota Multiphasic Personality Inventory (McCreary and Jamison 1975) and the Middlesex Health Questionnaire (Wolkind and Forrest 1972). The typical finding is that chronic pain patients are anxious, depressed, preoccupied with their pain, convinced that they suffer from an occult illness, and demonstrative in complaining about their pain (Crown 1978; Hall *et al.* 1981; Pilowsky *et al.* 1979; Wolkind and Forrest 1976).

There are a number of points to be made about the meaning of these findings. First, the patients series in which the findings are reported are selected. They are selected initially in being seen at a specialist clinic where the processes of referral operate to select patients with difficult diagnostic problems and patients who are more demanding of medical services (Chapman *et al.* 1979; Wood and Badely 1980). They are often further selected in ways that are not always made explicit. Wolkind and Forrest (1972), for example, report data on 23 patients who were initially selected from 186 patients attending a specialist orthopaedic clinic. These 23 patients were those who were still attending the clinic 90 days after their first presentation. Thus the characteristics of published series may not be representative of all back pain patients seen in specialist clinics let alone those seen by their general practitioners.

Second, even if we are careful in specifying the population of patients in whom these characteristics are found it is unwise to draw the inference that psychological problems cause back ache in these patients. There are two alternative explanations that have not been discounted: (i) psychological disorders may be a consequence of chronic back pain and the way in which it is managed, and (ii) the development of chronic back pain may bring about psychological decompensation in predisposed individuals.

SOCIAL AND PSYCHOLOGICAL CONSEQUENCES OF BACK PAIN

It is generally agreed that persistent pain saps morale and causes depression. It is less often appreciated that the processes of referral, investigation, and treatment may also produce some of the characteristic behaviour of back pain patients. The treatment history of patients with chronic back pain has often been a series of dispiriting encounters with their medical attendants. They have undergone various investigations, progressively more invasive and uncomfortable, which fail to disclose a cause for their pain. They have submitted to many trials of different treatments that fail leaving them pessimistic about ever achieving relief. And they have been treated by a succession of perplexed doctors and surgeons who begin to doubt their veracity because they are in pain without apparent cause and because they have failed to respond as expected to their doctor's best therapeutic efforts. It would be a sturdy individual who endured these experiences without displaying some of the depression, pessimism, and resentment encountered in patients with chronic back pain. Any predisposition to neurosis would understandably accentuate the psychological symptoms presented.

Scepticism about the veracity of the patient's complaints of pain is a special problem for the patient with chronic back pain. If there are no signs of an evident cause and the patient appears to be dramatic in the way he complains of pain, those in contact with him may believe that he protests too much. When, as often happens, the patient becomes aware of the scepticism of those around him, he often tries to convince them that his pain is severe. The efforts to convince the sceptics take the form of exaggerated wincing, the adoption of an atypical gait; behaviour which is largely counterproductive in that it serves to reinforce scepticism because it is observed only in the presence of others (Wiener 1975a). The patient's behaviour is dismissed as acting which it is in part but the pain it betokens is not thereby rendered less painful.

Back pain patients may also be regarded as 'complainers' because their behaviour compares unfavourably with that of the other patients in pain (e.g. arthritics or ankylosing spondylitics). More importantly, their pain behaviour compares unfavourably with the behaviour of those who care for them. Many nurses, for example, experience back pain as an occupational hazard but continue to work. Their attitude towards the complaining back pain patient may consequently be less than sympathetic. Others whose experience of back pain is limited to acute aches that abate with time may find it difficult to comprehend that back pain can be severe and that it can continue for months on end. The back pain sufferer is a victim of our habit of treating chronic pain in the way we often treat acute pain: as something to be borne with minimal complaint because of the reasonable expectation that it will be gone in a number of days (Hackett 1971).

Paradoxically, the patient who does not complain or who is matter of fact in describing his pain, may also be suspected of simulating because he does not complain enough. We assume that anyone who is in severe pain should be writhing, sweating, and complaining loudly. Anyone who does not cannot be in severe pain. Hackett has criticized this type of reasoning, arguing from personal experience that:

'When pain lasts days and weeks it either produces stress sufficient to exhaust, deplete and eventually destroy the victim or adaptation occurs. Somehow, either through pallation from drugs or distraction or stoicism or a combination of these and more, the individual manages to endure his pain and carry unlimited social transactions despite it. He usually can manage to appear untroubled, to be in no pain. However, it would be just as much a mistake to regard such an individual as exemplifying *la belle indifference* as it would to view him as a malingerer . . .' (Hackett 1971, p. 132).

A back pain patient's veracity is especially likely to be suspected if he or she is involved in legal action to obtain compensation for an injury that provoked the pain, or if he or she is in receipt of compensation for injury. Conventional medical wisdom has it that patients involved in compensation claims will not respond to treatment until a settlement has been achieved and that once settlement has been achieved the disorder will remit without treatment. One might expect that beliefs with such wide currency would be well substantiated by evidence. There is some evidence in their favour (e.g. Krusen and Ford 1958) but there is also evidence to the contrary. Cox (1979), for example, has reported considerable success in treating back pain compensation cases. Of 45 patients put through a didactic exercise programme 27 were restored to gainful employment, and all were still employed at 12-month follow up. All patients had been in receipt of compensation and 39 were involved in litigation. Good information on the effect of settlement on back pain is not available but similar studies of the effect of settlement on head injuries have failed to disclose any improvement (Merskey 1978, p. 90f). Given our comparative ignorance of these matters research on the effects of compensation on the course of back pain deserves priority; a higher priority than attempts to discover an 'objective' measure of pain to serve as a touchstone to separate 'real' pain from base malingering.

None of what I have said in criticism of the conventional wisdom implies that malingerers do not exist. What I am contesting is the implicit assumption that simulated pain is so common that doctors should be suspicious of it in any patient whose pain lacks an evident cause and who has some conceivable motive for complaining of pain. Here the comments of the French neurosurgeon Leriche are especially pertinent:

'. . . I have been able to collect many examples of simulated pain. And yet I am convinced that almost always, those who suffer quite as much as they say they do, and that if we assess their pain with the degree of attention it deserves, they suffer, indeed, more than we can imagine. There is only one pain that is easy to bear: it is the pain of others' (Leriche 1939, p. 25).

BACK PAIN IN GENERAL PRACTICE

Cases of back pain in which psychological causes operate are likely to be rare in general practice. That does not mean, however, that back pain in general practice is not responsible for a fair degree of distress and suffering. Back pain is a common but underrated cause of everyday misery and unhappiness as the community survey by Nagi *et al.* (1973) showed. The general practitioner is in a good position to minimize the misery of patients with persistent back pain by providing good information early in the course of the illness.

One of the most important pieces of information for the non-specific back pain sufferer is information about prognosis. As was the case with rheumatoid arthritis it is not always easy to be precise. What can be given is the reassurance that although the condition is likely to recur it is not necessarily progressive. Fear of becoming crippled and a burden to others may be a predominate fear in the case of an individual who believes he is destined for a wheelchair. The prospect of eventual pain relief is also important for the patient who may begin to believe that he will never be free of pain.

Patients who have been investigated without any cause being discovered need special consideration. They should be told that the failure to discover a cause (e.g. an obtruding intervertebral disc) does not mean that there is no cause for their pain; only that the investigative methods available have not disclosed one. It is especially important to explain the meaning of diagnostic findings because patients often have their own ideas about what is wrong (e.g. a nerve is trapped) and about how it can be resolved (e.g. by 'freeing' the trapped nerve). When the patient is not offered the treatment he thinks appropriate and no explanation is offered he may believe that his doctor is arbitrarily withholding treatment. It is wise to offer such explanations early because unrelieved patients may later become less accepting of their doctor's advice. Patients may not always find these explanations acceptable, one reason being the disparity between the well-publicized technological achievements of some medical specialities and the 'empirical' status of many of the symptomatic treatments used for back pain. As one patient asked : 'You can transplant hearts. Why can't you fix my back?'

Having disabused patients of any expectation that their problem can be easily fixed it is important that they not be left feeling that nothing can be done. The necessity for a broad therapeutic approach needs to be impressed upon the patient (Hart 1974). Explanations should be given of the role played by exercise, physiotherapy, modification of posture, ergonomic modification of furniture and workplace, and simple analgesics. Compliance with such a regime is more likely if the rationale of each component has been explained to the patient.

It is good policy to involve the patient's spouse in the process of patient education from the beginning. The spouse's involvement may defuse some of the potential areas of misunderstanding and disagreement that are likely to arise if the nature of the patient's condition is not known to his or her family.

Good information and sympathetic attention will not always prevent the patient from becoming depressed, anxious, irritable, or angry. As in rheumatoid arthritis, transient experiences of this sort are to be expected and can be managed by sympathetic discussion of their causes. If these symptoms become protracted and a characteristic feature of the patient's adaptation to his back pain, discussion may not be enough. In the case of depression, the most likely problem, antidepressant medication may be necessary if the depression becomes incapacitating. If antidepressants are prescribed they ought to be given in conjunction with normal management because the patient may believe that the treatment of his depression implies that the doctor no longer takes his pain complaints seriously.

When depression is unusually severe or unresponsive to antidepressants, or other psychiatric symptoms present serious obstacles to proper management, a psychiatric referral is appropriate. If referral is decided upon it is important that the reasons for it are discussed with the patient. When such an effort is not made and the patient's prior consent or agreement is not obtained, the psychiatrist often finds himself with an angry, resentful, and hurt patient who feels that he has been abandoned by his doctor. It is important too that referral be made earlier rather than later. If referral is deferred until all forms of medical management have been tried and found wanting the relationship between the patient and his referring doctor may have irretrievably broken down. These are not the most propitious of circumstances for the psychiatrist to attempt treatment in. Early referral in conjunction with a good explanation and the continuation of conventional medical treatment may do much to preserve good relations between patients and their doctors.

CONCLUSIONS

Three things emerge from the discussion of the psychological aspects of rheumatoid arthritis and chronic back pain: (i) the importance of providing the patient with the best available information about the causes, treatment, and prognosis of his illness; (ii) the necessity for more research on the natural history of both disorders, including the personal and social problems that each disorder poses for those afflicted by it; (iii) the desirability of changing lay and professional prejudices about the characteristics and behaviour of those who suffer from these disorders.

ACKNOWLEDGMENT

The preparation of this chapter was undertaken while the author was in receipt of an Applied Health Science Fellowship from the National Health and Medical Research Council of Australia.

REFERENCES

Alexander, F. Z. (1950). *Psychosomatic medicine.* Norton, New York.

Birkmayer, W., Neumayer, E., and Riederer, P. (1973). Masked depression in old age. In *Masked depression* (ed. P. Kielholz). Hans Huber, Berne.

Breuer, J. and Freud, S. (1955). *Studies on hysteria,* standard edition (ed. J. Strachey). Hogarth Press, London. (Originally published 1895).

Brooks, B. M. (1974). An investigation of factors affecting the use of buses by both elderly and ambulant persons. Transport and Road Research Contract Report. British Leyland UK Ltd.

Buchanan, W. W. (1978). Clinical features of rheumatoid arthritis. In *Copeman's textbook of the rheumatic diseases* (ed. J. T. Scott). Churchill Livingstone, London.

Bury, M. R. and Wood, P. H. N. (1978). Sociological perspectives on disablement. *Int. rehab. Med.* **1**, 25–32.

Carpenter, J. O. and Davis, L. J. (1976). Medical recommendations – followed or ignored? Factors influencing compliance in arthritis. *Archs phys. Med. Rehab.* **57**, 241–6.

Chamberlain, M. A., Buchanan, J. M., and Hanks, H. (1979). The arthritic in an urban environment. *Ann. rheum. Dis.* **38**, 51–6.

Chapman, C. R., Sola, A., and Bonica, J. J. (1979). Illness behaviour and depression compared in pain centre and private practice patients. *Pain* **6**, 1–7.

Cobb, S. (1959). Contained hostility in rheumatoid arthritis. *Arthritis rheum.* **2**, 419–21.

Cox. B. G. (1979). Chronic pain relief and employment restoration levels in the worker's compensation recipient. Paper presented American Pain Society Annual Meeting, San Diego.

Crown, J. M. and Crown, S. (1973). The relationship between personality and the presence of rheumatoid factor in early rheumatoid disease. *Scand. J. Rheum.* **2**, 123–30.

Crown, S. (1978). Psychological aspects of low back pain. *Rheum. Rehab.* **17**, 114–24.

——, Crown, J., and Fleming, A. (1975). Aspects of the psychology and epidemiology of rheumatoid disease. *Psychol. Med.* **5**, 291–9.

Dick, W. C. (1978). Drug treatment of rheumatoid arthritis. In *Copeman's textbook of the rheumatic diseases* (ed. J. T. Scott). Churchill Livingstone, London.

Egbert, L. D., Battit, G. E., Welch, C. E., and Bartlett, M. K. (1964). Reduction of post-operative pain by encouragement and instruction of patients. *New Engl. J. Med.* **270**, 825–7.

Empire Rheumatism Council (1950). A controlled investigation into the aetiology and clinical features of rheumatoid arthritis. *Br. med. J.* April, 799–805.

Feinstein, A. R. (1967). *Clinical judgement.* Williams and Wilkins, Baltimore.

Fordyce, W. E. (1976). *Behavioral methods for chronic pain and illness.* Mosby, St. Louis.

Forrest, A. J. and Wolkind, S. N. (1974). Masked depression in men with low back pain. *Rheum. Rehab.* **13**, 148–53.

Friedson, E. (1961). *Patients' views of medical practice.* Russell Sage Foundation, Philadelphia.

Gardiner, B. M. (1980). Psychological aspects of rheumatoid arthritis. *Psychol. Med.* **10**, 159–63.

Geertsen, H. R., Gray, R. M., and Ward, J. R. (1973). Patient compliance within the context of seeking medical care for arthritis. *J. chron. Dis.* **26**, 689–98.

Hackett, T. P. (1971). Pain and prejudice: why do we doubt that the patient is in pain? *Med. Times* **99**, 130–41.

Hall, W. D., Hayward, L. D., and Chapman, C. R. (1981). On 'pain and laterality'. *Pain,* in press.

Halliday, J. L. (1937). Psychological factors in rheumatism. *Br. med. J.* January, 213–17.

Harris, A. I. (1971). *Handicapped and impaired in Great Britain.* Part 1. Her Majesty's Stationary Office, London.

Hart, F. D. The control of pain in the rheumatic disorders. In *The treatment of chronic pain* (ed. F. D. Hart). MTP, Lancaster.

Hoffman, A. L. (1974). Psychological factors associated with rheumatoid arthritis. *Nursing Res.* **23**, 218–34.

Holmes, T. H. and Wolff, H. G. (1925). Life situations, emotions and back ache. *Psychosom. Med.* **14**, 18.

Jayson, M. I. V. (1978). Back pain, spondylitis and disc disorders. In *Copeman's textbook of the rheumatic diseases* (ed. J. T. Scott), Churchill Livingstone, London.

Kielholz, P. (1973). Psychosomatic aspects of depressive illness – masked depression and somatic equivalents. In *Masked depression* (ed. P. Kielholz). Hans Huber, Berne.

Krusen, E. M. and Ford, D. E. (1958). Compensation factor in low back pain. *J. Am. med. Ass.* **166**, 1128–33.

Kuipers, R. K. W. (1962). Imipramine in the treatment of rheumatic patients. *Acta rheum. scand.* **8**, 45–9.

Leriche, R. (1939). *The surgery of pain.* Williams and Wilkins, Baltimore.

Ley, P. (1977). Psychological studies of doctor–patient communication. In *Contributions to medical psychology,* Vol. 1 (ed. S. Rachman). Pergamon Press, London.

McCreary, C. and Jamison, K. (1975). The chronic pain patient. In *Consultation–liason psychiatry* (ed. R. O. Pasnau). Grune and Stratton, New York.

McDaniel, J. W. (1976). *Physical disability and human behaviour.* Pergamon Press, London.

Markson, E. W. (1971). Patient semiology of a disease. *Social Sci. Med.* 5, 159–67.

Merskey, H. (1978). *The analysis of hysteria.* Ballière Tindall, London.

—— and Spear, F. G. (1967). *Pain: psychological and psychiatric aspects.* Ballière Tindall, London.

Meyerowitz, S., Jacox, R. F., and Hess, D. W. (1978). Monozygotic twins discordant for rheumatoid arthritis: a genetic, clinical and psychological study of 8 sets. *Arthritis Rheum.* 11, 1–12.

Moos, R. H. (1964). Personality factors associated with rheumatoid arthritis: a review. *J. chron. Dis.* 17, 41–55.

—— and Solomon, G. F. (1964). Personality correlates of the rapidity of progression of rheumatoid arthritis. *Annls rheum. Dis.* 252, 145–53.

Nagi, S. Z., Riley, L. E., and Newby, L. G. (1973). A social epidemiology of back pain in a general population. *J. chron. Dis.* 26, 769–79.

Pilowsky, I., Chapman, C. R., and Bonica, J. J.' (1977). Pain, depression and illness behaviour in a pain clinic population. *Pain* 4, 183–92.

Popper, K. R. (1945). *The open society and its enemies.* 2 Vols. Routledge and Kegan Paul, London.

Rimon, R. (1969). A psychosomatic approach to rheumatoid arthritis. *Acta rheum. scand.* Suppl. 13.

Scott, W. A. M. The relief of pain with an antidepressant in arthritis. *Practitioner* 202, 802–6.

Shontz, F. C. (1975). *The psychological aspects of physical illness and disability.* Macmillan, New York.

Sternbach, R. A. (1974). *Pain patients: traits and treatment.* Academic Press, New York.

——, Wolf, S., Murphy, R., and Akeson, W. H. (1973). Traits of back pain patients: the low back 'loser'. *Psychosomatics* 74, 226–9.

Stimson, G. V. (1974). Obeying doctor's orders: a view from the other side. *Social Sci. Med.* 8, 97–104.

Terenius, L. (1979). Endorphins in chronic pain. In *Advances in pain research and therapy,* Vol. 3 (ed. J. J. Bonica). Raven Press, New York.

Traut, E. F. and Passarelli, E. W. (1957). Placebos in the treatment of rheumatoid arthritis and other conditions. *Ann. rheum. Dis.* 16, 18–21.

Wiener, C. L. (1975a). Pain assessment on an orthopaedic ward. *Nursing Outlook* 23, 508–16.

—— (1975b). The burden of arthritis: tolerating the uncertainty. *Social Sci. Med.* 9, 97–104.

Wolff, B. B. (1971). Current psychosocial concepts in rheumatoid arthritis. *Bull. rheum. Dis.* 22, 656–61.

Wolkind, S. N. and Forrest, A. J. (1972), Low back pain: a psychiatric investigation. *Postgrad. Med.* 48, 76–9.

Wood, P. H. N. (1978). Epidemiology of rheumatic disorders. In *Copeman's textbook of the rheumatic diseases* (ed. J. T. Scott). Churchill Livingstone, London.

—— and Badeley, L. (1980). Back pain in the community. In *The lumbar spine and back pain* (ed. M. I. V. Jayson). Sector, London.

Wright, V. and Owen, S. (1976). The effect of rheumatoid arthritis on the social situation of housewives. *Rheum. Rehab.* **15**, 156–60.

3 Interpretation and differential diagnosis of limb pain and arthritis

M. J. Dodd, C. Leon, and W. Carson Dick

GENERAL INTRODUCTION

The time-honoured division of renal, hepatic, or cardiac disease into 'acute' or 'chronic' has proven its value in clinical medicine. Similarly, it may assist the busy doctor to think in terms of 'locomotor' failure and to subdivide this into 'acute' and 'chronic'* Fig. 3.1. The patient with locomotor failure presents to the general practitioner with symptoms which are disrupting his function as an individual. The symptoms will be focused upon pain, stiffness, functional disability, swelling, or deformity of joints. A problem-orientated approach to diagnosis and management is particularly applicable in locomotor failure in that it leads to the construction of a management plan which is likely to be more appropriate to the patient's needs and more realistic in terms of the physician's ability to intervene in the disease process. A diagnostic label is helpful in the understanding of the likely pathology and prognosis of the disease but is of more restricted value in the assessment of a patient as an individual and of the likely interaction of that individual with his or her disease.

One of the most important early challenges in the management of a patient with chronic inflammatory joint disease is assessment and it is impossible to construct a therapeutic goal without first having obtained a thorough assessment of the patient's complaints, of his or her character and likely response to these complaints and in particular of the degree of reversibility of the process. So much of the management of chronic inflammatory joint disease is subjective that it is only by appreciation of the patient's own assessment of his needs that the doctor can achieve any understanding of the problem as the patient perceives it as opposed to the more usual definition of a problem in purely pathological or pathophysiological terms. As an example of this, pain is likely to be an early symptom and also a complaint to which the patient gives a high priority. Persisting pain, whatever the cause, is of itself destructive and produces chronic anxiety and depression. Adequate explanation of the process which produces

---

*The term 'locomotor' clearly directs attention to the lower limb and the purist may employ the words 'musculoskeletal' failure to embrace the upper limb and to describe the state in which any limb or part of a limb fails to meet the needs of the patient. In this chapter we have elected to use the more convenient term locomotor failure to encompass both and we have concentrated upon articular causes. Neurological and circulatory disorders which produce locomotor failure are covered in detail in other chapters.

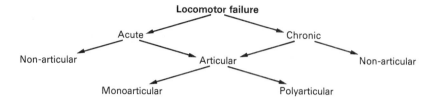

Fig. 3.1.

the pain and clear description of the possible modalities available to treat it may go a long way towards mitigating the feeling of helplessness which is such a large component of the patient's anxiety and depression.

Full descriptions of the subjective, semi-objective, and objective assessment methods available in chronic inflammatory joint disease are available elsewhere and it is our purpose at this stage merely to record how important these are to monitoring the course of any patient with chronic arthritis. It is worth emphasizing that a busy general practitioner, within the time available to him in the context of extended consultations, can quite easily record a numerical or adjectival pain scale, an articular index of joint tenderness, a haemoglobin, ESR, rheumatoid factor titre, and basic X-ray results. These, in the context of a patient in whom due primary regard has first been given to clinical history, the past history, the family history and the clinical examination, are the essential ingredients required and are well within the facilities available to most general practitioners today. It is also relevant at this juncture to point out that the patient's expectations may be very different from those of his physician. The doctor's objectives may well be relief of pain, restoration of function, and return to gainful employment, whereas the patient's objectives may quite easily be relief of pain, some restoration of function, and the acceptance of a long-term invalid status. This is the basis of the 'learned helplessness' concept developed by Seligman. Explanation, advice, encouragement, and the formulation of realistic therapeutic goals are far more important to the doctor and to the patient with chronic arthritis than are the words which encompass what are presently termed diagnostic labels. Common sense and compassion are of more importance than serum complement and copper and the most important single consideration is the need for the physician to project to the patient an attitude of interest and confidence rather than indifference and insecurity. Sadly, even today it is too often the case that the latter attitude prevails to the considerable detriment of the patient's future management, whether by that physician or by any other physician. It is far more difficult to correct an erroneous impression given by a previous medical attendant than it is to inculcate the proper principles of management right from the beginning.

One consideration which is looming larger and larger on our present horizon is the need to provide a 'diagnostic' label and the need to be aware that patients will obtain information from the popular press, the media, and, perhaps worst of all, other patients. Very often this sort of informal instruction is not only erroneous but even dangerously misleading and it is very much the function of a good general practitioner to arm the patient with knowledge about his disease in anticipation of just this very problem.

It is also worth emphasizing right at the beginning that most of the disorders covered in this chapter are long-standing and subject to remissions and exacerbations, a consideration too often omitted from our concept of disability today. Most of our ideas of disability stem from the last war and even the official forms provided display the bias towards stable and non-progressive as opposed to fluctuating problems. It is very difficult to explain to officialdom that a person with, for example, rheumatoid arthritis may be perfectly fit to work 10 months in a year but may have an unpredictable period of incapacity for the other two months in every 12. A consequence of this fluctuation is the need to retain balance at all times. Thus, neither physician nor patient should become too elated during a remission and similarly, it is important not to allow a temporary exacerbation to become a positive feedback system of pain, further disability, depression, further pain which can be very difficult to interrupt. Similarly, this sort of disease makes nonsense of routine return outpatient appointments whether to the general practitioner or to the hospital. The patients should have access to their physicians when they require them and not simply because their three-month appointment is due.

Finally, the fluctuation in disease activity underlines the absolute importance of not only the initial assessment but continual re-assessment. The management of the great majority of patients with chronic arthritis should be undertaken as close to the home and work environment as possible and should be centred on the interested general practitioner. The greatest danger to be avoided is what is now called the 'collusion of anonymity' whereby the patient is seen at regular intervals by a different junior hospital doctor on each occasion precluding sensible decision making and without anybody accepting responsibility for those decisions. Contrast this with a situation where the patient's management is centred on the interested local general practitioner with information on how to assess effectively, efficiently, and quickly and, most important of all, how to record and to retrieve the data generated. The general practitioner should then be in a position of dialogue with his consultant for points of general principle and this provides the optimum outcome for the patient, general practitioner, consultant, and all concerned. It is also important to emphasize that at different times in any patient's history there will be the need to consider help from other sources such as physiotherapy, nursing, occupational therapy, orthopaedic surgery, opthalmology, and disablement resettlement officers. It is absolutely essential that the co-ordination of all this should be firmly in the hands of the general practitioner to obviate the obvious disaster of a patient being shuttled from department to department undergoing piecemeal necrosis.

These then are the sort of general principles which spring to mind when faced with the problem of interpretation and differential diagnosis of acute or chronic locomotor failure based on articular pathology.

ACUTE LOCOMOTOR FAILURE (ALF)

ALF presents as pain, stiffness, swelling, or dysfunction of a limb or part of a limb. The first challenge is whether or not the seat of the problem is intra-articular and this is achieved in a vast majority of instances by history and clinical examination. It is true to say that if you have not reached a diagnosis at that stage then you are unlikely to do so later. Laboratory tests should be viewed as confirmatory or adjunctive and it is not possible to overemphasize the central importance of simple history coupled with a precise, carefully conducted clinical examination. It is becoming more and more important today to emphasize the central role of past history and of family history in the chronic inflammatory joint diseases and, as noted above, consideration of the socio-economic position of the patient is more important in this than in any other field.

In the differentation of intra- and extra-articular lesions the simple rules are easily applied. If there is restriction of passive movement in all directions then the problem is, without doubt, intra-articular. Similarly, swelling or enlargement arising from the joint margin provides reliable, objective evidence of an intra-articular pathology. If, on the other hand, dysfunction is produced in one direction of movement only, then it is much more likely that the diagnosis will be based on extra-articular pathology. Swelling from the joint margin will be due to synovial hypertrophy or synovial effusion or to bony enlargement or to a combination of all of these. Synovial hypertrophy or synovitis has a boggy or doughy feel and can be shown to encompass most of the joint. The normal outlines and skin contours around the joint may be distorted and synovial effusion may be detected by palpation. In the shoulder an effusion may be stroked from one compartment to the other and this may be clearly audible using the stethoscope. In the knee, in similar manner, the presence of a small amount of fluid may be detected by stroking out of one compartment into another and noting the bulge appearing on the opposite side of the joint margin. A larger effusion in the knee will allow the examiner to ballott or tap the patella against the femoral condyles. Synovial hypertrophy and effusion often co-exist making differentiation difficult at times.

Bony enlargement, on the other hand, is firm and fixed and usually irregular. A common example is the enlargement of the distal interphalangeal joints (DIP) of the fingers found in osteo-arthritis (Heberden's nodes) or of the proximal interphalangeal joints in the same disease (Bouchard's nodes).

In contrast to articular problems, peri-articular disease is confined to the outline of the involved structures and to the direction of movement of the structures involved. Examples of this are the synovitis affecting the tendon sheaths of the fingers or thumb in which case the patient may complain of a painful wrist but examination will show that the pain or tenderness is localized over the tendon sheath and not over the joint margin. It will be increased as the tendon is moved through its sheath. Certain enthesopathies (enthesis, attachment of tendon to bone – Fig. 3.2) may present as a painful joint but examination will show that pain is localized not on the joint margin but slightly divorced from it. Examples of this are the medial and lateral epicondylitis comprising golfer's and tennis elbow, plantar fasciitis and Achilles' tendonitis presenting as a painful heel, or the multiplicity of lesions which may occur around the shoulder. In most cases the nearby joint will retain a short range of pain-free passive movement but sharp pain will be induced when the damaged enthesis is subjected to traction by any appropriate manoeuvre. This highlights the importance of having a consistent and coherent approach to examination of the peripheral joints. Inspection should be followed by active movements to detect whether or not one particular structure is involved and then if these are pain-free, passive movements should be undertaken thereafter. Considerable amounts of information may be obtained on passive movement beyond the normal range of active movement when the 'end feel' of the articulation may be assessed and this in itself may produce the patient's pain where active movements had not done so.

The site of maximum tenderness is one of the most valuable clinical signs (Fig. 3.2). It is obvious that if a child complains of a painful knee but that the site of maximum tenderness is over bone, the possibility of an osteomyelitis looms large, a differential diagnosis which it is disastrous to ignore.

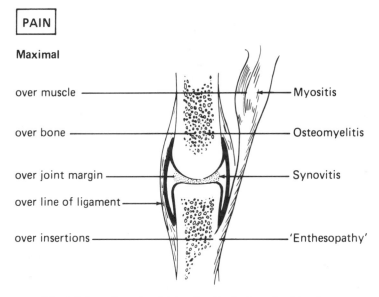

Fig. 3.2. Location of point (or direction) of maximal tenderness.

Non-articular disease may also present as acute locomotor failure. Only a careful history and examination of the musculoskeletal structures will enable differentiation between articular and peri-articular from more distant structures as the cause for the complaint. For example a variety of nerve entrapments including carpal tunnel syndrome, tarsal tunnel syndrome, thoracic outlet syndrome, and herniated disc may mimic many locomotor disorders because of pain radiating to the joint. The trap usually is pain being felt in a distal joint which is itself normal, deriving from pain in a more proximal articulation which is the seat of the disease. Table 3.1 lists various peri-articular and non-articular problems often localized symptomatically to a particular joint.

Other patients may complain of more widespread joint, bone and muscle pains which, on examination, have no clear-cut articular or periarticular basis. The diagnosis of these myalgic syndromes are often difficult but they may be caused by systemic diseases such as polymyalgia rheumatica, cranial arteritis, polymyositis, hypothyroidism, or metabolic bone disease such as osteomalacia. These conditions may be overlooked unless consciously sought and may require

Table 3.1. *Periarticular and non-articular conditions felt as if in a particular joint*

Joint	Periarticular condition	Non-articular disease
Hand	synovitis of flexor and extensor tendon sheaths	carpal tunnel syndrome
Wrist	tenosynovitis of extensor tendon sheaths tenosynovitis of thumb extensor and abductor tendons dorsal ganglia	carpal tunnel syndrome cervical spondylosis
Elbow	epicondylitis (golfers' and tennis elbows) olecranon bursitis	carpal tunnel syndrome (pain referred up the arm) cervical spondylosis myocardial ischaemia
Shoulder	calcific tendonitis subdeltoid bursitis subacromial bursitis bicepital tendonitis rotator cuff tears	pancoast tumour visceral disease (e.g. gall bladder, peritonitis) pleuritic chest pain myocardial ischaemia cervical disc disease
Hip	trochanteric bursitis	lumbar spondylosis ischaemic vascular disease pelvic inflammatory disease renal colic
Knee	bursitis – prepatella (housewives' knee) – infrapatella tendonitis collateral ligament injuries	lumbar spondylosis osteochondritis of tibial tubercle (Osgood–Schlatter) osteochondritis of patella poles
Ankle	achilles tendonitis achilles bursitis	erythema nodosum osteochondritis of the navicular
Foot	painful heel pad plantar fasciitis tenosynovitis	tarsal tunnel syndrome osteochondritis of metatarsal heads interdigital neurona

estimation of the sedimentation rate, muscle enzymes, thyroid function, and bone chemistry to begin to differentiate the various disorders. More specialized procedures such as temporal artery biopsy or muscle and bone biopsy are then needed to come to a firm diagnosis.

ACUTE LOCOMOTOR FAILURE PRESENTING AS MONOARTHRITIS

The patient who presents with a single swollen joint is described as suffering from monoarticular arthritis. Involvement of two or three separate joints indicates an oligo or pauciarticular problem. Since monoarticular disease may progress to polyarticular disease and polyarticular arthritis may start in a single joint, the diagnostic possibilities that this raises must always be kept in mind when the patient first presents (Table 3.2). The diseases which most often present as a monoarticular problem are listed in Table 3.3

In determining the possible cause of a monoarticular arthritis, having excluded periarticular or non-articular disease, a number of factors may help in arriving at a diagnosis.

Table 3.2. *Polyarticular disease which may present as acute monoarthritis*

Psoriatic arthritis
Rheumatoid arthritis
Rheumatic fever
Polyarticular juvenile chronic arthritis
Gonococcal septic arthritis
Rubella arthritis
Enteropathic arthritis
Sarcoid arthritis

Age of the patient

Monoarticular arthritis presenting in a child should always raise the possibility of trauma, haemophilia, or sepsis. Clues that a pyogenic arthritis may be present are suggested by a pyrexial child, a hot swollen joint which is very painful on attempted movement, a focus for primary infection such as skin abrasions or systemic infection or an acute arthritis developing in an immunosuppressed child. If you even consider a joint to be septic, aspiration and culture of the synovial fluid becomes mandatory.

Musculoskeletal symptoms are common with many viral infections in children. Complaints are transient but in some cases arthritis is prominent. Mumps may present as a self-limiting oligoarthritis of large joints although acute polyarthritis can occur. The diagnosis is suggested by the associated parotid gland inflammation which may precede the joint symptoms by 7–10 days.

Rheumatic fever, which is commonly a polyarthropathy, may present in a mono- or oligoarticular fashion. Arthritis is the most frequent manifestation of rheumatic fever and since the onset is acute with fever and joint pain, septic arthritis is often considered to be the cause of the problem and must be excluded. The arthritis, which is essentially a non-destructive process, often involves large joints such as knees, ankles, wrists, and elbows and the pattern is of a peculiar flitting character whereby one joint settles before another is involved. Inflammation in each joint lasts only 2–3 days and swelling without pain is rare. Clues to the diagnosis comes from a history of previous streptococcal infection and the other characteristic systemic features such as carditis, skin rash, development of small subcutaneous nodules, the prompt response of the fever to salicylates, and a raising ASO titre. Rheumatic fever is now a rare disease in Great Britain and other developed nations but is still commonly found in less developed parts of the world.

Table 3.3. *Usual causes of acute monoarthritis*

Septic arthritis
Gout
Pyrophosphate arthropathy (pseudogout)
Reiter's disease
Psoriatic arthritis
Enteropathic arthritis (ulcerative colitis and Crohn's)
Reactive arthritis
Ankylosing spondylitis
Trauma
Haemophilia
Haemoglobinopathies (e.g. sickle-cell anaemia)

Another possible cause of ALF presenting as a monoarthritis in childhood may be the first presentation of the pauci or oligoarticular form of juvenile chronic arthritis (JCA). This disease, often incorrectly referred to as 'Still's' disease, has been shown to be a heterogeneous condition comprising at least five distinct subgroups linked together by the presence of chronic synovitis and the absence of other identifiable diseases known to be associated with arthritis in childhood (see Chapter 4).

Pauciarticular JCA represents about 50–60 per cent of children with chronic arthritis. The distribution of the affected joints is asymmetrical and large joints tend to be affected with knees most commonly involved. Within the pauciarticular group are two subgroups which include that group, usually girls, who are seronegative for rheumatoid factor but seropositive for antinuclear antibodies and who have a high chance of developing chronic iridocyclitis. Early recognition is extremely important to preserve vision.

The other group tends to be boys presenting later in life often with associated back pain. These children are likely to develop sacroiliitis and progress to ankylosing spondylitis. Since by definition the disease is chronic, the progression of these children with acute monoarthritis to chronic disability supports the diagnosis.

Adults

Acute locomotor failure in an adult as a result of monoarthritis should always evoke the question 'is this a traumatic or a septic arthritis'. The latter diagnosis should not be missed and will suggest itself by the acute and rapid onset, marked deterioration in joint function with much pain on attempted passive movement of the joint in all directions. The patient is unwell with an associated systemic response to the infection. Examination may reveal a nearby portal of entry such as a skin abrasion, varicose ulcer, or a primary focus such as intercurrent pneumonia or even a suppurative otitis media. In other words, if you suspect the joint is septic, look for a possible primary site. The most important diagnostic procedures are blood and synovial fluid cultures. A septic arthritis supervening upon an already damaged joint, e.g. in a patient with rheumatoid arthritis particularly if on treatment with steroids may be associated with certain peculiar characteristics: there may be little to suggest infection on clinical examination and the usual constitutional signs of poly-morphonuclear leucocytes and pyrexia may be absent. Sometimes the only clue is the presence of somewhat disproportionate symptoms and signs requiring a high index of clinical suspicion.

An acute crystal-induced arthropathy (urate or calcium pyrophosphate) may mimic a pyogenic arthritis in many ways. The presentation is rapid; the joint is hot, red, and swollen; and movement is extremely painful. Features in favour of gout or acute pyrophosphate arthropathy are the relative wellbeing of the patient, often a history of recurrent similar short-lived attacks and the presentation of the particular joint. The overlying skin will be hot and dry as opposed to moist. Gout occurs more commonly in males and predominantly involves the first MTP joint of the foot and mid-tarsal joints although other joints are not immune.

Pyrophosphate arthropathy usually affects larger joints, especially the knees. Finger and toe joints are involved less often. Attacks may be brought on by trauma but in most instances there are no precipitating factors. A useful clue to acute pyrophosphate arthropathy is episodic attacks of monoarthritis on the background of a more chronic arthropathy such as osteoarthritis. In addition the patients are usually elderly and the disease is slightly more prevalent in women. Gout is uncommon in women and rare before the menopause.

ALF presenting as monoarthritis may occur as part of the seronegative spondarthropathies. These conditions which show a positive association with the HLA B27 antigen include ankylosing spondylitis, granulomatous bowel disease (Crohn's and ulcerative colitis), Reiter's disease, and psoriatic arthritis. The arthritis is asymmetrical and oligoarticular and tends to favour large joints particularly when associated with ankylosing spondylitis and Reiter's disease. Pauciarticular arthritis with 'sausage' swelling of one or more digits (denoting combined involvement of DIP and PIP joints) is also common, especially in psoriatic arthritis. These conditions usually progress to a more chronic arthritis with involvement of the sacroiliiac joints and the spine. Any patient presenting with an acute monoarthritis, where sepsis and trauma can be excluded, should be questioned as to the presence bowel disease, mouth ulcers, previous episodes of uveitis, urethritis or promiscuous sexual contact, psoriasis (don't forget the scalp, flexures, and nails), back pain suggestive of sacroiliitis (associated with stiffness, worse after rest and settles after exercise), and a family history of any of these conditions. Since the first symptom in about 10 per cent of patients with spondylitis is pain in the back radiating into the back of the thigh an acute lumbar disc prolapse may be suggested. Features in favour of spondylitis are the prominence of the pain or lumber stiffness after rest and especially in the mornings, the ability of the 'sciatica' to be worked off by activity and the lack of nerve-root involvement such as paraesthesiae.

Apart from the peripheral joints, examination of the spine may reveal an 'ironed-out appearance' with loss of the normal lumbar lordosis and lack of lumbar motion in all directions (in contrast to disc prolapse in which lateral flexion is usually spared). A raised ESR supports the inflammatory nature of the condition, and although radiology of the spine may be normal early in the course of the disease one may see blurring of the bone margins of the sacroiliac joints and straightening of the lumbar spine with loss of the normal lordotic curve.

Another conditions also having an association with the HLA B27 antigen and presenting as an acute monoarthritis is the arthritis occurring after gut infections with *Salmonella, Shigella, Yersinia enterocolitica,* and some forms of *Campylobacter.* This 'reactive' arthritis is an aseptic arthritis which presents 10–14 days after a recent enteritis. The knees and ankles are most commonly affected but usually asymmetrically. The severity and duration of the acute episode are extremely variable but it is possible that those possessing the HLA B27 haplotype have a more stormy course.

In all these conditions there may be associated inflammatory lesions of tendon sheaths and entheses resulting in pain which is localized to the periarticular structures such as Achilles' tendonitis, painful heels secondary to plantar fasciitis, periostitis or calcaneal spurs, and pain around large muscle insertions of the pelvic girdle.

In the elderly patient presenting with an acute monoarthritis any of the conditions mentioned may occur but after trauma, perhaps the most likely cause will be an acute crystal arthropathy caused by calcium pyrophosphate.

ACUTE LOCOMOTOR FAILURE PRESENTING AS POLYARTHRITIS

Acute inflammation of several joints is most often diagnosed as rheumatoid arthritis but 60–70 per cent of cases begin with a slow insidious onset of diffuse musculoskeletal pain and stiffness. It is only later that joint swelling becomes obvious to the patient who then brings this feature to the attention of the doctor. At this stage it is important to consider other causes of polyarthritis (Table 3.4) and again the age of the patient may suggest possible diagnoses.

Table 3.4. *Causes of acute polyarthritis*

Rheumatoid arthritis
Viral infections
  rubella
  infectious mononucleosis
  hepatitis
  mumps
Rheumatic fever
Juvenile chronic arthritis (polyarticular forms)
Psoriatic arthritis
Polyarticular gout
Systemic lupus erythematosus
Gonococcal arthritis
Haemostatic failure
Leukaemia

Acute polyarthritis in childhood

Musculoskeletal symptoms are common in many viral infections. In most cases complaints are transient with arthralgias rather than arthritis but in certain infections arthritis may be prominent.

Acute polyarthritis may follow or coincide with rubella and in some cases precede the usual syndrome of rash, adenopathy, fever, and malaise. Articular symptoms are characterized by transient and painful synovitis usually of the small joints of hands and feet but large joints such as knees or elbows may be involved. The arthropathy disappears within a few weeks and no progression to chronic disease occurs. As rubella often occurs without a rash it is important to consider it as a cause of a polyarthritis of unknown aetiology. A rising rubella titre confirms the clinical impression.

In infectious mononucleosis, frank polyarthritis is uncommon, but poly-arthralgias are more usual and the typical pharyngitis, lymphadenopathy and positive Paul Bunnel test confirm the diagnosis.

A symmetrical polyarthritis may precede viral hepatitis. The arthritis tends to affect the small joints of the hands and a particular feature of the condition is the degree of pain which is more severe than the amount of swelling would suggest. The arthritis usually disappears when the jaundice appears. In contrast to the two types of pauciarticular juvenile chronic arthritis already discussed, three forms of polyarticular disease may present as acute locomotor failure in children.

Systemic onset disease (Still's disease) is characterized by systemic features such as high fevers, rashes, lymphadenopathy, hepatosplenomegaly, pleurisy, and pericarditis. Any synovial joint may be involved but typically the small joints of the hands and feet are symmetrically affected. The patients are seronegative for rheumatoid factor (RF) and antinuclear factor and since by definition the disease is chronic, return of normal joint function is unlikely. The systemic features rarely last longer than 6–9 months.

The second subgroup is also seronegative for RF but differs from the systemic or Still's group in that systemic features are absent. An insidious onset is characteristic with mild pain and stiffness. It is not uncommon for the child to be brought to the doctor by the parents with the complaint that one particular limb or joint is being protected or not used and careful examination reveals more widespread polyarthritis.

The third subgroup of polyarticular juvenile arthritis resembles adult onset RA with positive tests for RF and a symmetrical, peripheral proliferative polyarthritis. The distinguishing feature of this subgroup is the rapidly progressive and destructive pattern of the disease. These children may develop the extra-articular complications which are common in adult RA. If the eye is affected the inflammatory lesions occur in the sclera and episclera rather than the uveal tract as in pauciarticular juvenile chronic arthritis.

ALF presenting as polyarthritis in adults

Rheumatoid arthritis is the prototype of this type of presentation. The insidious onset of the condition has already been emphasized. In the early stages the patient, more often female, will complain of stiffness which is particularly prominent on rising rather than pain and the hands and forefeet seem to be the most commonly affected parts. There may be an admission of feeling generally run down and out of sorts with fatigue, anorexia, and weight loss. In retrospect these constitutional symptoms may have preceded the articular manifestations by several months. Pain is usually first felt as metatarsalgia and at this stage apart from inducing pain by compression of the joints, nothing abnormal may be found.

Objective findings at first are subtle with filling in of the valleys between the knuckles when making a fist (denoting early synovitis around the metacarpophalangeal joints), a sensation of grating as the flexor tendons of the fingers in the palms are compressed by the examining finger and the patient's finger is passively flexed and extended (tenosynovitis may be one of the ealiest signs of RA). Slight thickening around the wrist suggesting synvovitis of the wrist or carpal joints may also be felt. Later, easily identified swollen joints may be seen, most commonly the MCP and PIP joints involved symmetrically. Inability to make a firm fist because of joint involvement may be prominent and full examination may reveal knee effusions and painful restricted shoulders or elbows.

Monoarticular rheumatoid arthritis cannot be diagnosed from examination alone regardless of which joint is involved.

A firm and definite diagnosis should only be made when, after the classical history and examination findings are recorded, rheumatoid factor and radiological erosions are identified. It may take time for these latter features to develop. The 'diagnosis' or diagnostic category appropriate in the intervening period is 'acute/chronic inflammatory, non-proliferative, non-destructive seronegative arthritis'.

There is no problem with the diagnosis in the later stages in the presence of the gross changes of RA with markedly restricted joint movement, ulnar deviation of the fingers, subluxed joints, nodules, and a history of pain and suffering spanning many months or years.

Psoriatic polyarticular arthritis may be confused with classical RA. Although it is usually asymmetrical, patients with psoriatic arthritis can present with a symmetrical peripheral and proliferative polyarthritis affecting the MCP and MTP joints, wrists, ankles, and knees. The presence of characteristic psoriatic skin lesions, nail pitting and onycholysis, and seronegativity suggest the diagnosis. The arthritis tends to be less extensive and more benign in course. Classically psoriatic arthritis affects the distal interphalangeal joints and reference has already been made to the oligoarticular forms and the possibility of an associated spondylitis.

Septic arthritis rarely presents as a polyarthritis except for gonococcal arthritis which may affect multiple joints and particularly large joints such as knees or ankles. There is usually a history of recent sexual exposure and in males urethritis may be present. Blistering, pustular skin lesions from which the gonococcus may be isolated are characteristic and should be looked for.

Another condition which may present as acute polyarthritis is sytemic lupus erythematosus (SLE). Although the clinical manifestations are protean, about 90 per cent of patients have arthritis or arthralgia. The articular symptoms most frequently involve the PIP joints with knees, wrists, and MCP joints also affected. The arthritis is symmetrical and apart from morning stiffness which is less prominent there are no clinical features to differentiate SLE from early RA at this stage. Articular symptoms may herald other signs of SLE and if the typical butterfly rash or more localized discoid eruption, non-scarring diffuse alopecia often occurring rapidly, photosensitivity or Raynaud's phenomenon are present, then the diagnosis is more likely. Confirmatory serological evidence in the form of positive tests for antinuclear factor and anti-DNA antibodies is necessary. As a general rule any young women presenting with polyarthralgias or polyarthritis, especially in the post-partum period which persist and increase in severity, should be considered seriously from the point of view either of RA or of SLE. As serological tests often take time to become positive regular follow up is necessary.

CHRONIC LOCOMOTOR FAILURE (CLF)

Most of the causes of arthritis presenting as acute locomotor failure are capable of producing chronic disability. The pattern of progression of the disease may be diagnostic in many cases as return of normal joint function is unlikely in such conditions as RA or psoriatic arthritis, variable in Reiter's disease or the peripheral features of ankylosing spondylitis but usual in arthritis associated with viral infections, rheumatic fever, gout, or a reactive arthritis.

The arthropathies most commonly presenting as chronic locomotor failure will be osteoarthritis (OA) or the insidious form of RA. The duration of the problem may be difficult to determine since the onset may have gone unnoticed by the patient and minor initial symptoms not brought to the doctor's attention. Again it is useful to consider CLF as a mono- or polyarticular problem once it has been established that the problem has persisted.

CHRONIC MONOARTHRITIS

This may or may not follow acute monoarthritis. Presentation will be with a story of many months or years of a painful, stiff, and sometimes swollen joint (Table 3.5). If the weight-bearing joints of knee or hip are involved osteoarthritis is most likely. It is important to enquire about previous trauma or surgery (e.g. a previous menisectomy) to the joint and there will be a long history of increasing disability with, in the initial stages, no objective physical signs. Later limitation of joint movement becomes obvious. Unlike an inflammatory arthropathy, tenderness on palpation is not prominent but pain occurs when the joint is moved. In the osteoarthritic knee a small effusion may be detected.

Table 3.5. *Causes of chronic monoarthritis*

Osteoarthritis
Pyrophosphate arthropathy
Any of the seronegative spondarthritides
Juvenile chronic (pauciarticular) arthritis
Gout
Neuropathic arthritis
Chronic infection (e.g. tuberculosis)

Certain patterns of joint involvement are characteristically found in OA. The joints commonly involved (the 'give away joints') are the carpometacarpal joints of the thumbs (rarely affected in RA) giving a 'squared' appearance at the radial side of the hand, the first MTP joint of the foot, the knee, and hip. These joints may be involved singularly or as part of polyarticular osteoarthritis. Most cases of monoarticular osteoarthritis occur in elderly patients and a particular problem in this group is OA of the hip. Pain is often associated with a limp early in the course of the disease and as hip pain may be referred to the buttock or knee, errors in diagnosis are common. In some patients all of the pain of hip disease is felt in the knee and when this joint is examined very little is found. If the hip is not examined the cause for the pain is missed. Clues to the diagnosis are limitation of hip movement, a short leg on the affected side and the hip may be held in external rotation, flexion, and adduction. If osteoarthritis is suspected radiology will confirm the diagnosis.

An uncommon but important cause for a chronic monoarthritis is chronic low-grade infection of the joint. Tuberculous arthritis may present in this way and since the onset is insidious with little constitutional upset, the joint may easily be considered as 'osteoarthritis'. If there is a previous history of TB and all other joints appear normal with no other evidence of osteoarthritis, the joint should be considered suspiciously. Primary osteoarthritis almost never affects wrists, elbows, shoulders, and ankles, although degenerative changes secondary to trauma may occur.

In teenagers or young adults a chronic painful knee may be due to chrondromalacia patellae. The cause of this condition is unknown but may reflect unnatural wear of the patella against femoral condyles due to malalignment of the patella. The characteristic physical sign is sharp pain when the patella is compressed firmly against the femoral condyles and the patient asked to contract the quadriceps. A small effusion may be found.

Of the inflammatory disorders any of the conditions comprising the seronegative spondarthritides group, particularly Reiter's disease, ankylosing spondylitis, or psoriatic arthritis may present with an insidious monoarthritis. Large joints are usually involved with the knee most often affected. If you suspect any of these conditions it is important to ask for and look for the other associated features already mentioned.

In children suspect the pauciarticular forms of juvenile chronic arthritis in those presenting with persistent pain and swelling of large joints.

CHRONIC POLYARTICULAR DISEASE

Table 3.6 lists the common causes of chronic polyarticular disease. This form of locomotor failure presents a different clinical problem from the acute form as often diagnosis is easier because the chronicity of the disease allows longer follow up. Thus a symmetrical peripheral and proliferative polyarthropathy which may or may not have started acutely but has progressed over the years to cause marked ulnar deviation of the fingers with subluxation of the metacarpophalangeal joints, flexion deformities of the fingers, elbows and knees and nodules over pressure points is almost certainly rheumatoid arthritis. At an earlier stage where the marked joint changes are absent differentiation from psoriatic arthritis may prove difficult but if the other features already referred to are dilligently searched for, the diagnosis may be made.

Table 3.6. *Causes of chronic polyarthritis*

Rheumatoid arthritis
Generalized osteoarthritis
Gout
Pyrophosphate arthropathy (pseudo gout)
Psoriatic arthritis
Reiter's disease
Ankylosing spondylitis
Systemic lupus erythematosus
Juvenile chronic (polyarticular) arthritis

Polyarticular generalized osteoarthritis with prominent soft tissue swelling and effusion in knees may also mimic rheumatoid arthritis. However, the bony enlargement of the distal and proximal interphalangeal joints, the involvement of the carpo-metacarpal joint in the thumb, the normal ESR, lack of rheumatoid factor and characteristic X-ray findings are diagnostic.

Occasionally it is difficult to differentiate chronic polyarticular gout from RA. The ESR may be raised adding to the problem. Certain features are more in keeping with chronic gouty arthritis than rheumatoid arthritis. It rarely occurs in premenopausal women, usually starts with recurrent attacks of acute monoarthritis particularly in the foot, there is often a positive family history or story of renal colic, and the description of nodules discharging white chalky material suggest gout. Careful examination shows that joint involvement is asymmetrical particularly in the hand and swellings are due to firm gouty tophi and not synovitis. A raised serum urate, negative rheumatoid factor and characteristic X-ray features with punched-out areas in the bones together with subcutaneous tophi provide additional support. Aspiration of a tophus or affected joint and the demonstration of the negatively birefringent, needle-shaped crystals of urate is diagnostic.

Apart from psoriatic arthritis chronic polyarticular disease occurring as part of the seronegative spondarthritides is uncommon. If it occurs the joints of the lower limbs are involved asymmetrically and the associated sacroiliitis and spondylitis together with the non-deforming nature of the disease enable differentiation from RA.

It must be accepted that not all causes of polyarthritis can be firmly established in the classical Oslerian sense and only on prolonged follow up may the characteristic disease pattern develop. Here again the problem of attitude becomes of extreme importance. It is dishonest to attempt to force the disease to fit the descriptions in outdated textbooks. If this is attempted it is likely that the diagnosis is wrong. This mental attitude must represent one of greatest setbacks for recognition of the truth. In the history of the rheumatic diseases the advances in the last decade have been made by the 'splitters'. Only by recognizing that ankylosing spondylitis and psoriatic arthritis were not simply inconvenient variants of rheumatoid arthritis has it been possible to acquire so much information about their clinical presentation, prognosis, and pathogenesis and this would have been entirely obscured had the crime of 'lumping' them all together carried the day.

Thus the challenge of the interpretation and differential diagnosis of locomotor failure due to articular disease is a fascinating one requiring all the skills of the clinician. The key may lie in the family history, in the past history, the patient's presenting complaints, the clinical examination, or simply the realization by the physician that something isn't quite right. This is a most exciting and challenging subject and the fact that a substantial proportion of patients attending a rheumatology clinic or GP's surgery cannot today, in all honesty, be categorized, may be looked on either as our present failure or as the major opportunity for further descriptive advances in the future.

4 Aches and pains in children and adults

Malcolm I. V. Jayson

INTRODUCTION

Of all the locomotor problems met in general practice aches and pains must surely be the most widespread and also the most taxing. So often the practitioner is confronted with the patient complaining of persistent and severe spinal and limb pains yet investigation reveals little amiss and therapy seems to be of no avail. The frustration experienced by the general practitioner is communicated to the patient and perhaps exacerbates the problem. Nevertheless, a number of organic conditions can present as aches and pains and sometimes may masquerade as non-organic illness. Awareness of the possible causes of aches and pains will minimize missing important diagnoses.

An understanding of the reasons that may underlie the complaint of aches and pains is essential. In this chapter we attempt to provide an approach to this problem emphasizing the salient clinical features and the principal conditions that may be recongized.

THE HISTORY AND CLINICAL FEATURES

Particular care must be taken with the patient who develops aches and pains for the first time. These symptoms, particularly if combined with ill health, weight loss, or loss of appetite, may represent the onset of neoplasia, infection such as tuberculosis, or some connective tissue disorder. The physician must make a systemic enquiry about general health and more specifically seek respiratory, cardiac, gastrointestinal, and neurological problems.

Aches and pains are rather loose words and patients should always be asked to describe the precise problem. What one person calls a pain another may regard as stiffness or cramp. The complaint of stiffness may indicate some neurological disorder such as Parkinson's disease.

It is characteristic of inflammatory rheumatic disorders and particularly rheumatoid arthritis and polymyalgia rheumatica that the symptoms are exacerbated by rest and relieved by exercise. Most patients will say that the symptoms, and particularly the stiffness, are at their worst when they first wake in the morning. They gradually ease up after anything from a few minutes to a few hours as they get going. The duration of this morning stiffness is a measure of the activity of the disease.

The mode of onset of the problem may be helpful. Aches and pains present for many years and perhaps exacerbated at times of emotional stress are less likely to be due to major underlying disease than the recent development of symptoms gradually becoming worse. The functional effects of the problem should also be sought. Not only will these indicate the magnitude of the problem but also they may suggest malingering or psychiatric causes.

The distribution of symptoms is important. Localized pain in one limb indicates some local pathology which may be amenable to local therapy. Obvious examples include tennis elbow, golfer's elbow, plantar fasciitis, De Quervain's tenosynovitis, etc.

The physical examination should include a general examination which may indicate the underlying diagnosis. By omitting this sooner or later important physical signs such as enlarged lymph glands, an enlarged thyroid, chest signs, or an abdominal mass will be missed. The more specific examination of the locomotor system should include the nervous system as well as examination of the muscles, tendons, and joints. Focal areas of tenderness should be sought as they may indicate areas amenable to treatment by local steroid injection.

Investigations should be performed particularly in patients developing generalized aches and pains for the first time. A limited number of screening investigations will usually exclude the principal problems. The tests should include the blood sedimentation rate or plasma viscosity the haemoglobin and white cell count, and a biochemical profile. A chest X-ray should be performed. Further investigations are only indicated as suggested by the clinical problem or if the preliminary screening tests are abnormal.

ACHES AND PAINS IN CHILDREN

It is interesting to appreciate that very young children do not recognize the symptom of pain. They cry because of hunger, and a desire for attention but the sensation of pain in response to some unpleasant stimulus is something that is learned during the formative years. The young child with arthritis may be disabled yet not complain of a significant pain problem. As he matures the sensation of pain becomes associated with unpleasant events. It may be learned as a form of operant conditioning by which the pain response such as tears and the appearance of misery evokes sympathy from parents and others and therefore gains attention. For this reason aches and pains do not appear as a significant problem in young children. However, from school-age onwards they are not uncommon and the general practitioner is frequently asked for help by frustrated parents.

It is always important to exclude organic disease. Many feverish illnesses produce aches and pains associated with the feeling of malaise and sometimes nausea and vomiting. These include influenza, and the pre-rash phase of many viral illnesses. The acute onset of symptoms suggests an infective aetiology. Most children rapidly recover with rest and simple analgesics.

Rheumatic fever fortunately is now rare in Britain. Malaise, a rash and flitting arthritis together with an elevated blood sedimentation rate and raised anti-streptolysin O titire indicate the diagnosis. Rather more frequent is Henoch–Schönlein purpura. This also may follow a streptococcal infection with associated malaise, aches and pains, and myalgia. The characteristic features are a purpuric rash in the skin, an abdominal upset sometimes with melaena, and arthritis. Leukaemia can present with limb pain. Suspicion should be alerted because of general ill health and anaemia perhaps with lymphadenopathy and splenomegaly and the diagnosis is usually made by a blood count and sternal marrow examination. X-ray of the limbs may show subperiosteal new bone formation. Scurvy may appear in a not dissimilar fashion but the dietary history and the social background together with broadened epiphyses on radiological examination will suggest the true diagnosis.

In some children aches and pains may herald the onset of chronic juvenile arthritis. This is not synonymous with juvenile rheumatoid arthritis as the latter is only one of several conditions making up the problem of chronic arthritis in children. Persistent inflammation usually in a number of joints associated with a raised blood sedimentation rate with exclusion of infective and other causes indicates the nature of the problem. Particular care must be taken in the younger girls with only one or two joints involved as they are at an increased risk of developing chronic iridocyclitis.

Finally many children are exposed to trauma and the history may not be available. Telltail signs of bruising should be sought but it is easy to be misled particularly in cases of non-accidental injury.

ACHES AND PAINS IN ADULTS

It is difficult to present any useful classification of the various causes of aches and pains in adults. They may be a feature of such a wide variety of conditions yet so commonly it is difficult to define any specific underlying cause. In this section we have listed some of the common problems frequently dismissed as aches and pains of no clinical significance.

Localized aches and pains

Some patients will identify a particular area as the source of the symptoms. Close and careful examination is required as this may well be the site of some underlying pathology of significance such as a primary or secondary neoplasm or an abscess. Careful inspection and palpation are required and if there is any doubt a radiograph of the part together with appropriate blood tests should be ordered.

Do remember that pain felt in one body part may be referred from elsewhere. The most obvious example is nerve-root pain referred from the spine. When there is pain in the shoulder or upper limb, the neck also must be examined and likewise the lumbar spine for the buttock and the lower limb. However, more peripheral examples of pain may occur. For example, pain arising in the hip is commonly felt in the thigh and knee alone. A cursory examination of the knee will reproduce the pain – because the hip is also bent during knee flexion when lying on a couch. Easily confused with this sort of pain is meralgia paraesthetica. In this condition the lateral cutaneous nerve of the thigh is compressed as it passes beneath the lateral border of the inguinal ligament. Pain, numbness, and parasthesiae are felt in the anterolateral aspect of the thigh. Pain can arise from the greater trochanter and again is easily confused with hip pain. The hips will move normally and freely but the pain is reproduced by pressure over the greater trochanter. Once recognized this condition is easily treated with a local steroid injection.

Entrapment neuropathies are another source of confusion. The carpal tunnel syndrome is due to compression of the median nerve beneath the flexor retinaculum on the palmar aspect of the wrist. The patient develops numbness and paraesthesiae which he may identify in the median nerve distribution in the hand. However, commonly patients of severe pain and tingling occurring in the whole hand which may radiate proximally up the forearem and upper arm even into the side of the neck. The give away to the diagnosis is that these symptoms occur predominantly during the night when they may wake the patient and are relieved by flexing and exercising the wrist and hand. Almost invariably the patients demonstrate this while they describe the problem. A careful neurological examination may reveal sensory loss and muscle weakness in the appropriate distribution.

An analogous problem may arise in the foot and is known as the tarsal tunnel syndrome. The posterior tibial nerve passing over the medial aspect of the calcaneum similarly may be compressed with pain and paraesthesiae in the sole of the foot which may spread proximally.

Many aches and pains are felt in the back and around the spine. There is an extremely complex cross-innervation in the back so that the site of the symptoms is a very poor guide to the source of the problem. So often it is impossible to make a precise diagnosis although it is clear that it is of mechanical cause. The best label here is 'non-specific back pain' (Chapter 9). A poor posture, particularly if standing in one position for long periods, too soft a bed, a chair with inadequate lumbar support contribute to this problem.

In some patients it is possible to palpate exquisitely sensitive nodules in the back which seem to be the source of the symptoms. These are the so called 'fibrositic' nodules which have been described for many years but whose nature is still not understood. The use of the term 'fibrositis' is wrong as there is no evidence of inflammation of fibrous tissue. They seem to be more frequent in patients with degenerative disease of the spine. Commonly they will respond to local physiotherapy treatment or a carefully placed local steroid injection.

Ankylosing spondylitis often presents with aches and pains in the back with, in the early stages, little to find abnormal on examination and often a normal

ESR and X-ray of the sacro-iliac joints. A history of the insidious development of aching and stiffness in the back, aggravated by rest, relieved by exercise and at its worst first thing in the morning, developing in a young person is highly suggestive of this disorder.

A localized area of tenderness can occur at the site of insertion of a ligament to bone – the so called 'enthesis'. Although enthesopathies are characteristic of ankylosing spondylitis and related disorders, commonly they occur on their own. Perhaps the most frequent site of enthesopathy is at the lateral epicondyle of the elbow. This is the so-called tennis elbow with severe pain and tenderness felt in the elbow and radiating down into the forearm and particularly occurring during use of the muscles of the forearm and hand. These symptoms are reproduced by pressure over the lateral epicondyle and by making a fist. An analogous problem occurs at the medial epicondyle and is known as a golfer's elbow. Pain may occur at the insertion of the plantar ligaments at the undersurface of the calcaneum and is known as plantar fasciitis and the Achilles' tendon at the back of the heel and is known as Achilles' tendonitis. These conditions may respond to carefully placed steroid injections.

Pain may arise within one or more joints. The source of symptoms is usually obvious but confusion can arise in monarthritis. This may be due to mechanical disorders of the joints such as a torn meniscus, osteochondritis dissecans, degenerative joint disease or inflammatory arthritis. One common problem is pain in the hand for which careful examination may reveal osteoarthritis of the carpo–metacarpal joint of the thumb.

Around certain joints there are bursae that may become swollen and painful. This is usually due to repeated minor trauma but they may be involved by rheumatoid arthritis or other inflammatory joint diseases. As examples there are the prepatallar and infrapatellar bursae in front of and below the knee and the olecranon bursa over the extensor aspect of the elbow.

In the lower limbs pain arising from varicose veins may be confused with locomotor problems. In overweight people and particularly women painful fatty lumps are not uncommon. This condition is called paniculitis but the evidence that there is actual inflammation of fat remains very doubtful.

GENERALIZED ACHES AND PAINS IN ADULTS

In some patients these symptoms appear as part of general ill-health and indicate some underlying disease. Patients with neoplasms, reticuloses, or blood diseases may all present in this way and the diagnosis may be suspected by the alert physician when a previously healthy person develops these symptoms for no obvious cause.

Aches and pains may be the prodromal phase of connective tissue disease. In particular polymyalgia rheumatica commonly presents with these symptoms. This condition usually occurs in the elderly and is more common in females than males. It starts with limb-girdle aches and pains with stiffness and some weakness and it is characteristic that these symptoms are worst first thing in a morning and gradually ease up after a variable period of time. Sometimes a low-grade synovitis perhaps in the hands or knees occurs. Some patients only develop tiredness and ill-health perhaps with weight loss. So frequently the patient with polymyalgia rheumatica merely thinks that she has become old and is so regarded by her family and physician. The diagnosis is easily missed unless it is considered. It can be confirmed by the finding of an elevated blood sedimentation rate. It is important to recognize this condition not only because it responds so well to treatment with steroids but because if left untreated it may be complicated by temporal arteritis with retinal vasculitis leading to blindness and by cranial arteritis leading to neurological complications.

The other inflammatory connective tissue diseases such as rheumatoid arthritis, dermatomyositis, systemic lupus erythematosus, and rarer syndromes may present with aches and pains. Tell-tale features include the recent onset of symptoms and the presence of specific physical features such as arthritis or rash. The diagnosis is confirmed by the appropriate blood tests. Some patients develop recurrent synovial effusions and rheumatic symptoms without any diagnostic feature. This is called 'palindromic rheumatism'. It may be the precursor of rheumatoid arthritis.

There is an interesting group of patients with excessively flexible joints who suffer from multiple joint pains. They are often labelled as non-organic as there is nothing to find abnormal on examination. Indeed the physician may be struck by the extreme hypermobility of the joints. These patients may be double jointed and can flex forwards when asked to touch their toes and put the palms of their hands flat on the floor. They can extend the thumb backwards to touch the forearm with similar features in other joints. Some patients show features of Marfan's syndrome with a high-arched palate, arachnoidactyly, and the span being greater than the height. It seems likely that the joint pains arise from the capsule being overstretched due to the extreme joint flexibility. These patients easily damage their joints and are prone to develop osteoarthritis in later life.

Aches and pains frequently accompany infection. The most obvious example is influenza but equally they may occur with other viral infections. Rubella and post-rubella vaccination arthralgias can be very painful and in their more severe forms may be confused with rheumatoid arthritis. Infective hepatitis can present without jaundice but with arthralgia and aches and pains. Low-grade chronic infections such as tuberculosis also can present in this way.

In metabolic bone disease there may be aches and pains. This includes both hypo- and hypercalcaemia. Paget's disease particularly when widespread may be associated with generalized bone pain. In some patients acute exacerbations may occur with a fracture. This often occurs in osteoporotic subjects when acute back pain develops with the progressive development of kyphosis and with multiple crush fractures and loss of height. Myelomatosis similarly may cause bone pain with acute exacerbations should a fracture occur.

Hypothyroidism and hyperthyroidism may both present with aches and pains. The tell-tale signs of thyroid deficiency or excess are usually present yet are easily missed. The former is characterized by lethargy, malaise, and a typical facial appearance. The delayed relaxation of the reflexes is a helpful diagnostic sign. In thyrotoxicosis weight loss, finger tremor, and other features may be helpful. Any suspicion should lead to the appropriate thyroid function tests. Again both conditions respond remarkably well to treatment.

Certain patients taking the contraceptive pill or barbiturates develop aches and pains particularly in the limbs. The causes of these symptoms are uncertain. The only way to be sure that a drug is responsible is to try withdrawing it and see if the symptoms are relieved.

PSYCHOGENIC ACHES AND PAINS

The counsel of perfection demands this diagnosis should be made not only when organic diseases have been excluded but also when appropriate psychological causes have been identified. Frequently we see patients disabled with limb or trunk symptoms in whom we cannot identify anything amiss or whose symptoms seem out of all proportion to the severity of the problems identified. In many patients there is clearly an element of depression or anxiety and it seems likely that this is responsible for the symptoms. Commonly, however, the patient states that the depression or anxiety is the result of a persistent disability which the physician has been unable to cure.

There are several mechanisms by which psychological problems can give rise to aches and pains of this sort. The first and perhaps the simplest is that of the malingerer, the person who pretends he has pain in order to avoid something unpleasant or to obtain some benefit. Aches and pains of this sort are used to avoid an unpleasant situation at work or to enable him to watch the local football derby. A related problem is that of compensation neurosis. This seems to be a subconscious effort by the patient to exaggerate problems in order to obtain some kind of social or financial benefit. Whether this can be classed as malingering or a natural human reaction to adversity is difficult to decide.

Some patients learn pain behaviour. The outward expressions of pain, that is complaining, groaning, wincing, etc. evoke sympathy from relatives and attendants and provides some kind of 'reward' to the subject. This is known as operant conditioning and seems to re-inforce pain behaviour and exacerbates and perpetuates the problem. In other patients, pain is associated with a chronic anxiety state. Patients with an anxiety state develop abnormal contractions of the muscles of the trunk. This particularly occurs in the neck and produces severe spasm of the cervical muscles. Biofeedback may be particularly helpful in this kind of problem. Aches and pains may also be an outward manifestation of depression. Instead of the characteristic features of this disorder it may appear disguised as bodily symptoms of this sort.

It can be difficult to recognize these psychological problems and it is far too easy to dismiss them as being of no importance. Except for malingering, the pain in these disorders is every bit as real as that arising from organic disease. Indeed the symptoms experienced by the patient may be far worse. Sympathetic and constructive help is required in order to provide appropriate relief.

In some patients the trunk and limb pains may be associated with a disordered sleep rhythm. In one study, a group of patients with fibrositis with pain, stiffness, and physical exhaustion, an abnormality of sleep rhythm was detected with abnormal sleep EEGs. Barbiturates did not restore a normal sleep pattern. Many of these patients obtain relief of symptoms together with restoration of natural sleep rhythms with tricyclic antidepressants.

IDIOPATHIC

Despite the wide range of conditions listed there are many patients with recurrent aches and pains in whom it is impossible to make a precise diagnosis. It is necessary to keep an open mind and to reassess the patient from time to time as the underlying problem may only become apparent during long-term follow up.

SUMMARY

A vast number of different conditions can underlie the complaint of aches and pains. A careful history and physical examination with certain laboratory tests will usually point to the right diagnosis. In this chapter I have described the more common problems but it is not possible to be exhaustive as virtually every medical condition can be associated with this problem.

Psychological problems frequently present with aches and pains. Even though they may have no organic cause, they are still real for the patient and demand relief.

5 Claudication and peripheral circulatory disorders

R. W. Marcuson

INTRODUCTION

Locomotor disability due to circulatory disorder is common and is due to acute or chronic problems in the arterial, venous, or lymphatic components of the circulation. The prime symptoms and signs of circulatory disorder are pain, swelling, discoloration, and ulceration. Whilst there is frequent overlap, patients often present with a dominant symptom and the problem is to assign the true cause. History taking and clinical examination are paramount in this context and may be time consuming.

Table 5.1 shows the circulatory disorders to be discussed in this chapter classified according to the dominant presenting symptom.

SYMPTOMS AND SIGNS

PAIN

There are four main vascular causes of acute onset of lower limb pain which should be considered. Severe pain of sudden onset, associated with paralysis, pallor, and absent peripheral pulses indicates *acute arterial ischaemia*. The term 'acute ischaemia' implies that the limb is dying and unless the circulation is restored within a few hours amputation will inevitably be the end result. Similarly *deep vein thrombosis* may present acutely with calf pain but usually with muscle tenderness and calf swelling although this may be minimal in the early stages. The peripheral pulses are palpable if not obscured by oedema. *Superficial thrombophlebitis* and *acute ascending lymphangitis* are usually easy to diagnose by virtue of the physical signs.

Persistent foot pain at rest, particularly at night, with dependent rubor and absent peripheral pulses, associated with rapid onset calf pain on walking suggests *critical ischaemia*. This term implies that the limb is viable although any fall in perfusion will probably lead to the loss of the limb (Fig. 5.1).

Intermittent claudication is a common and well-described syndrome usually affecting the calf and forcing cessation of walking due to pain. Claudication may affect the thigh or buttock with high arterial occlusions, but all true claudicants share the same characteristic of a well-defined exercise load giving rise to pain, relieved by rest, and recurring after repeat exercise. There are other causes of pain which may mimic claudication (see below).

Table 5.1. *A classification of circulatory disorders of the lower limb by predominant symptom*

Pain
    acute-onset pain
        (i) acute ischaemia
        (ii) acute deep-vein thrombosis
        (iii) acute superficial thrombophlebitis
        (iv) ascending lymphangitis
    rest pain of critical ischaemia
    pain on exercise
        (i) intermittent claudication
        (ii) pseudo claudication
Swelling
    acute-onset swelling
      acute deep-vein thrombosis
    chronic swelling
      chronic venous insufficiency
      lymphoedoema
Discoloration
    ischaemic changes
    pigmentation of chronic venous disease
Ulceration
    venous ulcer
    ischaemic ulcer
    diabetic ulcer

SWELLING

Acute-onset unilateral limb swelling should be considered as evidence of a *deep vein thrombosis* until that condition has been excluded by appropriate investigation. The extent of the swelling varies with the extent of the thrombosis. The degree of pain varies from discomfort to such severity that it may prevent walking or even putting the foot to the ground.

Chronic lower limb swelling, not of systemic aetiology, may be due either to *chronic venous insufficiency* or *lymphoedema*. A past history of deep venous thrombosis, post-partum white leg, or overt varicose disease clearly suggests chronic venous insufficiency especially in the presence of chronic *pigmentation* and *gravitational ulceration*. Sudden deterioration in such limbs always raises the possibility of a further thrombotic episode and should be investigated. The absence of pointers towards venous disease and a brawny nature of the oedema suggest *lymphoedema* which is the result of functionally deficient lymphatics. It is often associated with *recurrent cellulitis*.

DISCOLORATION

The most serious is that associated with chronic arterial insufficiency where the toes may become a dusky reddish-purple hue especially on dependancy. Elevation produces a bloodless foot especially if combined with calf exercise

(Fig. 5.2). The sign of delayed venous filling ultimately leading to rubor from stagnant hypoxia is an important index of the severity of the ischaemia. Paroxysmal changes in colour may occur in Raynaud's phenomenon but are more usually found in the hands. The pigmentation associated with chronic venous insufficiency has already been mentioned.

ULCERATION

The most frequent cause is chronic venous insufficiency. These ulcers are usually easy to diagnose in view of their medial situation, surrounding pigmentation, and associated venous abnormalities. Ulceration due to arterial insufficiency usually results from minor trauma and poor healing. The inner and outer sides of the heads of the first and fifth metatarsal are common sites. The neuropathic ulceration associated with diabetes is usually found on the sole of the foot under the metatarsal heads.

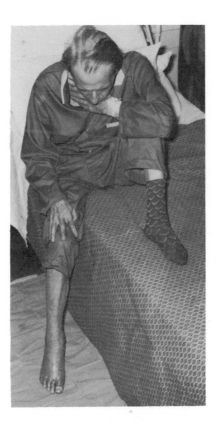

Fig. 5.1. Critical ischaemia – typical posture with leg hanging out of bed to relieve pain. Note oedema, skin dystrophy, and ischaemic ulcer.

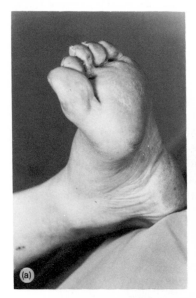

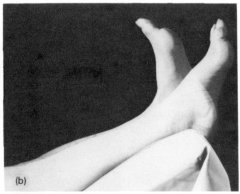

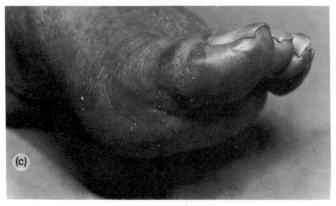

Fig. 5.2. (a) Pallor on elevation. (b) Exercise in elevation. (c) Rubor on dependency.

ARTERIAL PROBLEMS

ACUTE ISCHAEMIA

Clinical picture

There should be little difficulty in diagnosing an acutely ischaemic limb. The onset is usually sudden and the condition is characterized by the classical five Ps: pain, pallor, paralysis, parasthesiae, and pulselessness. They are all present in varying degrees. The pain is distal and predominantly affects the foot and calf. It is usually severe, unrelenting, and progressive. Relief implies opening up of the circulation by onward migration of the embolus or by dilation of the collateral vessels both of which are rare occurrences. Paraesthesiae are the manifestation of partial ischaemic neurological damage which may also produce other dysaesthetic symptoms and signs before anaesthesia develops in a sock or stocking type of distribution. Paralysis of the calf and anterior compartment muscles may be accompanied by tenderness on palpation. This weakness is easily demonstrated although loss of toe movements may develop earlier and should be sought. The limb shows a waxy bloodlessness, i.e. pallor, which is particularly obvious on elevation. Absent pulses define the location of the level of the occlusion as well as supporting the diagnosis. The symptoms and signs are obviously more severe if the block is at the aortic or iliac level rather than in the femoral, popliteal, or tibial vessels.

INDICATORS OF THE CAUSE

The differentiation between embolism and acute thrombosis is of importance but must not delay hospital admission or definitive treatment. Embolism is nowadays the rarer cause and is suggested by a previous history of mitral valve disease, atrial fibrillation, or recent myocardial infarction. Acute thrombosis is most often secondary to established atherosclerotic disease and is suggested by a previous history of claudication, the presence of ischaemic dystrophy of the feet and a pulse deficit or bruits on the contralateral limb. It is always worthwhile excluding an acutely thrombosed popliteal aneurysm by appropriate careful palpation of the popliteal fossa and remembering that the condition is usually bilateral. Thrombosis of the bypass graft, especially in the early post-operative period, may also produce acute ischaemia.

Management

The diagnosis of acute ischaemia implies that the limb will not survive if successful treatment cannot be undertaken and thus emergency admission to hospital is required. If any delay is anticipated, after consultation with the duty surgeon, an intravenous injection of 5–10 000 I.U. heparin may be given with the aim of preventing extension of the embolus or thrombosis by propagation.

The arteries of an acutely ischaemic limb should be explored. Local, regional, or general anaesthetic is employed as appropriate. Thrombo-embolus may be retrieved from the aortic bifurcation down to the popliteal trifurcation and below. The results of embolectomy are good especially if undertaken within six hours but those for thrombecomy are rather less impressive. Sometimes removal of an acutely formed thrombus from the common femoral bifurcation in the groin above an established superficial femoral artery block in the thigh saves the limb by opening up the profunda collateral system but more usually irreversible thrombosis has occurred in the tibial vessels so that the limb is lost. For the same reason femoro-distal bypass in the acute situation is similarly unrewarding. By contrast disobliteration of an acutely thrombosed bypass graft is often a worthwhile exercise.

CHRONIC ISCHAEMIA

Critical ischaemia

This concept has been discussed above and demands early investigatin with a view to bypass surgery for limb salvage. It should be appreciated that the mortality for amputation is greater than that for bypass surgery below or outside the peritoneal cavity. The principles of assessment and investigation are the same as those for claudication.

Claudication

This diagnosis is usually straightforward. The description of a tight cramping calf muscle pain on exercise relieved by rest and recurring after a further similar amount of exercise, associated with absent peripheral pulses, is well known and unlikely to be confused with other conditions. Minor variations in this pattern are not unusual. The pain may radiate to the knee or thigh and may develop after shorter distances on walking up inclines. Occasionally patients are able to 'walk through' their claudication pain. Buttock and thigh claudication indicate aorto-iliac disease but are usually associated with the commoner calf claudication. If the latter pain is absent the diagnosis is less sure. Isolated gluteal or buttock claudication due to internal iliac artery disease is rather unusual.

Difficulty in diagnosis may occur when the symptoms are typical but the pulses are present, where the symptoms are suggestive but atypical and in true claudicants who also have co-incidental diseases affecting gait.

True claudication can occur with palpable distal pulses although the pedal pulses are usually reduced in volume even at rest. After exercise there develops a proximal muscle 'steal' effect leading to greatly reduced perfusion of the distal muscles. Re-examination after exercise to the limit of tolerance in such cases will now reveal absence of the pedal pulses. Very rarely the cause is a metabolic disorder of muscle – McArdle's syndrome.

The more common problem is with atypical pain suggestive of true claudication usually associated with difficulty in feeling the pedal pulses. Venous claudication occurs as a possible sequel to old extensive deep vein thrombosis and gives rise to a bursting rather than a cramping calf pain. Peripheral pulses may be obscured by oedema. Spinal claudication is a condition associated with stenosis of the lumbar canal and nerve-root irritation on exercise. Thus the lower limb pain is related to exercise and may force the patient to stop walking. There may be demonstrable neurological signs after exercise but the peripheral pulses are usually palpable. The symptoms are usually relieved by flexing the spine forwards. Sometimes osteoarthritis of the hip or knee and various neurological problems affecting gait can superficially mimic true claudication.

Co-incidental orthopaedic, neurological, or venous disease in the presence of chronic arterial disease may obviously present a problem in terms of assessing the cause of the symptoms. In all of these situations careful clinical re-evaluation may elucidate the problem but Doppler ultrasound studies both at rest and after an exercise test are invaluable.

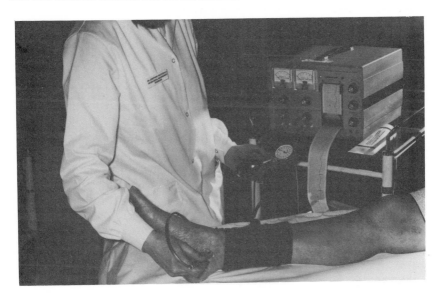

Fig. 5.3. Ankle pressure measurement by the Doppler ultrasound technique.

The probe (Fig. 5.3) is placed over a pedal artery as a blood-flow detector and with a cuff at the ankle the systolic pressure at this site can be measured. A qualitative estimate of flow can be obtained from the wave form. The ankle pressure is expressed as a fraction of the brachial systolic pressure to give an index which is around 1.0 is normals. Claudicants fall in the 0.4–0.7 range and critical ischaemia is found below 0.3. Serial ankle pressure measurements after exercise on a treadmill or bicycle show a fall proportional in extent and duration to severity of the arterial disease. A normal resting trace and no fall after exercise effectively excludes significant arterial problems.

Operative or conservative management?

Conservative measures are indicated for all patients with chronic ischaemia but surgery is relatively rarely undertaken. It is now established beyond reasonable doubt that cigarette smoking is detrimental to patients with symptomatic peripheral vascular disease. Not only is there a lower ultimate incidence of amputation in those who give up smoking but also those subjected to bypass surgery fare considerably worse if they continue to smoke. The implications are obvious.

Many patients are treated hypertensives on presentation and some are discovered to be hypertensive for the first time. Adequate control is important but may cause a deterioration of ischaemic symptoms by reduction of the perfusion pressure. Sympathetic beta-blockade should be advised with caution, especially if high doses of these drugs are required because of the potentially undesirable effect of uninhibited alpha activity on the limb. In such cases combined alpha-beta-blockers or other forms of treatment should be considered. The same caveat applies to the management of angina.

Urine should be tested for sugar and, if discovered, diabetes should of course be treated. Signs of hyperlipidaemia should also be looked for and fasting lipid estimations carried out in younger patients. Appropriate dietetic advice should be given.

Regular exercise is to be encouraged and many reports testify to the benefits of walking programmes. Particular care of the feet is very important because ischaemic digital ulceration frequently follows minor trauma. Regular bathing; careful drying and dusting with powder, and attention to shoes, socks, and stockings are important. Care of the nails is particularly important and whilst patients should be encouraged to attend a chiropodist it is only fair that the chiropodist should be made fully aware of the ischaemic problem. There is an unavoidable element of risk which has to be accepted.

In broad terms bypass surgery is indicated for claudicators who cannot manage their job, and for critical ischaemia or rest pain. It is only these patients for whom arteriography is indicated. Exclusions are of course made at this stage by virtue of general ill-health but it is worth while considering that the mortality of a successful bypass is probably less than that of a major limb amputation.

Claudication

Some guidelines are valuable to assess the severity of the claudication. Estimates by the patient of their claudication distance are notoriously inaccurate. An attempt should be made to judge the distance to the onset of calf pain – when patients frequently stop – and the maximum walking distance. Does inability to work really mean difficulty in getting to work? Is the pain so bad that the patient stops for a rest before crossing a road to be sure he can cross in one 'go'? In this context it should be accepted that housewives tolerate claudication poorly by virtue of their work in the home. Where very short maximum walking distances are claimed direct observation is of value. Doppler studies are also very helpful as indicated above.

In the absence of critical ischaemia a period of conservative management is advised for all patients on initial presentation. A sudden onset or deterioration of claudication distance is often followed by improvement as the collateral circulation opens up. This should be encouraged by complete cessation of smoking and regular exercise. There is no place for vasodilators whose action is more profound on the normal vasculature than the potential collaterals and therefore tend to 'steal' blood away from the ischaemic limb. The indications for newer drugs with activity on vaso-active peptides, red cell deformability, and calcium metabolism such as Praxilene, Trental, and Stugeron remain undefined and controversial.

Over succeeding weeks the collateral circulation develops and the claudication usually becomes 'stable'. In many cases the symptoms are minor and totally acceptable. A few fail to make such progress or even deteriorate and a decision must now be made with regard to angiography. It is emphasized that angiography presupposes that bypass surgery of some sort will be undertaken if a favourable anatomical distribution of the block is demonstrated and the patient should understand this before the angiogram is performed.

ANGIOGRAPHY AND SURGERY FOR CRITICAL ISCHAEMIA AND CLAUDICATION

Most angiograms are performed by a retrograde catheterization technique through the femoral artery under basal sedation or light general anaesthesia. Where the groin pulses are impalpable a translumbar or occasionally a trans-axillary approach is required.

Localized aortic or iliac stenosis can successfully be treated by endarterectomy. Extensive or total occlusions are treated by aorto-bifemoral bypass with a Dacron bifurcation graft – a major procedure usually involving abdominal and bilateral groin incisions. In the absence of femoral artery disease the results are good for many years.

Femoral artery disease is still best treated by reversed saphenous vein femoro-popliteal bypass grafts. Limb salvage procedures often require grafts to more distal vessels – the tibial or occasionally peroneal arteries. Where the saphenous vein is not suitable, various prosthetic materials, e.g. Dacron, Goretex, and Umbilical Vein (Dardik biograft) may be used. The long-term patency rates are about 50 per cent at two years for claudication and rather less for limb salvage procedures.

Late graft occlusion does not inevitably lead to amputation and limb salvage rates are usually rather higher than graft patency rates.

Angiography may reveal multisegment disease. In the presence of aorto-iliac and femoral disease a two-stage procedure may be undertaken if proximal reconstructions fail to relieve the symptoms. Occasionally a combined procedure, aorto-iliac and femoral bypass, is indicated.

A proportion of cases show extensive disease which is not amenable to bypass surgery. This is particularly true when there are multiple stenoses and occlusions of the tibial and peroneal arteries. Here there is a place for lumbar sympathectomy which, in the older patient, can easily be carried out by phenol injection under radiological control. It is not usually carried out for claudicants who may suffer a significant deterioration in their walking distance due to shunting of the blood from muscle to skin, but reserved for those with early ischaemic lesions and rest pain.

There remains of course a sizeable proportion of patients with ischaemic disease in whom amputation is required. The indication is uncontrollable rest pain with or without gangrene. Whilst a decision to accept amputation is frequently difficult not only for the patient but also for the surgeon the tendency should be to come to an early decision wherever possible. Long periods in bed contemplating life as an amputee accompanied by pain and sleeplessness despite analgesics are potent recipes for depression, muscle wasting, and flexion deformity of hip and knee. An optimistic but sensitive approach is required. Team work involving the medical and nursing staff, physiotherapist, occupational therapist, social worker and the early involvement of the limb-fitting surgeon is vital in securing the best outcome.

Modern tendency is toward below rather than through or above the knee amputations. There is a high incidence of bilateral amputation, perhaps between 10 and 20 per cent of unilateral amputees having to face a second-side procedure, and thus bilateral above knee amputations should be avoided if at all possible.

Stump problems are not uncommon. Many are due to oedema of the stump and corresponding difficulty with the socket of the artificial limb. Amputees may present with stump pain – the causes of which include ischaemia of the stump, local inflammatory problems and bony problems which must be sought by direct examination of the stump. Pain due to amputation neuroma or 'phantom limb' can often be helped by newer techniques available in pain clinics.

VENOUS PROBLEMS

SUPERFICIAL THROMBOPHLEBITIS

This condition is diagnosed by the tender blue–purple or red inflamed cord of superficial vein palpable in the thigh or calf. It may arise *de novo*, but is frequently associated with varicose veins. Precipitating factors include pregnancy, operation, trauma, long flights in aeroplanes, and even sclerotherapy undertaken to treat the varicose veins. Rarely an occult carcinoma may present in this way. Management is by analgesics, with or without anti-inflammatory agents, but not antibiotics, and supportive bandaging. Early mobilization is to be encouraged. There are two particular features to be watched for: (i) The development of an associated deep-vein thrombosis. Suspicious symptoms are deep muscle pain, tenderness and oedema. A specialist opinion should be sought to decide if anticoagulation is indicated (see below). (ii) The development of proximal extension especially toward the groin. In this instance operative sapheno-femoral disconnection is indicated to prevent further migration of the clot into the femoral vein with the attendant risks of pulmonary embolism.

Superficial phlebitis in patients with varicose veins is an indication for definitive treatment once the acute episode has settled – often, but not always, the varicosities deteriorate following recanalization of the thrombus and patients may be subject to recurrent attacks of phlebitis. Should such patients be required to undergo other forms of surgery serious consideration should be given to subcutaneous heparin prophylaxis against peri-operative deep-vein thrombosis. Similar prophylaxis is given for the definitive varicose vein operation.

ACUTE DEEP-VEIN THROMBOSIS

Clinical features

The symptoms of acute-onset calf pain and swelling should alert the doctor to the possibility of this diagnosis. The signs of unilateral calf tenderness, a positive Homans' sign (elicited gently) and oedema are strongly supportive findings. Oedema may be minimal in the initial stage and extension to involve the upper calf usually indicates involvement of the tibial or even the popliteal veins. The swelling is more extensive with ilio-femoral thrombosis when the thigh and even the buttock may be swollen and tenderness may be present on palpation over the femoral vein in the thigh. Venous ultrasound is a valuable technique for demonstrating occlusive thrombus extending to the thigh but is less valuable for calf thrombosis.

The common cause of diagnostic dilemma is the inability to palpate the pedal pulses, usually because of oedema, and especially when there is any discoloration of the toes. This raises the possibility of an ischaemic episode. A torn calf muscle or ruptured plantaris tendon should be diagnosed with circumspection in the absence of an appropriate history of trauma. Any calf injury to skin, soft tissue, muscle or bone may be complicated at any stage by a deep-vein thrombosis. Synovitis of the knee in inflammatory arthritis may lead to joint rupture. The synovial fluid leaks into the calf producing acute pain and swelling. This can mimic a deep-venous thrombosis very closely.

Management

The acute danger lies in the risk of pulmonary embolism – untreated the thrombus may extend proximally by propagation to give a long tail of loosely attached clot adherent only distally. This situation may be demonstrated by ascending venography (Fig. 5.4) from which the risk is clearly apparent. The long-term risk is that propagation leads to a wider involvement of the venous system. Although the natural course of the condition includes a variable amount of recanalization of the veins this process destroys or deforms the valves and the one way system they support. After many years this may lead to chronic venous insufficiency which is discussed below.

In general patients with a clinical diagnosis of deep-vein thrombosis should be admitted to hospital for anticoagulation with heparin. Ascending venography

Fig. 5.4. Ascending venogram showing adherent thrombus in the popliteal vein with loosely attached propagated clot in the femoral vein.

should be carried out to confirm the diagnosis and more especially to define the patient in the very high risk group with a floating loose thrombus. Where no sign of deep-vein thrombosis is demonstrated a correctly fitting elastic support stocking is prescribed and heparin may be discontinued.

Where loosely attached clot is demonstrated there is a strong case for performing a ligation of the superficial femoral vein just distal to the profunda vein, to 'lock in' the clot. There is no apparent increase in the incidence of chronic venous insufficiency after this manoeuvre because the superficial femoral and distal vein valves will have been destroyed by the initial thrombotic process and the potential for venous return still exists via the profunda system.

Heparin may be given by infusion for a period of several days with conversion to an oral anticoagulant and stabilization before discharge. Oral anticoagulants are maintained for at least three months or longer if the patient does not achieve full mobility at this time. Supportive graduated elastic hose should be worn.

CHRONIC VENOUS INSUFFICIENCY

The main features of this syndrome, also known as the post-phlebitic limb, are varicose veins, usually secondary to previous overt or occult deep vein thrombosis, oedema, pigmentation, induration (liposclerosis), ulceration in the lower medial or lateral 'gaiter' area of the calf, and rarely venous claudication. The essential pathophysiology is impairment of venous return due to deep or perforator vein valvular incompetence leading to inefficiency of the calf muscle pump which propels blood proximally. Not all features of the syndrome occur at one time but over the years all may develop.

The symptoms vary from those of ill-defined aches and pains in the calf with foot and ankle swelling at the end of the day, to severe incapacity due to painful venous ulceration accompanied by an offensive discharge due to secondary infection. A tendency to equinus deformity further limits ambulation. Painful thickening of the subcateous fat of the medial calf (liposclerosis) may be found with the varicosities lying in deep gutters or channels at this site. Chronic and extensive oedema may also occur and can be associated with a bursting pain on exercise often apparently made worse by wearing strong elastic hose – this is venous claudication.

Treatment may be problematical. Obesity is commonly present and all will benefit from appropriate dietary advice. Postural drainage of oedema by elevation of the limbs at night with a pillow under the mattress or bricks at the *foot* of the bed should be combined if possible with an hour of elevation in the middle of the day. Well-fitting graduated elastic hose, to be donned before getting out of bed, should be supplied.

When varicosities are present patients should be considered individually for treatment either by sclerotherapy or appropriate surgery. It is important to confirm the patency of the deep system at least by ultrasound but also by venography when there is any indication of venous claudication.

Much has been written on the management of venous ulceration. Only a brief outline will be given here. The aim should be to achieve healing if possible while the patient is ambulant by a combination of local dressings, preferably Eusol and paraffin or simple paraffin tulle, compressive bandaging and periods of elevation. Sitting with the legs dependent or standing for long periods should be avoided. When associated with superficial varicose veins and incompetent perforator veins, surgery should be considered after the ulcer has been healed. More severe cases frequently require a period of intensive local dressings with continued postural drainage. It is probably realistic to accept that a small minority of patients achieve a sort of symbiosis with their ulcer and provided local dressings and bandages keep them pain free and control offensive discharge, minimal distress is experienced.

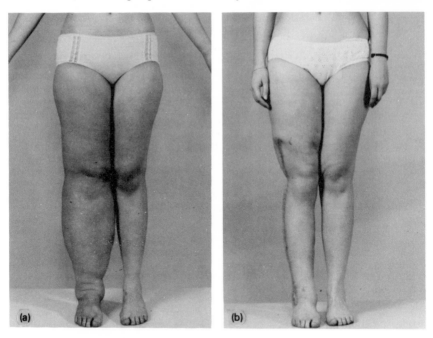

Fig. 5.5. Chronic lymphoedema: (a) before; (b) after operation. (Author's case referred to Mr Noel Thompson for operation.)

A few patients benefit from excision of the ulcer, skin grafting, and subfascial ligation of the perforating veins. All operative procedures on such patients carry the risk of a further deep-vein thrombosis and should have appropriate prophylaxis.

LYMPHATIC PROBLEMS

ACUTE LYMPHANGITIS

This is an unusual condition classically associated with streptococcal infection. The bright red lines of the inflamed lymphatics make the diagnosis straightforward. The site of the original infection should be sought. Tender groin lymphadenopathy is common. Active antibiotic treatment is indicated. Simple cases usually respond to high doses of penicillin but when associated with ischaemic ulcer a broad-spectrum antibiotic combined with an agent active against anaerobes is indicated.

CHRONIC LYMPHOEDEMA

An unusual cause of chronic leg swelling especially in adolescent females is anatomical or functional deficiency of the lower-limb lymphatics. Extensive disease leads to unsightly swelling of the lower limb and even difficulty with walking. Attacks of streptococcal cellulitis are characteristic and may be the presenting episode after which swelling of the limb deteriorates.

The differentiation from the oedema of venous disease is usually made by the age of onset, the absence of symptoms or signs of varicose veins or history of deep-vein thrombosis, and a tendency to brawny rather than pitting oedema.

Lesser degrees of swelling are treated by elastic support and posture, but gross problems can be treated by operative reduction (Fig. 5.5).

I have tried to offer a rationale of the pathology of common clinical vascular problems affecting the lower limbs. as they may present to the general practitioner, and hope this may be a guide to choosing the moment for consultation with hospital colleagues, as well as being an instruction in general management.

6 Locomotor disability in industry

J. A. D. Anderson

INTRODUCTION

Control of infection and improved nutrition have changed morbidity patterns
in Western Countries to focus attention on non-communicable diseases while
increased expectation of life means a higher proportion of the population
survives to adulthood. The diseases and injuries which result in locomotor
disability are essentially non-communicable now that the threat of paralytic
poliomeyitis has receded; furthermore, these conditions constitute important
components in the disability spectrum of those of working age.

Changes in occupational patterns are also relevant in bringing about an
increased awareness of the importance of locomotor disability in the working
population. Trends which started with the Industrial Revolution have become
more pronounced in recent years as industry has become concentrated to an
even greater extent in large units where workers carry out repetitive tasks. The
effects of these changes have been to accentuate the importance of disabilities
which may reduce the speed at which work can be performed. In addition the
era when the cost of sickness to industry was measured only in terms of
financial payments to affected individuals in the form of insurance benefit or
compensation has passed. The real cost of absence today must include some
allowance for the reorganization of staff and reduced productivity of a team
deprived of one member; furthermore, a fall in production of one component
can result in reduced production of the whole – a problem particularly important
when skilled workers are involved in the manufacture of a complex end product.
It is in these circumstances that peer-group comparisons of productivity tend to
be made and where failure to keep pace by one member can affect output and
hence bonus payments for others in a production team.

Another change affecting industrial disability generally and which is particularly relevant to general practitioners has been the introduction of social
insurance both nationally and privately. Today both sickness absence and
premature retirement can result in claims on such insurances and detailed
records have to be kept and supporting certificates issued. Most medical certificates issued to support claims for sickness benefit and industrial disability are
initiated by general practitioners who are only too well aware that the problem
is a substantial one. Thus, in Britain, the loss to industry from rheumatic
diseases alone amounts to some 37 million person days per annum (Wood and
McLeish 1974). Add to this the loss ascribed to an abridged list of other
conditions likely to cause locomotor problems such as strokes, multiple sclerosis,

and industrial injuries affecting the trunk and lower limbs and the loss to industry from locomotor disability is likely to be in the region of about 40 million days per annum, constituting 13 per cent of the total loss from sickness and injury. The proportions are similar in males and females (see Fig. 6.1); this reflects the fact that injuries though more common in males than females only account for a very small proportion of total absence (males 8 per cent; females 3 per cent), whereas there is more sickness absence among females for rheumatic diseases and multiple sclerosis. As might be expected the loss is greater in heavy industry than among those doing sedentary work as is evidenced by the fact that sickness absence from such causes accounts for 15 per cent of the loss among foundry workers and 4 per cent among their counterparts in commerce and insurance.

General practitioners themselves in a combined study conducted jointly by the Office of Population Censuses and Surveys (1974) and the Royal College of General Practitioners found that 14 per cent of their annual consultations for men aged 15 to 64 years observed over a period of 12 months were for the above group of diseases. As with insurance certificates rheumatic diseases accounted for the lion's share (77 per cent). In an earlier study (General Register

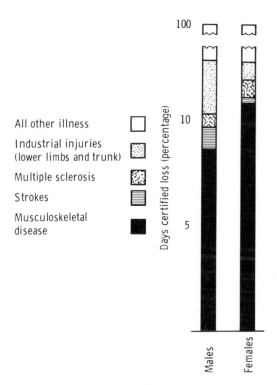

Fig. 6.1. Days lost from locomotor disability (percentage from annual insurance certificates).

Office 1960) it was reported that though the annual consultation rate for cerebrovascular accidents was marginally higher in Social Class I (20 per 1000 registered patients) than in Social Class V (14) the trend was markedly reversed for rheumatic diseases where there were 227 consultations per 1000 registered patients in Social Class V against 72 for those in Social Class I. These findings from general practice are in keeping with the evidence from insurance certificates that the impact of diseases causing locomotor disability is greater in manual workers than in those doing lighter work.

STUDIES IN INDUSTRY

Information such as the foregoing makes it tempting to suggest links between locomotor disorders and the strain and trauma of manual work. However, there are difficulties since no one checks the diagnostic labels used by doctors whether in describing their own workload or when writing insurance certificates. Furthermore, non-specific terms may mask those diseases indicative of locomotor disability either because of uncertainty on the part of the doctor or through a desire to protect the patient from knowledge of the condition – particularly if the prognosis is unfavourable. If more accurate information is to be obtained special surveys have to be conducted among workers themselves and these, though helpful in getting things into perspective, are difficult to perform accurately and are often time consuming.

HANDICAP

When considering disablement in any circumstances, but particularly in relation to industry, a distinction must be drawn between impairment (sometimes also called abnormality or malfunction), disability, and handicap as it is the prevention of the last named that is important in the context of locomotor disorders. To take a simple illustration, a man with shortening of one leg may be said to be abnormal but this alone need not limit activity. It may be only when the impairment is associated with osteoarthrosis of the hip or secondary disc degeneration from compensatory scoliosis that disability develops. Even if such disability is present and can be demonstrated clinically by, say, limitation of activity, the extent to which the condition is handicapping will depend on the requirements of that individual. In an industrial setting such requirements depend on the demand of the work to be performed. Thus, if the affected person is a miner working in a confined space at the coal-face then the handicap may well be total; it may be much less if he is a surface worker in the same colliery while if he is a banker, lawyer, or doctor the handicap, as far as occupation is concerned, could well be negligible.

In relation to rheumatic disorders gross abnormalities or even severe disabilities need not of themselves cause significant handicaps on the other hand, pain, whether or not accompanied by impairment and disability is usually handicapping – the problem is that pain is difficult to measure objectively. In relation to neurological causes of locomotor disability (including cerebrovascular catastrophies) the problem is more that of muscular weakness or inco-ordination than of pain. However, both pain and weakness are more than likely to lead to greater handicap and hence more sickness absence in those engaged in heavy manual work than among office staff.

CHOICE OF JOB

Prolonged or repeated sickness absence may force a man to change his job. However, caution is needed in interpreting information about such changes particularly as a person with locomotor disability fulfils the criteria of medical respectability discussed by Parsons (1951) when considering the 'sick role'. A label implying locomotor disability may be used in preference to what may be more factually accurate where the real reason for change is either social or mental inadequacy.

Doubts about the true relationship between locomotor disabilities and change of job are illustrated by a study in depth of manual workers (Anderson 1971) in which 201 were identified as claiming that rheumatic complaints were interfering with their work and, indeed, 30 of them stated that they had changed their jobs because of these disorders. These were matched by age and occupation including site and type of work with an equal number who had

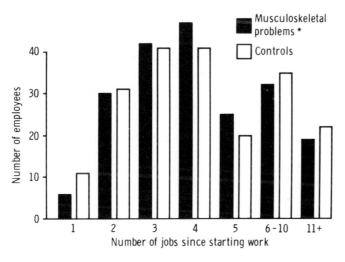

* ie. Musculoskeletal disease associated with one or more of the following:
a. Hospital admission   b. Sickness absence 4/52   c. Change of job

Fig. 6.2. Total number of jobs since starting work (201 matched pairs of dockyard employees).

never had rheumatic symptoms at any time. The total number of job changes since leaving school for each individual was obtained by means of a complete work history given in an interview with a social worker. Figure 6.2 shows a close similarity between the two groups in respect of their work histories and this gives cause for reasonable doubt that factors other than rheumatic complaints may well have contributed to the changes in occupation regardless of declarations made by those being studied.

Looking at the same problem from another angle a study was carried out on men registered as disabled and unemployed for three months or more (Anderson *et al.* 1963). Rheumatic complaints were detected in 56 (74 per cent) of the 76 men examined and they were the main cause of disability in 13 (17 per cent). However, there was little suggestion that rheumatic complaints alone played a dominant part in prolonged unemployment. Indeed, among the 41 men in the group who had been off work for at least two years there were 30 who had rheumatic complaints (i.e. 54 per cent of the 56 men with such complaints) compared with 11 (i.e. 55 per cent of the remaining 20) who had no such complaints.

AETIOLOGY

Occupational factors in the aetiology of traumatic lesions such as sprains, fractures, and head injuries sustained at work are self-evident; similarly toxic neuritis and traumatic synovitis affecting bursae, joints, or tendons can often be shown to have reasonably obvious links with working conditions. Locomotor diseases such as 'miner's beat knee', 'policeman's heel', and 'weaver's bottom' all have occupations incorporated in them as do toxic neuropathies such as 'hatter's shakes' and 'file-cutter's paralysis'. Such lesions are commonly, though not exclusively, associated with the occupation in their designation while 'march fracture' affecting footslogging postmen and soldiers alike can also have a clear occupational link. Other physical hazards include increased atmospheric pressure, which can cause both 'diver's bends' and bilateral avascular necrosis of the femoral heads (McCallum *et al.* 1954), while climatic extremes besides being the triggering mechanism for such disabling conditions as Raynaud's disease are associated with increased sickness absence due to arthritis and rheumatism without causing an increase in the prevalence of those affected.

Many diseases including some listed above have had their occupational links identified by the careful observations of doctors, many of whom were in general practice adjacent to the industry where the hazard responsible for the disability was a prominent feature. Furthermore, the recognition of such a hazard was in itself the factor which led in many cases to the elimination of the disease from among the work-force. It is natural, therefore, that thinking doctors should be mindful of the possibility that systemic diseases of unknown aetiology, but which have long-term consequences as far as locomotor disability is concerned, may be precipitated by an industrial hazard. The aetiological backgrounds of systemic diseases such as multiple sclerosis, muscular dystrophy, rheumatoid arthritis, and ankylosing spondylitis which can all cause locomotor disability in their advanced forms are still poorly understood; equally debatable are some of the risk factors associated with arterial changes which are the precursors of strokes. At present, however, no-one has been able to indicate any clearcut relationship between occupational conditions and the onset of such precursors to locomotor disability. The consensus of opinion remains that there is no proven connection between the aetiology of these conditions in working situations. However, if some factor relating to employment such as undue muscular effort, psychological stress, or even climatic extremes contributes to systemic upset, thereby affecting the reaction between infections and immune responses, then the possibility that this might precipitate the onset of clinical manifestations of such systemic disorders cannot be excluded.

In more specific terms the accepted view that chronic trauma can and often does produce osteoarthrosis had led to the concept that prolonged heavy work may do likewise. On the other hand, observers since the time of Hippocrates have associated osteoarthrosis with the elderly – a view that is maintained today, particularly in respect of generalized osteoarthrosis where it is believed that there may be a genetic predisposition (Kellgren *et al.* 1963). Thus, the relative importance of environmental conditions including those at work and genes may be difficult to assess when those near the end of their working lives claim that heavy manual work has been a causative factor in the development of degenerative changes which affect their locomotor capacity.

Survey Units supported by The Arthritis and Rheumatism Council and based at Manchester (Epidemiological Survey Unit), Edinburgh (Industrial Survey Unit), and Guy's Hospital, London (Occupational Health Survey Unit) have shown high prevalence rates for osteoarthrosis and also for degenerative changes of the spinal column among heavy manual groups such as foundry workers and coalminers. Those engaged in lighter industries (e.g. cotton operatives, electrical engineers, and office workers) have correspondingly low rates for these degenerative conditions. Age-specific prevalence rates suggest that the tendency to develop degenerative changes is accelerated among employees in heavier manual trades leading to the appearance of such changes earlier in life than is found among those engaged in lighter work even though ultimately – by the age of normal retirement – the overall prevalence rate is reasonably consistent (Lawrence 1977). However, jute stackers in Dundee regularly carry 200 lb bales on their backs without any apparent increase in complaints – unless, that is, those with problems have been obliged to seek alternative employment from an early age.

A further difficulty which arises when trying to relate any disease to occupational factors is that not all workers engaged in the same designated occupation do tasks which involve the same physical effort, nor do they adopt the same posture. If working conditions are suspected of being causal then they have to be examined individually using ergonomic principles and this is a time-consuming exercise.

A simple job analysis using a graded scale for such factors as effort by the *back, arms, hands,* and *legs,* and also the *posture* adopted at work and the *site* in terms of climatic conditions (*BAHLPS*) can be used to assess work patterns. In one such study it was found that 44 per cent out of 164 workers with jobs graded minimal on all four assessments of back, arms, hands, and legs had rheumatic complaints including back pain, and that 43 per cent out of 462 in the heavier grades on all four assessments were similarly affected. However, after allowing for age, increased effort by the back, arms, and legs significantly increased the likelihood of rheumatic complaints in the form of degenerative disc disease and generalized osteoarthrosis. Prolonged stooping and awkward posture were also shown to have a predisposing effect on the reported onset of low back pain.

The ergonomic assessments in the *BAHLPS* system are fairly simple and more sophistication is clearly needed. Furthermore, it is generally accepted that the presence of an observer at the work bench inevitably results in the observed worker acting atypically. Accordingly, research teams are currently experimenting with continuous electronic observations over an eight-hour shift (Sweetman *et al.* 1980). Modern equipment enables a recording device to be worn at work without difficulty causing minimal disruption of work patterns. A complementary method requires the worker to swallow a capsule containing a minute radio transmitter and this enables variations in pressure within the gut to be recorded (Davis *et al.* 1977); the pressure waves detected are believed to be related to pressure waves exerted on the lumbar discs and facetal joints by the strain of lifting heavy objects (Nachemson 1980).

By computerizing the records from devices such as these it is hoped to identify those tasks within a job specification which appear to be associated with the greatest effort by trunk and limbs and with excessive or awkward posture. Arguably the elimination of such procedures from the job specification could reduce the risk of developing locomotor disability on the one hand or render the job suitable for placement of those with an established problem.

MEDICO-LEGAL ASPECTS

Inability to prove a causal relationship between environmental hazards and the onset of a chronic disease with locomotor consequences may also mean that it is difficult to refute authoratively an allegation that one or more incidents at work have contributed to the onset of locomotor disability when circumstantial evidence suggests this. In such cases medical, managerial, and occupational considerations can be at variance in relation to patients with such complaints since conditions affecting the locomotor system as a whole are renowned for the divergent views which are presented by medical experts, employers, and patients.

There are few manual workers who have not injured their lower limbs or backs at one time or another while lifting or humping cumbersome loads. In addition, many sustain local trauma through blows or falls in the course of a working day. Office workers also suffer such setbacks but generally to a lesser extent, while those whose jobs include driving vehicles and who are involved in road accidents are very likely to receive injuries which can lead to residual pain, weakness, or loss of co-ordination.

Under these circumstances and given the vagaries of human memory in relation to the insidious onset of many of the diseases associated with locomotor disability it is hardly surprising that symptoms may appear to date from an injury at work. Indeed, the first clinical contact with such patients may leave in the minds of both doctor and patient a strong association between the onset of symptoms and such an injury without there necessarily being any intent to deceive since even a minimal episode may serve to focus attention on the legs or back. Such a patient thus becomes more aware of symptoms which have been largely or completely ignored up to the time of the injury. Superficial injuries and soft-tissue bruising may mask co-existing neuromuscular disease while ligamentous strain near a joint or in the back may serve to camouflage monarticular arthritis or ankylosing spondylitis.

Following such an accident a period of absence from work and even enforced immobility may be desirable – even inevitable – if the injury has led to a fracture. Delay in recovery may merge imperceptibly into the onset of a chronic disease when it becomes difficult to convince the patient to say nothing of his legal advisers and Trade Union supporters that the injury concerned did not 'cause' the disabling disease.

Clearly this is a difficult equation to solve not least because injuries at work are common whereas diseases such as multiple sclerosis, ankylosing spondylitis, and monarticular arthritis are comparatively rare. However, the possible existence of underlying disease in patients who are slow to recover from simple occupational trauma should be borne in mind particularly in relation to back injuries masking the onset of ankylosing spondylitis in young male patients.

Conversely it can be argued that failure to encourage a patient to return to work as soon as possible is, in effect, contributing to the resultant handicap. Unfortunately, such a policy may occasionally backfire through the erroneous view of patient, relatives, or employer that returning to work too soon has contributed to the ultimate state of locomotor disability.

PREVENTION

The implications of these and other similar findings in relation to working conditions can be considered under the heading of primary and secondary prevention.

PRIMARY PREVENTION (INCLUDING BOTH ENVIRONMENTAL AND PERSONAL COMPONENTS)

Dangerous machinery, falling objects, slippery floors, and obstructions to traffic lanes at places of work are all recognized hazards and most efficiently run industries take adequate steps to reduce such hazards to a minimum. Where such accidents are likely protective headgear and other clothing are required to be worn usually with the threat that failure to do so will bring instant dismissal.

Clearly, doctors should encourage their patients who work in potentially dangerous situations to follow all the required safety regulations. However, there are grey areas such as manual exertion and weight lifting since it has been shown that prolonged heavy work is correlated with increased chronic back pain and osteoarthrosis. The introduction of statutory maxima for weight and size of loads might help in this respect. It would also seem possible that in future prolonged stooping or other awkward posture may come to be discouraged by Works Safety Committees in the same way as happens today for the hazards listed above.

It would be naive to think that these aims can be achieved quickly in Britain, particularly as the new Health and Safety at Work Act (1974) does not lay down any weight limits. Obviously it will be necessary to produce more evidence – perhaps through experiments and observations with continuous recording devices. Nevertheless there is probably sufficient evidence available at the present time to suggest that education about lifting methods is worth advocating as a personal preventive measure not only to counter strained backs but also dropped packages and slipping injuries.

Most people know the right and wrong way to lift but how can they be persuaded to do the right thing? One method of educating those most at risk is to display posters and pamphlets for guidance either on the shop floor or in leisure rooms and some of these are available from The Back Pain Association; much of this advice is good but in the past some have been rather impracticable and this can bring the rest into ridicule. As an example Fig. 6.3 shows one pair

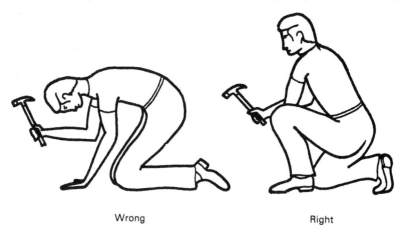

Wrong   Right

Fig. 6.3.

of sketches in such a pamphlet which might or might not help to reduce the incidence of back pain but could well lead to an increase in injuries of the left hand among those seeking to wield a hammer in this position.

It is interesting that health education is still taken on trust. Thus, the Department of Employment has refused to fund a controlled trial to assess sickness absence from musculoskeletal causes before and after a period of instruction in handling had been given to a group of factories even though such a study was recommended by a Working Party set up by a Committee of the Department to consider the problem.

A more positive approach has been shown by a chain of supermarkets seeking advice on which is the least stressful method of checking out purchases as far as cash-desk assistants are concerned. An extension of the continuous recording method is being used to try and measure the amount of asymmetrical bending and twisting associated with the various types of cash desk and transfer methods from trolley to purchaser. It is to be hoped that the success of this process of predetermination will be the subject of subsequent evaluative assessment.

In the domiciliary setting Do-It-Yourself experts combine the roles of manager, foreman, employee, and safety officer with dangerous equipment, precarious positioning, awkward posture, and excessive weight lifting being everyday occurrences. Advice on preventive measures from a visiting member of the Primary Care Team may be regarded by some as unwarranted interference in personal liberty but nevertheless every chance should be taken to offer timely advice about obviously hazardous situations whenever these are observed.

EARLY DETECTION

Screening examinations such as routine serology for rheumatoid factor or X-ray of hip joints to show early osteoarthrosis could, in theory, be used as secondary preventive measures to supplement pre-employment examinations in an occupational setting. However, such techniques are unimpressive since they fulfil few of the criteria which are prerequisites of useful screening procedures. In particular their levels of sensitivity (i.e. the proportion of affected people who are graded positive by the test) and specificity (i.e. the proportion of unaffected people who are graded negative by the test) are far too low. Furthermore, current therapies to help those with potential problems revealed by such tests are not very effective.

Another such test is the use of HLA 27 as an early detector of ankylosing spondylitis in young males. This is notoriously unreliable and can be likened in this respect to the suggestion currently gaining popularity in the United States that schoolchildren should have their backs examined routinely so that those with identifiable scoliosis can be treated with a Milwaukee brace – on the same principle as dental plates and wires are used widely to remedy occlusion defects at the present time.

Another example of the absurdity of such screening procedures is to be found in a report of a series of pre-employment special X-rays on 4103 potential employees in the United States. Abnormalities discovered by this method led to 1181 (29 per cent) being classed as unacceptable and it was later shown that this significantly reduced the amount of litigation in respect of long-term back disabilities against the firm adopting this procedure (Kosiak *et al.* 1968). However, it does not seem very practical to reject nearly a third of the work force without being much more specific about which abnormalities put potential employees at risk.

Routine investigation of cholesterol levels in an attempt to control atheromatous plaques and reduce the incidence of strokes are also of questionable value. However, advice to those who are overweight and routine estimations of blood pressure particularly in middle-aged males may be more helpful and such measures are gaining popularity in the field of secondary prevention particularly now that there are reasonably effective hypotensive agents. However, there remains the problem of ensuring accurate and consistently reliable readings with sphygmomanometers. In the setting of general practice it is also worth bearing in mind that some authorities (Wilson and Ross Russell 1977) take the view that flitting blindness (amaurosis fugax) and transient ischaemic attacks, which may well manifest themselves in a working situation, are to be regarded as imminent precursors of strokes which may be prevented by timely surgery. Urgent investigation of these cases is therefore recommended even though those suffering such phenomena at work and their work-mates may tend to dismiss the episodes as trivial.

MANAGEMENT AND REHABILITATION

Selection of the right therapy to reduce chronicity and prevent recurrence may prove a more practical possibility than either primary or secondary prevention as far as most locomotor disabilities in industry are concerned. However, this is often much more difficult to achieve than at first appears.

For minor complaints, recovery is often spontaneous and complete but for chronic conditions it is generally accepted that treatment of the physical component may be largely palliative. Accordingly, it is necessary to consider management less from a curative point of view than a rehabilitative one and in this context the occupational setting is of critical importance.

Some industries have established their own Rehabilitation Units with special workshops to encourage disabled employees to readjust more rapidly. One is at Longbridge and another in Luton also associated with the motor industry where reports indicate that duration of absence from several conditions including injuries and some locomotor disorders can be reduced by this means.

Uncertainties abound as far as employment is concerned for those with chronic locomotor disabilities and only a few firms have enough resources to absorb all their own handicapped workers let alone employees from elsewhere.

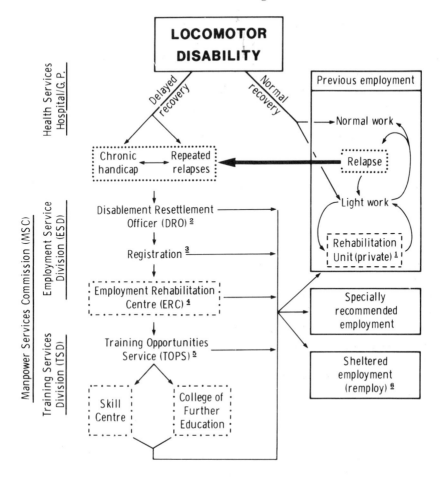

Fig. 6.4. Diagram illustrating the help available to those with locomotor disability.

1. Rehabilitation unit (private): administered by a firm for its own employees though occasionally a number of firms share such a resource as part of a privately financed occupational health service.

2. Disablement resettlement officer (DRO): special officers of the ESD with special local knowledge of the job opportunities for the handicapped; also providing a range of services for employers prepared to accept handicapped staff.

3. Registration placement of patients on the *register* maintained by ESD provides protection for those whose handicaps prevent them from performing certain tasks or who have prolonged and/or repeated periods of sickness absence.

4. Employment rehabilitation centre (ERC): used to assess the potential of those requiring a change of job; has both clerical and industrial sections. Also helpful in reaclimatizing those with erratic work attendance because of low back pain; emphasis on rhythm in working routine by setting standards within functional capacity and insistence on compliance. Doctor and gymnasium available so remedial programmes started elsewhere can be continued. NB (a) Only 26 such Centres in the United Kingdom. This may pose travelling problems or necessitate temporary absence from

home. (b) If the proportion of depressed patients among attending clientele is unduly high this may be detrimental to those with low back pain.

5. Training opportunities scheme (TOPS): Courses arranged by TSD held at either one of 60 *skill centres* or at a College of Further Education. Not exclusively for handicapped but those officially *registered* stand good chance of acceptance, particularly if supported by prior assessment at ERC.

6. Sheltered workshops (REMPLOY): government maintained to provide employment for severely handicapped, particularly those with below average production speeds. Limited number of able-bodied supervisors may also carry out complicated/hazardous tasks.

Accordingly, it is usually wise to persuade those with such problems to invoke the help of statutory agencies set up for this purpose by the Manpower Services Commission (MSC) established in Britain in 1974. The MSC is divided into two components concerned with resettlement and retraining known respectively as Employment Service Division (ESD) and Training Services Division (TSD). Both Divisions have special responsibility for those who are handicapped as a result of injury or chronic disease from any cause.

The services are represented diagrammatically in Fig. 6.4 together with the accompanying notes. Many of those affected through injury and also those with locomotor disability from other causes recover in less than three months and return to their own employment, perhaps doing a spell of light work or, where available, attending a special Rehabilitation Unit of their firm. Those with delayed recovery should be referred for discussions with the disablement resettlement officer of the ESD (contact being made through the nearest Job Centre, though a few hospitals have such an officer seconded to work in hospital). Patients with chronic disability may be *registered* with the ESD. This is regarded by some patients and clinicians as permanent though the process is easily reversible; there is also a feeling by some that they have been 'branded'. However, the advantages outweigh the disadvantages and currently include:

1. Obligation by larger firms to include a percentage of *Registered* persons in their work force.*

2. Insurance companies generally include *registered* persons in their Employer's Liability Policy on the same terms as able-bodied persons since the inference is that account has been taken of functional capacity when allocating tasks.

3. Help may be given with fares and/or transport to and from work to selected persons on *Register*.

4. Special equipment to obtain or maintain employment may be provided (on loan) from public funds.

Registration usually makes it easier to obtain placement at an Employment Rehabilitation Centre and also to undertake one of the options for retraining (at a Skill Centre or a College of Further Education) offered under the Training Opportunities Scheme (TOPS) of the ESD.

*The MSC is currently reviewing registration as well as the quota system.

Very rarely it may be necessary to seek employment in a sheltered workshop for a patient with severe locomotor disability though the need for such a referral will be lessened if the affected person has been encouraged from the outset to see continuing occupational activity as an integral part of the resettlement programme. However, where the underlying disease is progressive, when deformities are far advanced before a rehabilitation programme is instituted, or if the situation has been mishandled then a sheltered workshop may be the only hope of employment. Vacancies are scarce and delays may be long; another important disadvantage is that such workshops like the Employment Rehabilitation Centres described in Fig. 6.4 are limited in number and this may result in an awkward journey or even a requirement to move house. There is thus an added incentive to use maximum pressure to achieve retraining for suitable employment in the open market supported, if necessary, by inclusion on the ESD *Register.*

It is generally agreed that the earlier these services are involved once a decision has been made to use them the better the opportunity for successful rehabilitation. On the other hand, advising people to change jobs cannot be undertaken lightly particularly in view of the economic consequences which usually ensue when such a change is made. Furthermore, facile views about a job change as an answer to the problem of locomotor disability may be misleading and counter productive.

Taylor and Fairie (1968) have commented that there is a lack of understanding by doctors and managers and indeed by the disabled themselves about occupational aspects of rehabilitation. They have suggested that a suitable educational programme is required to correct this misunderstanding since it is unlikely to be successful in patients with chronic locomotor disability unless those responsible for their care can be persuaded that the services offered are useful. This in turn depends among other things on adequate and convincing evaluative studies in this field. Unfortunately, the demarcation zone between the Department of Health and Social Security and the Department of Employment enables both these bodies to evade the issues thus perpetuating in United Kingdom the rather unsatisfactory *status quo.*

CONCLUSION

Locomotor disabilities are numerically among the most important causes of absence from work and long-term or permanent disability in Britain. Their effects are formidable in terms of loss of earnings, and those affected make heavy demands on medical services and social insurance. The extent to which occupational hazards contribute to aetiology by causing permanent changes is difficult to assess and more studies need to be made in relation to those who have retired prematurely, changed their jobs or been off work for long periods. Cohort (logitudinal) studies, though costly in manpower and time, could probably shed more light on facts about workers, their locomotor disabilities and the handicapping effects of such disabilities in relation to jobs which might affect the long-term prognosis.

Primary prevention of accidents both in industry and in the home should receive more attention though success seems to be out of reach at the present time particularly in relation to long-term disabilities arising from chronic rheumatic and neuromuscular diseases. Secondary prevention or early detection is of limited use in respect of progressive rheumatic or neuromuscular diseases since it is comparatively rare that treatment is successful in achieving complete cure. However, the possibility that screening for hypertension could reduce the incidence of strokes presents interesting opportunities.

There remains, therefore, a large number of patients requiring rehabilitation and this problem seems likely to increase rather than decrease. Closer liaison should be established between the health services and industry by such methods as attaching DROs. not only to hospitals but also to health centres. The appointment of specialists in rehabilitation with responsibility for both hospital-based patients and those who have either never been to hospital or who have been discharged and are struggling to survive in the community. The present Health Authority structure, bridging as it does the hospital and the community health services, should be able to help in this way. On the other hand the tendency to concentrate resources in institutions of high technology and the separation of supporting social services under the aegis of local government militates against collaborative action. It is depressing also that central government has still not seen fit to include occupational health as an integral part of the re-organised health services in Britain. Furthermore, only limited resources are being deployed in evaluating methods of caring for those disabled by loco-motor disabilities; notable exceptions are drug trials sponsored by pharmaceutical manufacturers where the lure of future profit appears to justify costly evaluation.

REFERENCES

Anderson, J. A. D. (1971). Rheumatism in industry: a review. *Br. J. indust. Med.* **28**, 103–21.

—— Duthie, J. R., and Moody, B. M. (1963). Rheumatic diseases affecting those registered as disabled. *Annls rheum. Dis.* **22**, 188–93.

Davis, P. R., Stubbs, D. A., and Ridd, J. E. (1977). Radio pills: their use in monitoring back stress. *J. med. Engng Technol.* **1**, 209–15.

General Register Office (1960). Morbidity statistics from general practice. Vol. II Occupation. Studies on Medical and Population Subjects, Vol. 14. HMSO, London.

Kellgren, J. H., Lawrence, J. S., and Bier, F. (1963). Genetic factors in generalized osteoarthrosis. *Annls rheum. Dis.* **22**, 237–55.

Kosiak, M., Aurelius, J. R., and Hartfiel, W. F. (1968). The low back problem. *J. occup. Med.* **10**, 588–93.

Lawrence, J. S. (1977). *Rheumatism in populations.* Heinemann, London.

McCallum, R. I., Stanger, J. K., Walder, D. N., and Paton, W. D. M. (1954). Avascular necrosis of the femoral heads in a compressed air worker. *J. Bone Jt Surg.* **36**, 606–11.

Nachemson, A. (1980). Lumbar intradiscal pressure. In *Lumbar spine and back pain,* 2nd edn (ed. M. I. V. Jayson) pp. 341–58. Pitman Medical, Tunbridge Wells, Kent.

Office of Population, Censuses and Surveys (1974). Morbidity statistics from general practice: second national study. Studies on Medical and Population Subjects, Vol. 26. HMSO, London.

Parsons, T. (1951). *The social system.* Free Press, Glencoe, Ill.

Sweetman, B. J., Page, S., McMaster, G. W., Ellam, S., and Anderson, J. A. D. (1980). EMG correlates of low back pain. In *ISAM 1979* (ed. F. D. Stott, E. B. Raftery, and L. Goulding). Academic Press, London.

Taylor, P. J. and Fairie, A. J. (1968). Chronic disability in men of middle age. *Br. J. prevent. social Med.* **22**, 183–92.

Wilson, L. A. and Ross Russell, R. W. (1977). *Br. med. J.* **ii**, 435–7.

Wood, P. H. N. and McLeish, C. L. (1974). Digest of data on the rheumatic diseases. *Annls rheum. Dis.* **33**, 93–105.

Section II

Upper limb problems

7 Pain in the neck and shoulder

D. M. Grennan

INTRODUCTION

Pains in the neck and shoulders are among the most common complaints to cause an adult to seek medical attention. In children the commonest causes of cervical pain are acute infections often associated with tonsillitis and a cervical lymphadenopathy and this chapter will presume that such causes have already been excluded. The diagnosis of most adults presenting with neck and shoulder pains is possible on the basis of a clinical history and examination alone and the management of many of the soft-tissue rheumatic disorders causing pain in these regions falls within the province of the general practitioner. A few of the local causes of cervical pain are serious and threaten neurological involvement and in them immediate hospital referral is indicated whilst neck and shoulder pains may also be symptomatic of a wide range of general medical disorders. This chapter will seek to provide a guide to the differential diagnosis and management of the various causes of neck and shoulder pains in adults.

CLINICAL HISTORY AND EXAMINATION

HISTORY

A good clinical history and examination will differentiate most of the causes of neck and shoulder pain shown in Table 7.1. Pain in these regions may occur in disorders of all the local cervical and shoulder structures as well as being referred from disorders affecting other structures which also have a cervical nerve-root innervation. This includes the serosal surfaces of the diaphragm and intra-thoracic structures. Factors to be elicited about the pain include its radiation, onset, diurnal or nocturnal variation, actions which precipitate or relieve it, its periodicity and whether it is associated with pain or other symptoms elsewhere. Many of the disorders causing pain in the neck and shoulders can be distinguished by a simple history alone. Most of the local cervical causes of such pain (Group A, Table 7.1) are made worse by movements of the neck whilst the local 'shoulder' disorders are made worse by movements of the shoulder and although they may produce pain which radiates down the arm this does not usually radiate into the neck.

Rotator cuff disorders of the shoulder often cause pain which radiates down to the deltoid insertion while such lesions are also often differentiated by a history of appearing during particular movements of the arm above the head.

Table 7.1. *Causes of neck and shoulder pain*

(A) **Local cervical disorders**
Cervical spondylosis
Acute disc protrusion
Cervical rib
Cervical involvement by generalized polyarthritis (RA, ankylosing spondylitis, Still's disease)
Acute wry neck
Traumatic and post-traumatic
Disc abscess
Tumours (vertebral bodies, spinal cord, nerve and nerve sheaths)

(B) **Local shoulder disorders**
Soft tissue, periarticular disorders (rotator cuff disorders, frozen shoulder)
Reflex sympathetic dystrophy
Arthritis of glenohumeral joints (polyarthritic process, osteoarthritis)
Arthritis of acromioclavicular joint
Fractures

(C) **Generalized disorders**
Generalized inflammatory disorders (RA, SLE, polymyalgia rheumatica, peripheral arthritis
  of ankylosing spondylitis)
Systemic infections (viral)
Neoplastic (metastases, multiple myeloma)
Endocrine (hypo- or hyperthyroidism)

(D) **Referred pain**
Peripheral nerve syndromes (carpal tunnel, ulnar neuritis)
Cardiovascular (myocardial ischaemia, dissecting aneurysm)
Pulmonary and pleural (basal lung infections, pulmonary infarct)
Abdominal (subphrenic abscess, perforated viscus)

(E) **Non-organic**
General fatigue
Psychiatric (depression, hysteria, personality disorders)

Radiation of pain below the elbow is not caused by rotator cuff lesions and is more likely to be symptomatic of a nerve-root disorder.

Knowledge of the periodicity of the pain is of help in that pain from most inflammatory or degenerative disorders tends to be intermittent whilst pain which is described as constant throughout day and night suggests either some form of bone neoplasm or that the pain aetiology is not wholly organic. Several of the soft tissue rheumatic disorders of the shoulder may be worse at night or when the patient lies on the affected side and shoulder pain which is bilateral and associated with marked early morning stiffness particularly if associated with thigh stiffness, suggests the possibility of polymyalgia rheumatica. A history of a recent fall should suggest the possibility of a fracture whilst a history of recent unaccustomed physical household activities such as decorating would help to explain the onset of a soft-tissue disorder. The presence of joint symptoms elsewhere suggests that the neck and shoulder symptoms may be part of some more generalized disease process (Group C, in Table 7.1). Features which indicate the possibility of serious underlying disease include fever, weight loss or suggestion of cervical cord involvement with symptoms such as weakness of arms or legs or recent loss of bladder or bowel sphincter control and would be reasons for early specialist referral. If history and subsequent examination findings are negative for the various local abnormalities shown in Table 7.1 under A and B then questions which might suggest a referred cause of pain (Group D) such as whether there was associated chest pain, abdominal symptoms, or paraesthesiae in the fingers, or dyspnoea should also be asked.

Lastly, particularly in patients with pain of obscure aetiology in whom the subsequent local and generalized examinations are negative, questions which might help to reveal an underlying depressive illness should also be posed. These will include questions about mood change, sleep disturbance, family problems, appetite, sex, memory, and concentration.

EXAMINATION

The physical examination extends the information available from the history. Simple examination of cervical spine and shoulder regions often defines the cause of symptoms. Neurological examination of the limbs is required in patients in whom a cervical disorder is suspected. Examination of other peripheral joints will help indicate any generalized articular process whilst full medical examination including respiratory, cardiovascular and abdominal examination is advisable in patients in whom a local cause of pain is not identified. Failure to find a local abnormality in neck or shoulder suggests that a disorder categorized under C, D. or E in Table 7.1 is present.

General observation should include note of factors such as whether the patient looks ill (suggesting malignancy or systemic disease process) or whether he is breathless or looks depressed (Fig. 7.1).

In order to examine the neck and shoulders the patient should be stripped to the waist. Inspection of the cervical spine may show tilting of the head and neck to one or other side, local swelling, or kyphosis. Careful palpation should be carried out to define areas of local tenderness and enlarged lymph nodes. The patient is asked to perform the movements of flexion, extension, left and right lateral flexion, and rotation and pain and limitation of the range of movements noted. This may be followed by careful passive movements assisted by the physician again noting any particular movements which produce pain.

The 'shoulder' is a complicated region and movements of the arm may be influenced by disorders of glenohumeral joint, acromioclavicular, and sternoclavicular joints and by disorders affecting the rotator cuff. Inspection of the region for muscle wasting, local swelling, and deformity is most important. This should be followed by palpation around the above joints for local tenderness and warmth, and to determine the nature of any swelling. Active movements of the arms should be examined commencing with abduction. Both arms are raised sideways to above the head with palms meeting, while from behind the physician fixes the inferior angle of the scapula and observes the degree to which scapular rotation and glenohumeral movement contribute to the full range of abduction. Normally the movement is accomplished in a

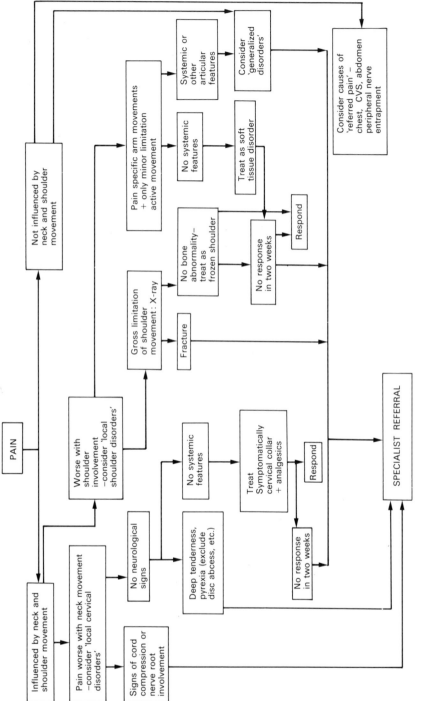

Fig. 7.1. Pain in the neck and shoulder: management guide.

smooth arc but if the glenohumeral joint is fixed abduction to as much as 60° may be produced by rotation of the scapula alone. The rotator cuff is formed by the tendons of supraspinatus, infraspinatus, teres minor, and subscapularis muscles in close apposition to the capsule of the glenohumeral joint. It is responsible for rotation of the humerus and via the supraspinatus component for initiation of abduction. The supraspinatus lies beneath the subacromical bursa and disorders of the bursa and of the tendon are clinically indistinguishable. Disorders of different parts of the rotator cuff may produce pain on particular active movements of the shoulder depending on the tendon predominantly involved, e.g. supraspinatus inflammation gives a painful arc on abduction between 60 and 100° as the clearance beneath the acromion for soft tissues reaches a minimum. Inability to initiate abduction other than by scapular rotation of a full range of abduction after passive assistance for 30° or so indicates rupture of the supraspinatus tendon. Active movements of internal and external rotation and flexion and extension are then observed. If any of the active movements are abnormal, passive movements should be carried out as many of the soft tissue peri-arthritic disorders of the shoulders limit active movement only whilst in true arthritis of the glenohumeral joint or in a frozen shoulder both active and passive movements are reduced (Fig. 7.2).

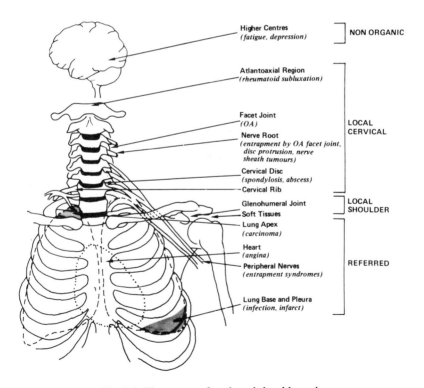

Fig. 7.2. The causes of neck and shoulder pain.

Neurological abnormalities found with cervical disorders most commonly consist of sensory and lower motor neuron signs in upper limbs and upper motor neuron signs and sometimes posterior column signs in lower limbs. Power of main muscle groups in both upper and lower limbs and muscle tone should be tested. Biceps (C5,6), triceps (C7,8) supinator (C7,8), knee (L3,4) and ankle (S1) jerks and plantar responses should be examined as should sensation of light touch, pin prick, vibration, and proprioception. In the case of pain radiating down the arm, the wrist and elbow should be examined for possible evidence of a carpal tunnel syndrome or of ulnar neuritis at the medial epicondyle.

Signs suggestive of subclavian artery compression as may occur with a cervical rib or other thoracic outlet syndrome should also be sought. Examination should include palpation of the radial pulses and observation of colour and warmth of fingers, hands, and forearms.

LOCAL CERVICAL CAUSES OF PAIN

CERVICAL SPONDYLOSIS

In primary generalized osteoarthritis, cervical spondylosis is associated with osteoarthritis of other joints which typically include the distal and proximal interphalangeal joints of the fingers, the carpometacarpal joints of the thumb, lumbar spine, hips, knees, and first metatarsophalangeal joints of the feet. Primary generalized osteoarthritis presents most commonly in women aged from 40–50 who often have a family history of the condition.

The symptoms of cervical spondylosis are pain or aching in the back of the neck and shoulders which may radiate down the arm or into the occipital region. There may be paraesthesiae in the fingers. Patients with cervical spondylosis may present with an acutely stiff neck which has been precipitated by trauma. On examination there is usually pain and restriction of movements of the cervical spine. There may be signs of nerve-root involvement in the arms. In a few patients with very severe disease, cervical cord compression may occur. Typical features include shooting pains in the upper limbs and pyramidal tract signs in lower limbs with weakness, spasticity, extensor plantar responses and bladder involvement.

The radiological signs of cervical spondylosis include loss of disc space with sclerosis and osteophytosis in the vertebral bodies and also in the facet joints. Some degree of degenerative change of the cervical spine is common in people over 40 year old and it is important not to attribute all pains in the neck to coincidental radiological abnormality. Features such as weight loss, general malaise, fever, or rapid progression of symptoms are a warning that cervical spondylosis may not be responsible for the patient's symptoms and the physician should seek some other cause such as an abscess. Nerve-root symptoms with paraesthesiae radiating down the arms due to cervical spondylosis should be differentiated from other causes such as acute disc protrusion, cervical ribs, and tumours compressing the nerve roots or brachial plexus. In addition peripheral nerve entrapment such as carpal tunnel syndrome and ulnar neuritis from trauma of the nerve at the medial epicondyle of the elbow may cause pain radiating up the arm to the shoulder and should be distinguishable by local examination. Sometimes nerve conduction tests may be helpful in the differential diagnosis of these syndromes.

In the first instance I usually immobilize the neck in a cervical collar. This will provide firm support and when combined with analgesics or low doses of non-steroidal anti-inflammatory drugs most acute stiff necks recover in a few days. Pain unrelieved by these measures may require more rigid immobilization of the cervical spine with a fitted plastazote collar. Manipulation is potentially hazardous in patients with degenerative changes of the cervical spine and should not be attempted. For more persistent symptoms a period of complete bed rest with only one pillow may relieve the problem. Some patients suffer severe pain and stiffness when they wake in the mornings. For them I advise sleeping on a firm supported bed with only one thick pillow and even wearing the collar during the night.

Most patients with pain caused by cervical spondylosis can be managed by their general practitioners. Persistent pain which is unrelieved by the above measures, the presence and advancement of neurological signs (either nerve root or the more serious signs of cervical cord compression) are indications for specialist referral. Patients with signs of cervical cord compression may require a neurosurgical opinion and in some cases decompression or spinal fusion may be necessary.

ACUTE DISC PROTRUSION

Acute disc protrusion may occur posteriorly when it bulges into the posterior longitudinal ligament or if larger may involve the spinal cord or may protrude posterolaterally to threaten the cervical nerve roots. C5,6 and C6,7 discs are most commonly involved. Disc protrusions in the cervical spine are less common than in the lumbar region and are often precipitated by sudden twisting movements or whiplash injuries of the neck. Patients present with the sudden onset of an extremely stiff painful neck. With nerve root involvement pain radiates down the arm to the fingers and the head is held stiffly to one side. Neck movements are restricted by pain and there may be nerve-root signs in the arms (e.g. loss of reflexes, muscles wasting, sensory loss). With posterior disc protrusions there may be pain alone or if massive there may be cervical cord compression with pyramidal tract and posterior column signs.

Diagnosis of a cervical disc protrusion is made on the basis of the history and X-rays of the cervical spine which may be normal or may show narrowing of the involved disc space. This loss of disc space may be related to previous disc disease rather than the recent episode. Treatment in the absence of cervical cord compression is with analgesics or anti-inflammatory drugs and firm support of the neck with a soft or plastazote collar. Complete bed rest may be required as before. If symptoms are not relieved by these simple measures, specialist referral is indicated and further treatment such as traction may be initiated. In most patients symptoms resolve in about eight weeks. This is thought to be the time taken for the prolapsed disc material to heal (by fibrosis). If there are signs of cervical cord involvement or of definite nerve root signs other than paraesthesiae at any stage, then hospital referral is also necessary. A myelogram may be required and some patients may come to surgery.

CERVICAL RIB (FIG. 7.1)

Cervical ribs are congenital bony or fibrous abnormalities formed at the level of C7 and are often asymptomatic. They may occasionally present as a cause of local swelling or pain. More usually when symptomatic they produce thoratic outlet syndromes with root signs and vascular symptoms in the arms caused by pressure on the brachial plexus and the subclavian artery. These symptoms are characteristically made worse by depressing the arm so increasing the pressure on the neurovascular structures. The features of this syndrome include pain and paraesthesiae radiating down the ulnar side of the arm to the finger, weakness, and muscle wasting of the arm and hand, and vascular symptoms in the hands including coldness, pallor, and Raynaud's phenomenon.

Cervical ribs causing the above symptoms have to be distinguished from other causes of nerve-root lesions (see cervical spondylosis). A lateral or oblique X-ray of the cervical spine may show up the abnormality. A similar problem may be caused by a fibrous band or by abnormalities of the scalene muscle, neither of which will show up on a straight X-ray. If a cervical-rib syndrome is suspected and the symptoms are severe, specialist referral is indicated with a view to surgical resection.

ACUTE WRY NECK SYNDROME

This describes the sudden onset of pain and stiffness in the necks of young adults which resolves spontaneously in about two weeks. No neurological symptoms or signs occur. The cause is not known but it is not due to acute disc protrusion. It seems likely that the syndrome may arise from a variety of causes including malalignment of the facet joints. X-rays of the cervical spine are normal apart from loss of cervical lordosis.

The treatment is similar to that of a minor acute disc protrusion. Manipulation has been used to treat some patients with this disc condition in the belief that the facet joints are being restored in their normal alignment. Such attempts should only be made by those experienced in this procedure and after exclusion of other causes of neck problems. An X-ray of the cervical spine is an essential preliminary to exclude other diagnoses. Failure of a so-called acute wry neck to settle within two weeks suggests more serious pathology and is an indication for hospital referral.

TRAUMATIC AND POST-TRAUMATIC DISORDERS

Discussion of the immediate treatment of acute spinal injuries is outside the scope of this chapter. The commonest form of trauma to cause persistent neck pain is a whiplash injury. This may damage muscles, ligaments, discs, and occasionally oesophagus, larynx, and sympathetic nerves. Immediately after a severe whiplash injury there may only be slight discomfort and severe symptoms are often delayed for 24 hours.

The investigation of neck pain following whiplash injury should include X-ray of cervical spine to exclude bone injury which if present would be an indication for immediate referral to an orthopaedic surgeon. Otherwise the initial treatment of symptoms is symptomatic with a firm cervical collar and analgesics or non-steroidal anti-inflammatory drugs. Severe neck pain made worse by extension of the neck and referred into the posterior scapular muscles across the shoulders is typical of laceration of the disc and anterior longitudinal ligament. If this is suspected or if pain from a whiplash injury persists for more than a few days despite symptomatic therapy then specialist referral is indicated. Compensation neurosis is a common cause of persistent pain after accidents. Sometimes the diagnosis is difficult as it is not possible to exclude organic damage in such patients.

GENERALIZED ARTHROPATHY

Cervical spinal involvement is common in generalized arthropathies such as rheumatoid arthritis, ankylosing spondylitis, or Still's disease. In rheumatoid arthritis abnormalities at both atlanto-axial and subaxial levels may occur. Twenty five per cent of hospital patients with rheumatoid arthritis are said to have anterior subluxation at the atlanto-axial joint. Severe neck pain occurs in a minority of patients only (Matthews 1974). Anterior atlanto-axial subluxation is best shown by a lateral X-ray of the cervical spine in flexion where the gap between the anterior arch of the atlas and the odontoid peg should not be more than 4 mm. Vertical subluxation is less common and associated with more severe local disease and on lateral X-ray the tip of the odontoid peg should not be higher than 3 mm above McGregor's line (drawn from the posterior edge of the hard palate to the lower margin of the occipital curve). Symptoms of cervical spine involvement in rheumatoid arthritis include pain in the head and neck which may radiate to the occiput and temple. Neurological involvement from cervical-cord compression may occur and is suggested by weakness in arms or legs which is out of proportion to the severity of the articular disease, shooting pains in the arms or by bladder or bowel disturbance.

In ankylosing spondylitis calcification and fusion of vertebral bodies and facet joints with consequent spinal ridigity may occur as elsewhere in the vertebral column.

Both the unstable rheumatoid cervical spine and the rigid brittle spine found in late ankylosing spondylitis are highly vulnerable to trauma and such patients should by advized regarding risk situations such as driving (head restraints) and getting into and out of the bath (hand rails to lessen risk of falling). Apart from the general management of arthritis treatment of cervical pain due to rheumatoid arthritis is with anti-inflammatory drugs and a soft collar or, if this fails, a firm plastazote collar. If there is any suspicion of neurological involvement such as those symptoms already discussed or signs such as brisk reflexes or extensor plantar responses then specialist referral is indicated and surgical stabilization of the spine may be necessary.

In patients with chronic rheumatic disorders there is a tendency to ascribe any pain which arises to the known arthritic process. If a sudden exacerbation of the neck pain of a patient with ankylosing spondylitis occurs it is essential to X-ray the cervical spine to exclude a recent fracture as there is a danger of cord compression. Specialist referral is indicated.

DISC ABSCESS

A disc abscess should be suspected if a child or young adult develops marked cervical pain associated with fever or other systemic upset. Both tuberculous and pyogenic abscesses may occur. It is important to diagnose and treat these as early as possible in order to prevent cervical-cord involvement. Patients presenting with a disc abscess usually complain of neck pain made worse by head movements in all directions. On examination the head is held stiffly with cervical muscle spasm and there is usually local bone tenderness. If the disease has progressed to destroy the disc and adjacent bone there may be a cervical kyphosis. The abscess may track along muscle planes in tuberculosis so that in addition to examining the limbs for neurological involvement it is important to look for abscesses or sinuses at the back of the neck and for evidence of septic foci or tuberculosis elsewhere in the body.

If a disc abscess is suspected urgent hospital referral is indicated. Lateral X-ray will show destruction of disc and adjacent bone. The ESR and white cell count are usually elevated. Treatment of will include appropriate antibiotic or antituberculosis therapy, immobilization of the spine, and sometimes surgical drainage and spinal fusion.

TUMOURS

Tumours including those arising from bone, meninges, nerves, and nerve sheaths may all cause pain in the neck and shoulder regions (Table 7.2). Both benign and malignant primary bone tumours may occur but metastases are

Table 7.2. *Neoplasms causing neck and shoulder pain*

Structure affected	
Bone	Primary (uncommon)
	metastases
Meninges	Meningioma
Extradural space	Lymphoma
Spinal cord	Glioma
Nerve and nerve sheath	Neurofibroma
Lung	Pancoast's syndrome
	Bone metastases

more common and will be discussed further on page 160. Carcinoma of the bronchus may also produce pain by direct extension from a carcinoma of the apex of the lung into the brachial plexus (Pancoast's tumour).

Neoplastic disease may be suspected as the cause if cervical pain is severe and unremitting during day and night or in the case of metastatic disease if other features such as weight loss or those of a primary neoplasm elsewhere are present. On examination there may be evidence of nerve-root involvement such as muscle wasting or weakness, sensory loss in a nerve-root distribution, or loss of reflexes. Pancoast's tumour may invade the sympathetic plexus in the neck and produce a unilateral Horner's syndrome. General examination for a primary tumour elsewhere should be carried out. Straight X-rays of cervical spines and humerus may show lytic areas indicating a bone tumour but often will be normal particularly with locally growing primary tumours such as neurofibromata. If metastases are suspected general investigations as discussed on page 167 are appropriate.

Patients who have features suggestive of a neoplasm should be referred to hospital. Although the prognosis of metastatic disease is poor certain slow-growing local tumours may be treated successfully by surgery whereas if left untreated produce progressive pain and neurological damage. Metastatic deposits may respond to radiotherapy, chemotherapy, or hormone treatment and sometimes surgery provides useful palliation.

LOCAL DISORDERS OF THE SHOULDER

The commonest causes of pain in this group are the soft-tissue rheumatic disorders.

SOFT-TISSUE RHEUMATIC DISORDERS

These are common, often self-limiting, and ill-understood disorders which affect the periarticular structure of the shoulder joint. They affect a high percentage of the otherwise normal population. A group of minor soft-tissue disorders include disorders of the rotator cuff and bicipital tendinitis. These cause pain on active movement of the arm only and usually do not require hospital referral. The more severe frozen shoulder causes limitation of both active and passive arm movements.

Supraspinatus tendinitis

This occurs in adults and may occur at any age from 30 years onwards. It is thought to be associated with degeneration and subsequent inflammation of the tendon associated with repeated stress as it is compressed between humeral head and the arch of acromion on abduction of the arm (Bland *et al.* 1977). Often symptoms are precipitated by sudden bursts of unaccustomed activity such as home decorating. A clinically indistinguishable syndrome is thought to be associated with subacromial bursitis whilst in other patients with similar symptoms supraspinatus calcification is present and crystal release may be responsible trigger inflammatory changes.

Supraspinatus tendinitis causes pain in the upper arm referred to the deltoid and associated with abduction of the arm. On examination there may be a painful arc of movement between 60 and 120° of abduction. There may be tenderness over the lateral aspect of the humeral head. When other parts of the rotator cuff are involved other specific movements of the arm will be painful.

X-rays of the shoulder are unnecessary in most individuals with mild rotator-cuff syndromes as they are usually normal. Sometimes sclerosis of the greater tuberosity is seen but it has no special significance. Calcification in the areas of the supraspinatus tendon may be significant. Although it is reported in 8 per cent of the normal population, one in three deposits are said to produce symptoms at some time (Bland *et al.* 1977). Calcification is often bilateral.

Most patients with symptoms of supraspinatus tendinitis can be managed by their general practitioners. With mild symptoms advice about avoiding the particular precipitating activity or acquiring new working habits is sufficient. More severe symptoms may require subacromial injection of local anaesthetic and steroid. Immobilization of the shoulder should be avoided as it may produce the more severe frozen shoulder syndrome.

Bicipital tendinitis

Tendinitis involving the long head of biceps tender may occur alone or in association with rotator-cuff disorders. It produces pain over the anterior aspect of the glenohumeral joint and is associated with tenderness in the bicipital groove. Flexion at the shoulder against resistance reproduces the pain. The onset of symptoms is often associated with overuse of the arm.

Treatment should include advice about avoiding overuse of the arm. Analgesics and anti-inflammatory drugs are helpful and local injection around the tendon sheath may be necessary for patients whose symptoms do not settle with rest. In view of the risk of spontaneous rupture of the long head of biceps tendon in association with inflammatory changes, common sense dictates that local steroid injections should be used sparingly and given cautiously, making certain that the injection is not into the tendon itself.

Frozen shoulder

This is thought to be associated with inflammation of the capsule of the glenohumeral joint and adjoining bursae. Patients complain of pain in the shoulder which is worse at night and which may radiate to the base of the neck and also down the arm into the hands. Arm movements in all directions are affected. Symptoms may start acutely and in some patients are associated with immobilization of the arm. Immobilization may be one reason why frozen shoulders occur following myocardial infarctions and cerebrovascular accidents. A frozen shoulder syndrome may also occur in and be a presenting feature of rheumatoid arthritis. Apparent bilateral frozen-shoulder symptoms in patients over the age of 60 years may be a presenting feature of polymyalgia rheumatica. Such patients should be asked about the presence or absence of other symptoms which might suggest this disorder (see p. 000).

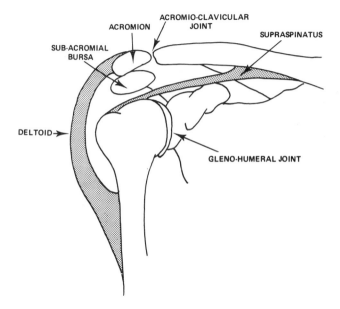

Fig. 7.3. Diagram of shoulder structure.

In patients with frozen-shoulder it is worthwhile arranging for X-ray of the shoulder and for blood count, ESR, and serological tests for rheumatoid factor. In a frozen shoulder X-rays are usually normal although there may be osteoporosis of the humeral head in severe cases. The ESR should be normal and if elevated implies that the symptoms are part of some generalized inflammatory process such as polymyalgia rheumatica or rheumatoid arthritis. There is a known association between frozen shoulder and diabetes so that it is worthwhile to test the urine for glycosuria.

Treatment of a frozen shoulder includes analgesics or anti-inflammatory drugs given symptomatically and graded mobilization exercises from a physiotherapist. Local injection of corticosteroids and local anaesthetic into areas of maximal tenderness may be helpful. Some established cases of frozen shoulders may be resistant to such simple measures and if this is the case or if there is a possibility that the frozen-shoulder syndrome is part of a more generalized disease process, then specialist referral is indicated. Occasionally manipulation may be indicated but this must be done with caution as there is a risk of tearing the capsule or fracturing bones. With adequate treatment and enthusiastic practice of mobilizing exercises, most frozen shoulders respond in six months.

Reflex sympathetic dystrophy (shoulder/hand syndrome)

This is a term given to the association of frozen-shoulder syndrome with pain in the hand and fingers and trophic skin and nail changes. It usually follows immobilization following illnesses such as myocardial infarction or cerebrovascular accidents. There is often swelling and tenderness in the hands and fingers. Although the precise mechanisms are obscure, both immobilization and a sympathetic mechanism may play a part in its pathogenesis. Treatment is with graded active mobilization exercises and anti-inflammatory drugs.

ARTHRITIS OF THE GLENOHUMERAL JOINT

The glenohumeral joint may be affected by a range of inflammatory and degenerative processes.

Inflammatory joint diseases

All the polyarthritides may affect the glenohumeral joint.

Osteoarthritis

Primary osteoarthritis of the glenohumeral joint is uncommon but is occasionally seen in the elderly and usually in association with long-standing degeneration of the rotator cuff.

Secondary osteoarthritis may occur following destruction of the articular surface by severe forms of inflammatory joint diseases. Less commonly it may follow loss of proprioceptive sensation in the neuropathic (Charcot) joint (e.g. diabetic neuropathy, tabes dorsalis, syringomyelia), metabolic disorders such as chondrocalcinosis and haemochromatosis, or avascular necrosis of the humeral head.

Septic arthritis

Septic arthritis may occur in the glenohumeral joint as part of a generalized septicaemia. It should be suspected if pain and stiffness in the shoulder occur in association with fever or signs of sepsis elsewhere. If sepsis is suspected immediate hospital referral is indicated where joint aspiration together with blood cultures can be carried out and appropriate measures adopted.

Established glenohumeral joint arthritis restricts both active and passive joint movements. There is often associated muscle wasting and local tenderness and sometimes an effusion detectable anteriorly over the head of the humerus. Crepitus may be felt if degenerative changes are present. X-ray of the joint is often helpful in documenting the degree of destructive change which has occurred and hence gaining an idea of the amount of improvement which may be expected from treatment.

Treatment of inflammatory arthritis of the glenohumeral joint is directed at the generalized inflammatory process with non-steroidal anti-inflammatory drugs or second line antirheumatic drugs, general supportive measures, patient education, and physiotherapy. Both intra-articular and peri-articular involvement of surrounding soft tissues and bursae may occur with rheumatoid arthritis and local steroid injections may help an acute exacerbation. Physiotherapy with a graded active exercise programme is used to maintain or regain a range of movement and combat muscle wasting. Hospital referral should be early rather than late in patients with rheumatoid involvement of the shoulder region as movement can be lost easily and muscle wasting may develop rapidly. In patients with severe destructive changes in the shoulder and secondary osteoarthritis the prognosis for functional improvement is poor. Replacement surgery of the glenohumeral joint has been tried in a few centres but no entirely satisfactory prosthesis is yet available and such surgery is still best regarded as at a developmental stage.

ARTHRITIS OF ACROMIOCLAVICULAR JOINT

Arthritis of the acromioclavicular joint may occur alone but often develops in many forms of generalized joint inflammation such as rheumatoid arthritis, with primary generalized osteoarthritis and osteoarthritis secondary to local trauma. Pain is felt over the joint and may also radiate to the base of the neck. Pain is worse on abduction of the arm past 100° and the joint is tender and may be swollen on palpation. Acromioclavicular arthritis responds well to local steroid injection. If the acromioclavicular pain is part of a generalized arthropathy then appropriate treatment with anti-inflammatory drugs is indicated.

FRACTURES

Most fractures of upper humerus or clavicle following trauma are easily recognized as the cause of pain in the shoulder region. However, when a pathological fracture occurs in the upper end of the humerus the diagnosis may not be obvious particularly if the patient is obese. If a fracture is suspected because of gross restriction of arm movement, X-ray of the region is essential and appropriate local measures can be instituted. If a pathological fracture has occurred orthopaedic referral is indicated for local treatment and investigation.

GENERALIZED DISORDERS

Neck and shoulder pain may occur as a result of a wide range of generalized disease processes (Table 7.1).

GENERALIZED INFLAMMATORY DISORDERS

Bilateral shoulder pain may be due to systemic disease processes such as rheumatoid arthritis, polymyalgia rheumatica, and less commonly connective tissue disorders such as systemic lupus erythematosus (SLE). When shoulder pains occur as part of an established disease the diagnosis is usually obvious but may be more difficult if they are presenting features of the disease. Suspicion arises when any of the following occur: sudden onset of bilateral 'rotator-cuff symptoms' in the elderly, association of symptoms with thigh stiffness, marked early morning stiffness, association with joint symptoms elsewhere, presence of fever, malaise, tiredness, or rash. Referral to a rheumatologist is indicated if there is any indication of a generalized inflammatory process or if screening tests such as the ESR are abnormal.

Polymyalgia rheumatica

This is an ill-understood inflammatory disorder which usually affects elderly people although occasionally it occurs in people in their 40s. Patients typically present with bilateral shoulder girdle and thigh pains which may start suddenly and are associated with early morning stiffness. On examination there may be limitation of active shoulder movements and in a minority of patients swelling of one or two peripheral joints occur. In up to 55 per cent of patients polymyalgia rheumatica may be associated with the more serious, giant-cell arteritis and up to 40 per cent of patients who present with polymyalgia rheumatica without clinical evidence of giant cell arteritis develop the latter condition on subsequent follow-up (Jones and Hazelman 1981). Polymyalgia rheumatica and giant-cell arteritis may be different manifestations of the same condition. Arteries which may be involved by giant-cell arteritis include the temporal arteries, intracranial vessels including the opthalmic arteries, and other vessels including the coronary arteries. Clinical features of giant-cell arteritis besides polymyalgia rheumatica include headache and scalp tenderness, pain in the temple on chewing, blurring and loss of vision of one or both eyes, myocardial ischaemia, cerebrovascular accidents, fever, weight loss, and anaemia. On examination there may be a tender, thickened temporal artery or other scalp vessel (Fig. 7.1) and if the retinal artery has been involved evidence of visual disturbance and retinal oedema or optic atrophy.

A diagnosis of polymyalgia rheumatica is made on the basis of the clinical features in association with an elevated ESR and the absence of other rheumatological causes of the symptoms such as rheumatoid arthritis. The differential diagnosis between polymyalgia rheumatica and rheumatoid arthritis in the elderly is not always easy and some patients are seen who have been treated for polymyalgia rheumatica for several months or even years before developing definite features of rheumatoid arthritis. The difficulty is increased by the occurrence in some polymyalgia patients of synovitis in the knee or other joints.

Specialist referral is appropriate for most patients in whom a diagnosis of polymyalgia rheumatica is entertained in order to avoid mis-diagnosing conditions in which corticosteroid therapy would be inappropriate (such as rheumatoid arthritis) and to recognize patients with clinical evidence of giant-cell arteritis in whom a higher dosage corticosteroid regime is necessary.

Patients with polymyalgia rheumatica who have no features of giant-cell arteritis should be treated with 10–20 mg of prednisolone or equivalent daily. Most patients with polymyalgia show a dramatic response to treatment within 48 hours. When the symptoms remit and the ESR falls, steroid therapy is gradually reduced over several months. The dosage is titrated to provide the minimum dose to control the symptoms and the ESR and usually most continue for several years.

Patients with visual symptoms from giant-cell arteritis are medical emergencies in whom immediate treatment with corticosteroids (40–60 mg prednisolone daily) is essential in order to decrease the risk of blindness. The diagnosis of giant-cell arteritis later can be confirmed in hospital by biopsy of an involved temporal artery. Treatment should not be delayed even for one day whilst awaiting the biopsy procedure as retinal artery occlusion will produce permanent blindness and may occur during the intervening period.

Rheumatoid arthritis

Shoulder symptoms may be a presenting feature of rheumatoid arthritis particularly in the elderly. More commonly they occur during the course of the disease and they are less liable to cause diagnostic difficulties. The presence of other articular pains or of peripheral joint swelling should be looked for in patients with shoulder symptoms. A negative serological test for rheumatoid factor early in the course of the disease does not exclude the diagnosis.

Treatment of the shoulder symptoms of rheumatoid arthritis include treatment of the general disease process with anti-inflammatory drugs or second line antirheumatic drugs as indicated, and local measures as previously discussed.

SYSTEMIC INFECTIONS

Neck and shoulder pains may arise as part of a generalized myalgia in a range of febrile, viral infections such as influenza. A stiff painful neck may also arise as part of bacterial and viral meningitides. The presence of fever and severe headaches and sometimes other persistent systemic features will usually suggest such a possibility. If meningitis is suspected medical referral is indicated for futher assessment including blood and CSF cultures.

NEOPLASTIC

Bone metastases arising from neoplasms in a variety of sites of which breast, prostate, lung, kidney, thyroid, and bowel are the commonest, may give rise to persistent bone pain. A metastatic deposit may present as a pathological fracture and should be suspected if pain is persistent throughout day or night or is associated with weight loss. X-ray of painful areas should be carried out. A normal X-ray does not totally exclude the presence of bone metastases and occasionally a bone scan will show deposits where X-rays have been normal. If a diagnosis of bone metastases is entertained, specialist referral in indicated to help in the diagnosis and further management. Further investigations which may be helpful include bloodcount and ESR, serum calcium, inorganic phosphate, alkaline phosphatase, acid phosphatase, protein electrophoresis, and examination of urine for Bence Jones protein and for k and λ-light chains (multiple myeloma). A definite diagnosis is of help in subsequent management of metastatic disease.

ENDOCRINE

A variety of endocrinological disorders including hypo- and hyperthyroidism and hypo-adrenalinism may cause musculoskeletal symptoms. Of these hypo-thyroidism is the commonest to cause neck and shoulder pains. Practitioners should look out for features suggestive of hypothyroidism such as weight gain, loss of energy, intolerance of cold, bradycardia, hair loss, sluggish tendon reflexes, and dry skin. Thyroid function tests (serum thyroxine and serum TSH) can be estimated in most routine laboratories. Such disorders are satisfying to diagnose as treatment is straightforward with replacement thyroxine.

REFERRED PAIN

Pain may be referred to the neck and shoulders from a variety of thoracic and abdominal structures and from peripheral nerve entrapment syndromes in the arms (Fig. 7.1).

PERIPHERAL NERVE ENTRAPMENT

Entrapment of the median nerve at the wrist where it lies beneath the transverse ligament (carpal tunnel syndrome) usually presents with pain and paraesthesiae in hand and fingers which are worse at night and may be associated with pain radiating into the upper arm and even into the side of the neck. The condition is easily confused with the various causes of nerve-root compression. On examination there may be tenderness of the median nerve at the wrist to percussion (Tinel's sign) and loss of sensation to light touch and pin prick in a median nerve distribution in hand and fingers. There may be wasting of the thenar eminence. Classically the little finger is spared but is affected if the superficial branch of the ulnar nerve at the wrist is also involved. The diagnosis of carpal tunnel syndrome can usually be made clinically. Occasionally it may be difficult to differentiate from other entrapment syndromes and from cervical causes of nerve-root symptoms in the upper limb. Nerve conduction tests are commonly used to demonstrate the presence and site of nerve compression.

A carpal tunnel syndrome may be secondary to pregnancy or disorders such as rheumatoid arthritis, any cause of fluid retention or hypothyroidism, or may be idiopathic. Treatment includes that of any primary disease state together with resting the wrist in a splint in a neutral position at night. If this fails carpal tunnel injection of a small volume of corticosteroid (less than 0.5 ml) may relieve symptoms, although relapse is common. Failure to respond to these measures or the presence of definite muscle wasting or sensory loss are indications for referral for surgical decompression of the median nerve and synovectomy of the wrist when appropriate in patients with rheumatoid disease.

CARDIOVASCULAR

Myocardial ischaemia

A history of pain radiating down the arm or into the neck which develops on exertion, is relieved by rest and usually but not always is associated with pain and tightness across the chest should suggest angina. Cardiovascular findings may be completely normal and a normal electrocardiograph does not exclude the diagnosis.

A diagnosis of myocardial infarction is suggested by more persistent pain which is not relieved by rest and which may be associated with other features such as breathlessness and sweating and signs of shock. An electrocardiograph usually reveals the typical ST wave changes although these may take some hours to develop.

Dissecting aneurysm

This is an uncommon cause of pain radiating into the neck and back. Pain is usually severe and of sudden onset and may be suspected if it proceeds down the back or into the abdomen or is associated with loss of radial pulses. Immediate hospital referral is indicated if the diagnosis is suspected.

PULMONARY AND PLEURAL

Basal lung infections and infarcts may cause pain felt in the shoulder as it shares a common nerve supply with the diaphragm. Respiratory system examination and chest X-ray should be included in the assessment of all patients with obscure shoulder pain particularly if features such as cough, expectoration of purulent sputum, and breathlessness are present.

ABDOMINAL

A subphrenic abscess may produce pain referred to the shoulder via irritation of the diaphram. The condition may be suspected in patients who have recently had abdominal surgery or who have other form of intra-abdominal sepsis such as cholecystitis or diverticulitis. Associated features include swinging pyrexia, malaise, and sometimes weight loss. If suspected hospital referral is indicated where further investigation will include fluoroscopic screening of diaphragmatic movements.

NON-ORGANIC

GENERAL FATIGUE

General fatigue and lack of sleep may result in a range of non-specific muscle aches and pains involving structures in neck, shoulders, and often lower back in particular. The term fibrositis is often used loosely for this type of non-specific muscle pain. The relationship between fatigue and soft-tissue pains has been studied experimentally. EEG examinations of patients with 'fibrositis' during sleep show a disturbance of the sleep pattern. Experimental deprivation of the non-rapid eye movement associated sleep component in normal subjects, lowers pain threshold (Smythe and Moldofsky 1978). This relationship between sleep deprivation and soft tissue rheumatism may be germane to the non-specific pains found in depressed patients (see below).

The presence of persistent neck and shoulder pains without obvious cause requires full examination and investigation to exclude particularly the generalized disorders described in Chapter 4. This will include local X-rays, blood screen and ESR, serum proteins and electrophoresis and serum calcium, inorganic phosphate and alkaline phosphatase estimation. A bone scan may be necessary to exclude bone metastases. Specialist referral is advisable if any of the screening blood tests is abnormal or if there remains doubt about the diagnosis.

PSYCHIATRIC

Depression

Non-specific neck and shoulder pain may be a presenting feature of a depressive illness in which the other features only become apparent on further questioning. Other features include mood disturbance, abnormality of sleep pattern, difficulty in getting to sleep, early morning wakening, constipation, loss of appetite, loss of libido, flat mood, and poverty of thought and expression. Sometimes it may be difficult to distinguish an endogenous depression from a mood disturbance occurring as a consequence of chronic pain. In the latter situation a reactive depression may exacerbate symptoms and treatment may be required for the mood disturbance as well as for the primary painful condition.

In patients with a primary depressive illness full medical assessment with screening blood tests including ESR should be carried out to exclude organic disease. A definite diagnosis of depression should be treated accordingly.

Hysteria

In this disorder the patient gains a clear advantage by his illness. This should be a positive diagnosis based on psychological assessment after exclusion of organic disease.

Personality disorders

These include hypochondriacs and cranks who appear to 'enjoy' a variety of recurrent non-specific aches and pains without apparent organic cause. After exclusion of organic disease such patients should be handled firmly and sympathetically and expensive and potentially toxic drugs withheld.

REFERENCES

Bland, J. H. *et al.* (1977). The painful shoulder. *Semin. Arthritis Rheum.* **1**, 21–47.

Jones, I. G. and Hazelman, R. L. (1981). Prognosis and management of polymyalgia rheumatica. *Ann. rheum. Dis.* **40**, 1–5.

Mathews, J. A. (1974). Atlanto-axial subluxation in rheumatoid arthritis. A 5 year follow-up. *Ann. rheum. Dis.* **33**, 526–31.

Smythe, H. A. and Moldofsky, H. (1978). Two contributions to understanding of the 'fibrositis' syndrome. *Bull. rheum. Dis.* **28**, 928–31.

8 Disorders of hand function

Jonathan Noble

INTRODUCTION

With his hand man may earn, caress, or condemn. The agent of art and labour is accordingly apportioned a large area of the cerebral cortex. Hand surgery is one of the most rapidly and solidly expanding sub-specialities of surgical practice today. But what in a busy practitioner's surgery has the hand surgeon to offer in coping with locomotor disability? Let us start with the practitioner's assessment of his or her patient's problem. The scope of hand problems can be approximately classified as follows:

Injury
Arthritis
Congenital abnormalities
Swellings
Inflammatory conditions
Contractures
Neurological disorders

Most patients with a hand problem will present with one or more of the following symptoms:

Pain
Paraesthesia
Weakness
Clumsiness
Stiffness
Contracture
Swelling
Inflammation
Injury

This oversimplication at least gives us an outline of a useful basis for examination of the patient.

EXAMINATION AND ASSESSMENT OF THE HAND

There can be no better example than the hand of a presenting part, the origin of whose problem lies elsewhere. Obviously it is impossible in general practice to examine the peripheral pulses, neck, chest wall, breasts, etc., of each and every hand problem but a mental check-list of lung, heart, breasts, neck, thoracic outlet, shoulder, elbow, radio-ulnar joints, and wrists is a wise safeguard in all but the most obvious problems. This author has within the last six months seen cervical spondylosis, cervical rib, cervical metastases, lung tumour, angina, hypocalcaemia, Raynaud's disease, frozen shoulder, and an old elbow fracture referred as hand problems (as indeed secondarily they were). Release of the collar, two or three leading clinical questions (chest pain, weight loss, other joints affected, etc.) the sleeve rolled up and a quick assessment of the range of painfree movement from the neck downwards takes but two or three minutes. The competent clinician may then decide whether he needs to look further, order tests, or merely prescribe reassurance (the most undervalued medicine today, provided it is used correctly and appropriately).

But how does one examine the complexities of the hand itself, after these essential preliminaries? The following is a quick, reasonably safe routine. It is not fully comprehensive, but it will pick out most conditions seen in general practice.

POSTURE AND ATTITUDE

The way the patient holds his hand on entering or leaving a room can be revealing. The homemade sling (Grandma's shawl usually) may have induced stiff, swollen fingers already. The 'unbearable pain' preventing a return to work may be accompanied by a bizarre posture, not consistent with the grime, bearing testimony to a weekend spent gardening or under the car. A 'spectacle sign' is a useful clue 'I haven't been able to do a thing for three months doctor' is hardly compatible with the telltale specking of home decorating, adorning the complainant's spectacle frame. I have seen dramatic improvements in manual dexterity as the all-important sick note is safely stored in its recipient's wallet or purse. The hand in which a bag, motor' bike helmet, toddler, or even newspaper is held are all useful indicators of function. Most is to be learned from the way the coat is removed, the shirt or blouse buttoned, or the tie tied. Difficulties in pinch grip and sensory perception, whether they are due to joint stiffness, deformity, pain or sensory deficit soon become apparent.

SKIN QUALITY

The skin too tells a thousand tales. The cold skin of vascular problems, especially Raynaud's, the dry skin of denervation (often accompanied by an increased pinkness) and the thinned skin of rheumatoid and/or steroids are all familiar. The absence of a normal sweat pattern is probably the most reliable sign of a nerve lesion. It should be sought early in the examination before the clinician's further manipulations smear the pattern. An invaluable adjunct is the 'fountain-pen cap sign' (alternatively the 'tea-spoon sign'). A smooth metal surface will run easily on dry skin, but friction is encountered as it 'binds' on a moist surface.

The same healthy scepticism must be applied to the skin as in the previous section. Not only is the honest (sometimes dishonest) grime of toil recognized, but so too are associated skin thickness and callosity.

Scars, scabs, and the features of skin disease must be apprehended. Always right and left should be compared and as experience and confidence in assessing the hand increase, so too does the time spent simply looking at the hand. Indeed in my practice 95 per cent of the diagnoses may be easily made on a history and inspection alone.

SWELLING

A generalized swelling of the hand must be differentiated from discrete swellings or lumps.

Oedema

Oedema of the hand is usually manifest predominantly on the dorsum. However, when it involves the volar aspect, as revealed by a loss of concavity of that surface, then the problem has become a potentially grave one.

Infection

Infection in the hand always presents with oedematous swelling, which tends to follow the anatomical compartments. Thus each digital tendon sheath, interdigital web space, should be inspected as well as the thenar and hypothenar eminences, along with triangular palmar space between them.

Discrete swellings

Discrete swellings about the hand and wrist are numerous. Most will be *ganglia*, usually on the dorsum, or at the mouths of fibrous flexor sheaths, at the level

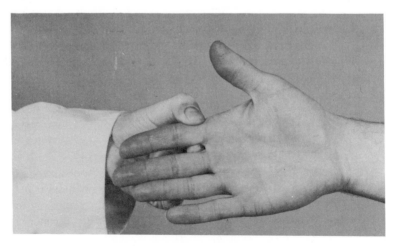

Fig. 8.1. The pinch test for flexor tendon synovitis.

of the distal palmar skin crease, or they will prove to be early harbingers of rheumatoid disease. Most typical in that respect are swelling or swellings of the metacarpophalangeal joints (MPs), or of the proximal interphalangeal joints (PIPs). Also common in rheumatoid are swelling, or thickening, following the line of the dorsal, extensor tendons, which so frequently may rupture in that disease. Flexor tendons too may rupture in the fibrous-flexor sheaths of the fingers, owing to the pressure of synovitis. This can be palpated at the mouths of these sheaths, at the level of the distal, palmar skin crease. It may also be felt by means of the *pinch test*, with which the examiner's finger and thumb approach the patient's finger from behind and roll the skin surfaces together. In the presence of synovitis there is a rubbery sensation between those two skin surfaces (Fig. 8.1).

NAILS

As every good haematologist knows the finger nails may reveal signs pathognomic of many conditions outwith the hand surgeon's area of expertise. Moreover examination of these structures will often betray much about the patient's psyche and the extent and cleanliness of his or her work.

VASCULATURE

We have already noted the hand's colour and temperature, but the radial and ulnar arteries must always be examined and the capillary bed blood flow assessed in the pulp and nail-beds of the fingers.

INNERVATION (FIG. 8.2)

Much may have been learned already by examining the sweat pattern of the skin. Many find it difficult to remember the nerve supply of the hand, but it is quite straightforward. *All those muscles which flex the wrist and/or fingers are supplied by the median nerve, except the ulnar carpal flexor (FCU) and the ulnar half of the flexor digitorum profundus (FDP)* and thus the flexors of the distal inter-phalangeal joints of ring and little finger, which are supplied by the ulnar nerve. As these muscles (FCU and ulnar FDP) are supplied well above the wrist a case of ulnar nerve pathology, whose presentation includes their weakness, is likely to arise at the elbow.

Thus the median is essentially the nerve of the flexor aspect of the forearm and similarly the ulnar nerve is the supply to the small muscles of the hand; that is the intrinsics (interossei and lumbricals), and the hypothenar group. However, the median nerve contributes by almost always supplying the thenar eminence – that is the opponens, the short flexor and the short abductor, of which the last is invariable. Observing abduction of the thumb, away from the palm and concurrently feeling for contraction of that muscle in the thenar

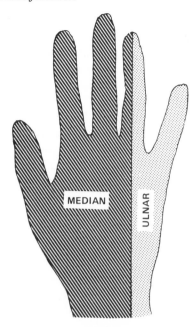

Fig. 8.2. The median and ulnar nerve distributions in the palmar aspect of the hand.

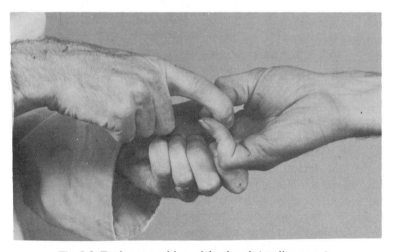

Fig. 8.3. Testing opposition of the thumb (median nerve).

group, is a more reliable test of median nerve integrity than the better known observation of opposition of thumb (circumduction) to the little finger (Fig. 8.3). Abduction of the index finger against resistance, accompanied by palpation of the belly of the first dorsal interosseous in the first web space is the simple 'acid test' for integrity of the ulnar nerve (Fig. 8.4).

The muscles whose tendons extend the wrist and MP joints, as well as both joints of the thumb are all innervated by the *radial nerve*, which has no significant distribution within the hand itself. An ability to extend the wrist or thumb tip against resistance is a good routine test for the radial nerve (Fig. 8.5).

The sensory contribution of the radial nerve is small, being confined largely to the skin overlying the anatomical snuff-box. If a line is drawn along the middle of the volar aspect of the ring finger to the level of the wrist then this classically separates the median territory on the radial aspect from the ulnar, to the ulnar side of that line (Fig. 8.2). There are considerable variations in this division, however. It is well to remember the dorsal branch of the ulnar nerve, which supplies a somewhat variable area on the ulnar aspect of the dorsum of the hand. This branch arises above the wrist and if its area is hypoaesthetic then this too may betray an ulnar nerve lesion at the elbow.

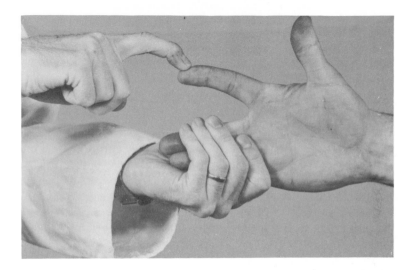

Fig. 8.4. Testing abduction of the index finger (ulnar nerve).

There are many pitfalls in examining neurology in the hand but there are perhaps two which are especially compelling.

1. The freshly lacerated nerve may on occasions not at first produce all (or even some) of those signs which the foregoing suggests it should. There is no obvious explanation for this peculiarity, which behoves the examiner to assess the potential severity of lacerations by their depth and by their situation. Thus all cases with deep cuts, however short, about the palm of the hand, or the volar aspects of wrist or fingers need careful e*X*tension, e*X*ploration, and e*X*cision of the wound (the three Xs).

2. The other word of caution relates to the myotomes and dermatomes (Fig. 8.6). This text is no place to recount the outflow of the cervical nerve roots.

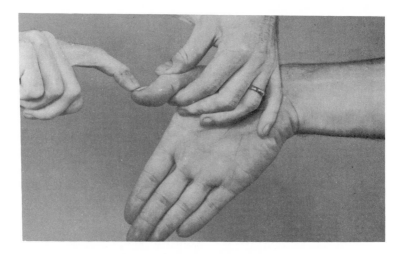

Fig. 8.5. Testing extension of the thumb (radial nerve).

Suffice it to say, for example that a C8/T1 lesion can have similarities to an ulnar nerve lesion and thus meticulous examination is essential. An area of hypoaesthesia in the ulnar nerve distribution which extends for several centimetres proximal to the wrist is now encroaching on the territory of the medial cutaneous nerve of the forearm. One lesion is always more probable than two and thus a T1 lesion is the greater likelihood, redirecting our attention to the neck, rather than wrist or elbow. Neurological deficits which do not fit into either pattern are liable to be peripheral neuropathy. Having a dermatome chart hanging in the surgery is a wise help. After all it is likely that 30–35 per cent of all your patients will have musculoskeletal problems.

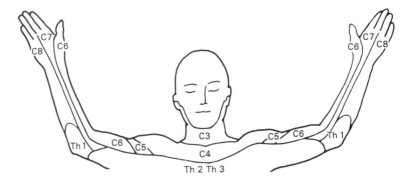

Fig. 8.6. The sensory distribution of the cervical and upper thoracic nerve roots.

TENDONS

Here too is a frequent ill-examined area. 'Make a fist, now straighten your fingers.' The patient obeys, there is no grossly obvious deficit. So all must be well. Such is the basis of many cases ultimately finding their way into medicolegal practice. Auscultation of the right pulmonary apex tells us little about the left lower lobe. Each pulmonary area is examined in turn, and so it should be with the tendons of the hand.

As the patient flexes his wrist the radial carpal flexor (*FCR*) (Fig. 8.7a) and the flexor carpi ulnaris (*FCU*) (Fig. 8.7b) can be palpated in turn (Fig. 8.8). By

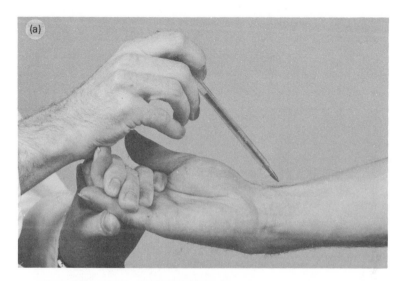

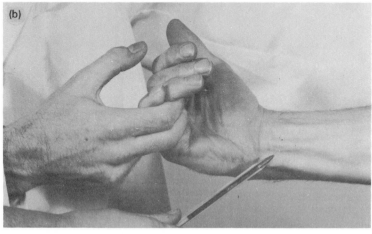

Fig. 8.7. Testing the flexor carpi radialis (a) and flexor carpi ulnaris (b).

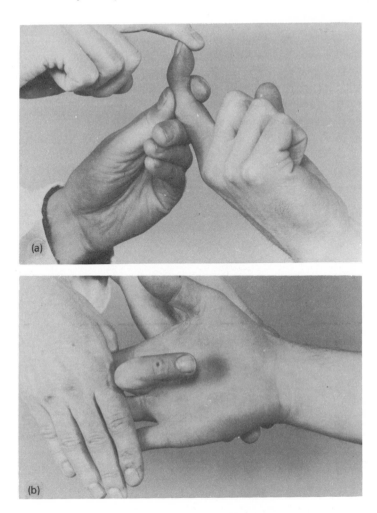

Fig. 8.8. Testing flexor digitorum profundus and flexor digitorum profundus and flexor digitorum superficialis.

holding the proximal phalanx of the thumb, flexion at the interphalangeal joint, due to *flexor pollicis longus* (FPL) may be assessed. Similarly by holding each intermediate phalanx of each finger in turn the integrity of the deep flexor (*Flexor digitorum profundus, FDP*) can be assessed (Fig. 8.8). The *flexor digitorum superficialis, (FDS)* is more difficult to understand. Its action can only be tested at the proximal interphalangeal, when the profundus action at the distal joint is abolished, by placing it on the stretch. Thus all the fingers, except the one being tested, are held fully extended and the patient is invited to flex the tested finger (Fig. 8.8b). If active flexion is observed at the proximal

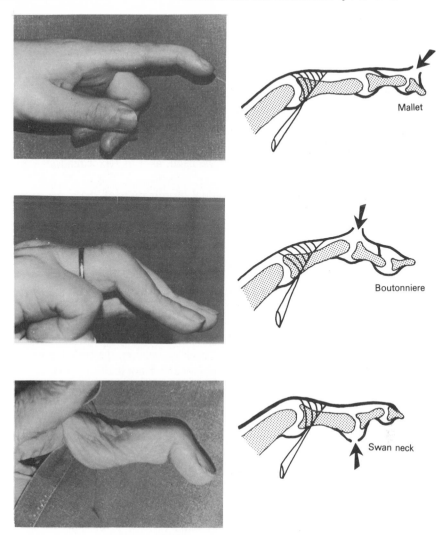

Fig. 8.9. (a) The mallet finger with rupture of the extensor tendon over the DIP joint. (b) The boutonniere deformity with rupture of the extensor tendon over the PIP joint. (c) The 'swan neck' finger with rupture of the volar plate over the PIP joint.

interphalangeal joint then FDS is in tact. Beware of flexion at the MP joint, this can be actively achieved by the intrinsics alone.

Extension is easier to assess. The key here is to examine each joint, starting with the wrist and going through each MP joint in turn, before ending with the interphalangeal joints of the thumb and fingers. It is important to examine these joints separately so that *boutonniere deformity* (Fig. 8.9b) of the proximal interphalangeal joint or *mallet deformity* of the distal joint is not overlooked (Fig. 8.9a).

It is always important to check the integrity of both flexor and extensor tendons in rheumatoid disease. Tenosynovitis always follows the line of these tendons. That a tendon has ruptured in one hand means that if tenosynovitis affects the corresponding tendon on the opposite side then rupture is highly likely.

It is important to realize that adherence or friction between firmly adducted digits may enable, for example, the ring and index finger to bring a tendon-less long finger fully down into the palm of the hand, as if its flexor tendons were in tact.

It is surprising how little time individual examination of these structures actually takes.

JOINTS

The principle here is much as it was with the tendons, *each joint must be examined in turn.* Firstly, however, one can learn much by briefly observing the overall hand and wrist flat, from both dorsal aspect and ventral aspects. Then it is useful to observe the contour of the MP joints with the fist clenched. These three observations will almost certainly reveal any deformity or swelling, such as that due to synovitis. Filling in of the 'valleys' between the 'mountains' of the metacarpal heads in the clenched fist is very important in this respect.

Another vital clinical point in assessing bones or joints in the hand is to seek any evidence of rotational deformity of the fingers by looking at their nails. In the absence of deformity the nails will all face the ceiling, when the bone is rotated they become oblique to it. Even mild rotational deformities become obvious in this way, once the fingers are flexed.

An important, but often overlooked, joint in the hand is the *carpometacarpal joint of the thumb*, which may be palpated at its base and a little distal to the wrist. This is quite a frequent site of pain, swelling, and tenderness due to either rheumatoid, or most typically osteo-arthrosis. The surfaces of this joint may be disrupted by a *Bennet's fracture* sometimes mistaken, like scaphoid fracture, for a Colle's fracture.

The metacarpophalangeal joint of the thumb is most typically injured in a fall on the outstretched hand, with the thumb in valgus. In such cases swelling tenderness and bruising may be elicited on the ulnar aspect of the joint. This is usually due to rupture of the sizeable ulnar collateral ligament, as a result of which the proximal phalanx can be pulled widely into valgus, whilst stabilizing the patient's metacarpal head and neck, between ones own index finger and thumb (Fig. 8.10).

The Z shaped thumb (Fig. 8.11) may be seen most typically in rheumatoid. It can take one of two forms. In one pain at the basal carpometacarpal joint results in an adduction deformity of that joint. This brings the thumb into the

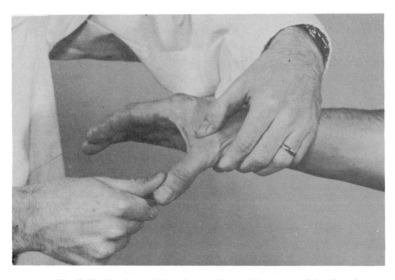

Fig. 8.10. Rupture of the ulnar collateral ligament of the thumb.

palm and to widen the decreased span between thumb and index finger tips the MP joint of the thumb hyperextends, often potentiated by synovitic damage to the volar plate (i.e. ligament) of that joint. Finally the thumb yet again tries to realign itself by hyperflexing at the interphalangeal joint. This may be secondary to (or be a cause of) rupture of the extensor pollicis longus tendon. The alternative pattern is that of hyperextension of the basal joint, flexion deformity of the MP joint, and a hyperextended interphalangeal joint, frequently associated with a ruptured flexor pollicis longus.

The best known *finger deformities* may be seen in the rheumatoid hand. If the central slip of the extensor mechanism, on the dorsum of the PIP joint is

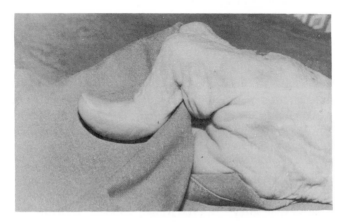

Fig. 8.11. The 'Z' shaped thumb of rheumatoid arthritis.

cut, ruptures or becomes attenuated then the proximal phalangeal condyles button-hole through, causing this joint to become flexed. The side slips of the extensor tendon, which extend the distal joint cause that joint to become hyperextended; the so-called *boutonniere deformity* (Fig. 8.9b). The exact reverse of this may be caused by the combination of rupture of the volar plate of the PIP joint, with tightness of the intrinsics. This causes a hyperextension of the PIP joint with flexion of the terminal joint, known as *swan-neck* (Fig. 8.9c) *deformity*. The flexion deformity of the distal joint alone, commonly seen after injury to that joint is a *mallet finger* (Fig. 8.9a) which in itself can lead to secondary swan-neck deformity, in that entire power of the extensor mechanism is centred upon the PIP joint. It is important when examining these features to ascertain whether they are still passively correctible, or if they are now fixed deformities.

CONCLUSION

The practitioner unfamiliar with the foregoing may be overawed by it and understandably so. It is nevertheless a relatively simple sequence of observations, for which no special equipment is required. The individual may well feel himself more inclined towards perinatal work, social psychiatry, or geriatric care in his community. However, to ignore locomotor disability is to ignore an estimated one-third of general practice. Within that third afflictions of the hand are the commonest of all, reflecting its great vulnerability and practicability. The hand which has been cut, crushed, whose nerve supply is compressed, or which is being ravaged by rheumatoid is the province of us all. It is opportune now to consider some specific groups of disorders and to indicate how they may be treated.

CLINICAL PATHOLOGY IN THE HAND

INFECTION

There is probably no better example than this of where the general practitioner's role has been of paramount importance. The incidence of hand infection as a surgical problem has declined over recent years. This can only be due to the prompt administration of antibiotics outwith the hospital service. This must be tempered with caution. Most hand infections, whether a paronychia, or one involving an anatomical space, will declare themselves as responsive or refractory to treatment within 24–36 hours of starting antibiotics. *Such treatment MUST be reviewed within 48 hours.* The obvious presence of pus, the persistance of erythema, the advent of localized, let alone generalized swelling, the signs of proximal spread (lymphangitis, lymphadenopathy), a systemic reaction (e.g. a fever, however mild), or evidence that flexor tendons within their sheaths are under pressure are signs that all is not well and that it is time for the patient to be admitted to hospital. Here repeated compression bandaging, early immobilization with elevation and then supervized mobilization may be all that is necessary, although surgical exploration and drainage will be urgently needed in some cases, such as a flexor sheath infection, which if neglected may readily result in amputation, as several well-known surgeons have learned to their own cost and loss.

INJURY

Swelling

Probably more important than any individual lesion to bone, joint, tendon, or even nerve is the problem of swelling. Swelling is oedema. The fluid of oedema becomes invaded by fibroblasts and a rapidly stiffening, sticky mess has been made, as if glue had been poured into the 'works' of a clock.

Our approach to this problem is much as it is with the infected hand. Diuretics have no significant part to play in treating oedema of the hand. *The cardinal sign that the problem is potentially grave is when the concavity of the palm of the hand has been lost.* A patient with oedema of the dorsum of the hand and inability to fully close the fist is set upon the same road. In such cases hospital referral is probably the wisest step in urban or semirural practice. However, this may be very difficult for the 'Tannochbrae' type of practice. *Now the general practitioner must learn to apply a 'boxing glove' compression bandage in the safe position of the hand.*

The ligaments of the small joints of the hand and fingers are at their maximum length and thus their least vulnerable (in terms of developing secondary flexion contractures), when the MP joints are flexed to 90° and the interphalangeal joints are in full extension. This position is most easily achieved with the wrist dorsiflexed (Fig. 8.12a). Now gauze swabs are layed between the fingers and a large resilient pad of either fluffed gauze (Fig. 8.12b), or better still best cotton wool is placed into the hollow before the palm and beneath the flexed fingers. This is now consolidated by winding 10 cm plaster wool around the wrist and palmar pad and then bringing that over the top of the fingers (Fig. 8.10b). This position can now be held by the application of a 10 cm crepe bandage, which helps to keep the MP joints at 90° and the fingers straight (Fig. 8.12c). The thumb should lie abducted, beneath the index finger and should be neither flexed nor extended. The dressing is a very comfortable one – if it is not, it is too tight. Finally it is mandatory that the tips of all five digits visibly emerge from the dressing (Fig. 8.12d). The patient must know that increased pain, paraesthesiae, numbness, thermal, and/or colour change of the digits is an indication for him to seek IMMEDIATE medical help. The swollen hand so bandaged should be elevated for at least 24 hours. This is most conveniently done by suspending it in a roller towel. In hospital a drip stand is usually used. In practice it is more difficult, but hat stands, bed posts, hooks in rafters are all mentioned. An alternative to suspension is to prop up the arm, flexed to 90° at

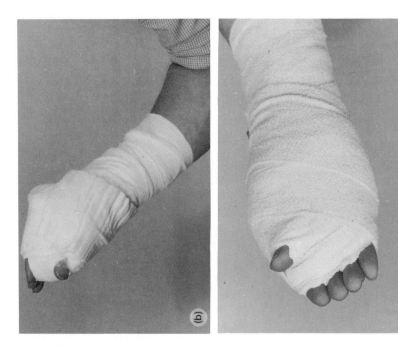

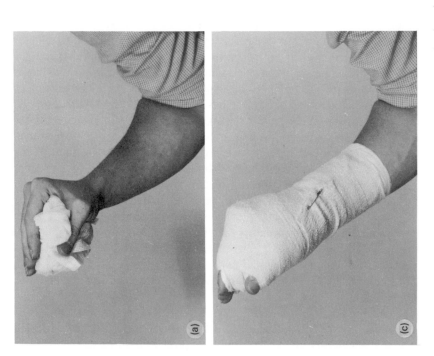

Fig. 8.12. Forming the 'boxing glove' compression bandage for the hand.

the elbow, with pillows and cushions. During the day the arm may be elevated by applying a sling which carries the hand almost up to the level of the opposite shoulder.

Having mentioned *slings,* it is important to recognize their ill-use (and they are both abused and grotesquely over-used) as a *cause* of swelling. Whereas there may be a cause for giving someone a sling for 24 *hours,* whilst a plaster dries out, there are thereafter almost NO indications for the use of a sling, beyond the doctor's observation of tradition and the patient's sense of entitle-ment. The nice patient not given a sling will even think that the poor, dear doctor, so busy, forgot, whereupon Aunt Tabatha's shawl will be pressed into service as a sling, only to encourage the development of stiff and swollen fingers, not to mention a frozen shoulder. Most slings should be slung out.

That this advice is given before even considering the cause of swelling is quite deliberate. But surely there may be an underlying fracture, some may ask. Affirmative as the answer may be, there can be no better example of treating the patient, rather than the X-ray.

Lacerations

It is in considering these injuries that the earlier rejoinder to appreciate some basic, simple, clinical anatomy is all important, as is the warning that each nerve and each tendon must be examined in turn. However, nerve injuries are highly treacherous, when examined shortly after their infliction. Although some of their sufferers may be drunk, drugged, or a little mad, this does not entirely explain the observation that a median nerve an hour after its division may *objectively* appear to still be functioning, which it certainly will not be the following day. Therefore any decision as to whether a laceration requires *formal exploration* under some form of anaesthesia and tourniquet, will often depend as much upon circumstantial, as upon direct evidence. The *two cardinal factors* here are the *situation and the depth of a wound.* Wounds in which tendons can be seen, wounds where the deep fascia has been divided and deep wounds overlying the position of vital structures all should be treated with such circum-spection. A few unnecessary, and trivial explorations are a small price to pay for preventing some of the difficulties of delayed nerve or tendon repair.

This is not to say that all nerve and tendon divisions should be repaired primarily. Nerve, especially digital nerve, repair like flexor tendon repair is highly skilled and difficult. The practitioner who does not have easy access to a hospital where they can be dealt with urgently is wise to attend, instead, to the primary management of the skin wound itself and it is to that, which we must now turn our attention.

The treatment of the wound itself can be summarized, a little didactically by a set of rules:

1. Suspicion of nerve or tendon injury, as described, should be dealt with as a hospital in-patient, whenever possible. If this is impossible then only the skin wound itself should be treated.

2. Contaminated and dirty surfaces, especially skin edges, should be excised.

3. The wound should only be sutured if that can be achieved without tension. Otherwise the excised, cleansed wound should be dressed and allowed to heal by secondary intention.

4. Wound excision and suture should not be attempted in wounds more than 12 hours old, some would say 6–8 hours.

5. The patient's antitetanus status must be established and brought up to date.

6. Antibiotics are given usually as a matter of local policy, but are generally advisable in cases of dirty, contaminated wounds.

7. Relieving incisions to facilitate wound closure are outwith the province of general practice, or most casualty departments.

8. 'Tagging' nerves or tendons, so they can be found more easily at a secondary procedure reveals good sense in firstly leaving a difficult repair to someone with more experience, but in itself is totally unnecessary. To tack the ends together to make it easier later verges on malpractice.

Amputations

These injuries have become fashionable, mainly by virtue of the great advances brought about by those specializing in *microsurgery*. But how can the practitioner help? Firstly he should have a rough knowledge of what is possible and where it may be achieved. For these purposes partial single-finger amputations are seldom if ever worth re-implanting. Multiple-finger amputations, even partial thumb amputations and certainly anything proximal to the digits is worth packing with ice in a plastic bag and sending with the patient to the nearest centre. The patient's proximal stump should have a clean dry dressing applied with a firm crepe bandage. Bleeding can almost inevitably be staunched by applying firm pressure dressings. Application of ligatures, haemostats, or tourniquets is potentially most counterproductive and should seldom if ever be attempted.

Most amputation work in the hand is thankfully at the much more mundane level of *crushed finger tips*. Here the general practitioner in a rural practice, or doing 'A and E' sessions will frequently be called upon to carry out primary management. In the thumb every effort should be made to preserve all possible length and hospital referral to this end may be advisable. With the finger the guiding principle is to secure primary wound healing. Macerated or avascular skin should be sacrificed and it is often necessary to resect some bone to close pink skin, likely to remain innervated, without tension.

Bone and joint injuries

The principles here are as follows:

(i) firstly treat and/or prevent oedema;

(ii) secondly once a fracture is diagnosed ask is it displaced or undisplaced and if present is that displacement significant?;

(iii) therefore is there any deformity evident clinically;

(iv) if a dislocated joint has been reduced is that joint now stable or unstable;

(v) is a fracture stable or unstable.

To review these principles in practice is outwith the scope of this chapter, but there follow a few common sources of errors or confusion.

1. There is a conviction that a fracture deserves a plaster, or at the very least a heavy bandage and of course a sling. Thus a very common fracture, such as that of the *fifth metacarpal neck,* is immobilized (and what is worse, so is the rest of the hand) by all manner of means. The problem, however, with these injuries is almost never one of fracture displacement or stability, it is one of soft tissue swelling, with consequent risk of subsequent stiffness.

2. Very few *metacarpal fractures* require surgical treatment. The common exceptions are those at the base of the first metacarpal, entering the joint (Bennet's fracture) and uncommonly those with displacement through the necks of the fourth or third metacarpal.

These injuries may require internal fixation or stabilization. If neglected the patient with an unreduced Bennet's fracture may rapidly develop osteo-arthritis of the thumb basal joint, with pain and dysfunction. Volar displacement (i.e. into the palmar aspect) of the third or fourth metacarpal heads may feel, to the patient, like 'having marbles in the gloves'. For a person whose work or recreation involves forming a grip around any object, albeit a screwdriver, or a tennis racket, this may be a disaster. What is more it can be extremely difficult to correct secondarily.

Fractures of the proximal and intermediate phalanges are much less forgiving. They have a significant tendency to develop both non-union and mal-union. Many of them may enter a joint and/or may be unstable and both may be indications for internal fixation. The most important clinical point is to judge not just whether the affected digit is deviated towards the ulnar or radial side of the hand, but if there is a rotational deformity as well. This can be most easily checked by remembering that when the fingers are flexed they should all point towards the scaphoid. An initial failure to apprehend these problems has a surprisingly high incidence of subsequent medicolegal difficulties. Angular or rotational mal-union results in fingers crossing or overlapping, whilst making a grip. Such difficulties have been known to end in amputation.

A failure to realize that a *dislocated joint may be unstable after reduction* may result in subsequent re-dislocation or painful instability. This is a common problem at the PIP joints, and less so at the MP joints. The worst offender is the rugby-player or watching doctor, who reduces a fresh (and thus not yet very painful) dislocation on the touchline. Such injuries always need a check X-ray and careful re-examination for ligamentous instability.

A similar problem which frequently causes subsequent dysfunction, with pain and difficulty in thumb opposition, is *the tear of the ulnar collateral ligament of the thumb.* Although classically associated with skiers or Swedish game-keepers it frequently occurs domestically from no more than a fall on the outstretched hand, with the thumb abducted. These injuries are very easy to treat surgically initially, and should always be referred. The physical signs have already been described.

Almost any of the arthritides may affect the hand. Although only rheumatoid arthritis and osteoarthritis will be described here, the principles are common to all these disorders.

Osteoarthritis

Osteoarthritis is not a common affliction of the hand, although it is frequently to be observed in the distal interphalangeal joints of middle-aged and elderly people. It may effect the wrist and require the use of splints, or even arthrodesis, after fracture or dislocations of the wrist or carpus. Some cases will settle with a period of splintage. Persistent cases should be referred for a specialist opinion.

The most troublesome joint for osteoarthritis in the hand is the *carpometacarpal joint at the base of the thumb*. This may be secondary to injury and Bennet's fracture has been mentioned in this respect, however it is usually idiopathic and is common in middle-aged women. Perhaps because of pain radiating into the thumb some of these patients are referred as cases of carpal tunnel compression. Initial treatment may include immobilization in a scaphoid type of plaster for a month, or injection of local anaesthetic and a steroid preparation. Once more the refractory case should be referred, as many are amenable to a variety of relatively simple surgical procedures. It is wise, before considering surgery, to assess the effect of plaster immobilization first, as a case not settling in a cast is unlikely to be a favourable candidate for surgery.

The practitioner's role in managing osteoarthritis of other joints of the hand is essentially one of prevention, with the prompt treatment of fractures or dislocations, as already outlined.

Rheumatoid arthritis

Many of the worst ravages of this disease are wrought upon the hand and an interested practitioner can do much to help current problems and to prevent future ones. The following is a review of some of the problems the general practitioner might expect to encounter. Their separation is to some extent artificial, as they interrelate considerably, as cause and effect.

Rheumatoid harbingers – there are several common and simple afflictions of the hand, for which rheumatoid may be the underlying pathology. These include carpal tunnel compression, trigger finger (or thumb), or the dorsal ganglion, which is not a ganglion, but is some form of synovitis. This all behoves the doctor to examine the patient a little more closely. Firstly we should consider some of the key symptoms of which the patient may complain.

1. *Pain* is usually a transient symptom in the hand and is often due to

synovitis, most commonly effecting the MP joints. It may be a more long lasting, troublesome symptom at the distal radio-ulnar and wrist joints.

Synovectomy is a useful and undervalued procedure, whether it is effected medically or surgically. It is particularly valuable in relieving pressure on tendons, which may ultimately rupture, but it is also useful in relieving pain in the wrist, distal radio-ulnar, MP, or PIP joints. At the distal radio-ulnar joint it is usually combined with excision of the distal end of the ulnar, in what is one of the most satisfactory of surgical procedures in the rheumatoid hand (Darrach's operation).

2. *Altered and/or diminished sensation* is a fairly common complaint and there is no doubt that microscopic neurovascular changes accompany rheumatoid disease. However, compression of median and ulnar nerves at elbow or wrist is quite a common problem. These problems do well if the pressure is relieved within three to six months. Longer periods of delay usually result in permanent neurological deficit. Thus referral for prompt surgical decompression is urged.

3. *A decrease in strength* may be just that. It can be due to the overall wasting associated with rheumatoid arthritis. It could be due to joint stiffness, with limited movement, or it may be due to tendon ruptures. Pain will always reduce strength.

4. *An inability to grip firmly* is one of the commonest symptoms. It may be due to any of the following, singly or in combination:

(i) decreased range of joint movement, especially at MP or PIP joints;

(ii) ruptured flexor tendons;

(iii) ulnar drift;

(v) tightness of the intrinsics.

Quite frequently all these factors combine to impede either the power (cylinder) grip of all four fingers or the more precise three-point or key grip between thumb and one or two finger tips.

An inability to oppose the thumb to some or even any of the fingers and a failure to make a full fist are common and serious disabilities consequent upon rheumatoid arthritis.

5. *Instability.* The patient may complain of difficulty in making thumb to index pinch, in that the thumb gives way. This is due to ligamentous laxity at either MP or interphalangeal joints, or quite often both. It is often assumed, as with the knee joint, that the ligaments are ruptured. This is usually incorrect. It is merely that there is such a loss of joint and bone structure that they become floppy.

6. *An inability to open the fist.* This is an unusual, but most depressing complaint, with which not only does the hand become almost functionless, but it also becomes unhygienic and macerated. The main cause is fixed flexion deformities of the MP and most especially of the PIP joints.

7. *An inability to straighten joints.* This is usually due to one of four causes:

(i) There may be a flexor, stenosing tenosynovitis (i.e. trigger finger).

(ii) *Progressive flexion deformity of the joint(s)* may occur. It is very important to intervene here before such deformity has become irrevocably fixed. Treatment may be splintage or surgical correction.

(iii) *Extensor tendon rupture.* Most typically this occurs on the dorsum of the wrist, due to extensor tenosynovitis often augmented by subluxation of the distal radio-ulnar joint. Tragically it is not uncommon that three or more tendons are allowed to rupture before an opinion is sought. Extensor lag in just one digit in rheumatoid implies advanced, but retrievable pathology. Persistent extensor tenosynovitis, in a case where there has been a previous extensor tendon rupture on the opposite side, is a very strong indication for extensor tenosynovectomy, before inevitable tendon rupture supervenes.

(iv) *Extensor tendon rupture* may also occur more distally, as explained in the descriptions of boutonniere and mallet deformities. Again it is so important to intervene before these deformities became progressive and fixed.

8. *Cosmesis.* I have been surprised by the many occasions upon which the patient seemed thrilled by what seemed to me to be an indifferent objective result. One was forced to conclude that an improvement in the hand's appearance was more gratifying than might have been predicted.

Arthritis mutilans

We have all seen the small, flabby, floppy, telescoped, functionless, but rarely painful hand. To attribute this disaster exclusively to a failure on someone's part to refer the patient for heroic hand surgery earlier is naive. Equally this author would be so blunt as to suggest that there may have been an element of neglect in these cases. Synovitis can be interrupted, tendon ruptures often can be prevented, progressive deformity may be arrested even reversed. It all takes time and there can be a large investment of effort and some disappointment for patient, surgeon and referring physician. It then seems apposite to conclude by considering in broad terms what is available to treat the rheumatoid hand and to prevent locomotor disability.

Treatment of the rheumatoid hand

1. *Drugs.* So often the recent hand afflictions may be part of an overall, generalized flare-up, for which salicylates may be insufficient and for which second- or third-line antirheumatoid drugs are indicated.

2. *Steroid injections.* These are warmly advocated by many skilled colleagues. This author has had to repair too many tendons and scrape too much cortisone out of peripheral nerves to be enthusiastic about their use for nerve compression or tenosynovitis. Alternatively the dangers of their use in joints seems to be much less than the danger to such joints of persistent synovitis.

3. *Splints* (Fig. 8.13). This is a complex subject, none more so than in the management of the rheumatoid hand. Basically splints are RESTING or ACTIVE. Active splints may be *static* or *dynamic* in that the patient wears the active splint during the active use of his hand. The static splint holds a position

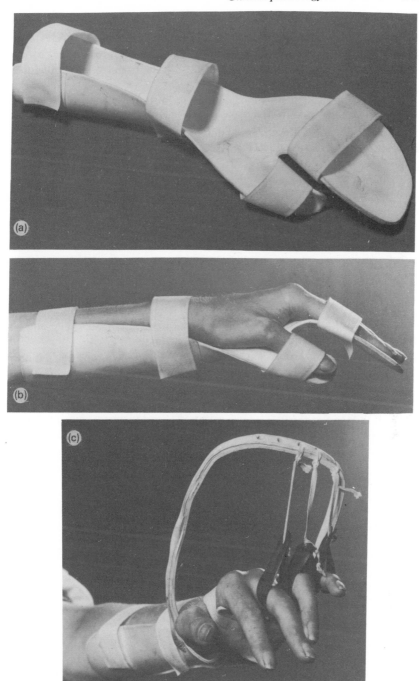

Fig. 8.13. (a and b) A static wrist splint for the rheumatoid hand. (c) A dynamic lively splint.

or prevents a deformity (Fig. 8.13a and b), whereas the dynamic one will actually assist one set of movements or counteract others (Fig. 8.13c). The effective prescription and supervision of splints must involve the patient, practitioner, rheumatologist (be he medical, surgical, or both), physiotherapist, and/or occupational therapist.

Several traditional splints are often ill-prescribed and some are fit for little other than showing as museum exhibits. The custom-made splint for a specific individual with a specific problem, manufactured in 20 minutes, on site, in light, clean, attractive materials is hard to beat by any modular commodity out of a box.

4. *Physiotherapy and occupational therapy.* The role of these two disciplines in accepting a watching brief for some of these patients and their splints can not be over stressed, nor can their contribution to a surgical campaign.

5. *Surgery.* Quite simply – what has the surgeon to offer? The following is a very brief account of what he may be called upon to do:

(i) *Synovectomy.* The role of this procedure in preventing extensor or flexor tendon rupture should not be underestimated. Its contribution to pain relief at several joints has been mentioned.

(ii) *Tendon repair.* This too has been outlined and these procedures are best undertaken when only one tendon has ruptured. Better still they and the ruptures are best prevented by controlling the synovitis.

(iii) *Nerve decompression.* The ulnar nerve at the elbow or the median nerve at the wrist are the best known examples.

(iv) *Intrinsic release.* Tight intrinsics are a very common component of much rheumatoid hand pathology. The easy test for this is to extend the MP joint fully and then to flex the PIP joint passively (Fig. 8.14). If this reveals an elastic resistance relieved by now flexing that MP joint then intrinsic tightness, usually a progressive condition is present. This can be simply released surgically.

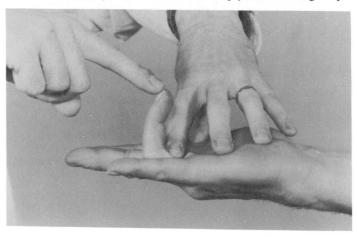

Fig. 8.14. Testing for tight intrinsics.

(v) *Correction of ulnar drift.* MP joint synovitis, ulnar dislocation of the extensor tendons off the tops of the MP joints, and intrinsic tightness all contribute to this deformity, which is usually progressive. Thus synovectomy, intrinsic release and centralization of those tendons is very worth while, before the ulnar drift becomes fixed and causes secondary dislocation of the MP joint.

(vi) *Soft tissue correction of finger deformities.* If still correctable or partially correctable deformities, such as boutonniere or swan neck, can be corrected by releases, repairs or transfers of tendons or tendon components.

(vii) *MP joint replacement.* The implantation of silicon rubber spacers into the resected MP joints (Swanson's arthroplasty) is one of the best-known operations for the rheumatoid hand. It is a useful procedure in abolishing pain and correcting gross deformities at these joints. The amount of movement obtained post-operatively is very variable, but this seldom matters much if there is good PIP joint movement and the indications for surgery were correct initially.

These implants (they are not truly joint replacements) are seldom indicated in the PIP joints or at the wrist joint. Silastic replacement of the lunate or trapezium is occasionally useful.

(viii) *Joint stabilization.* To stabilize a painful and/or flail rheumatoid joint is one of the best procedures, which we can offer. The painful, subluxed wrist, often confined to ugly, uncomfortable splints for long periods, is easily stabilized after resection of what remains of the joint surfaces by passing a heavy metal pin through the third metacarpal across the wrist and into the radius. Other joints, particularly those of the thumb, and the PIP joints of the fingers, are easily and conveniently stabilized by the insertion of stiff plastic pegs (Harrison–Nicholl). They are available in a variety of different angles, so that a joint may be stabilized in anything between 0 and 50°.

With deformed MP joints, flexion contractures of 80–100° at the PIP joints can render the hand almost functionless. Re-alignment, or silastic replacement of the MP joints and setting the finger PIP joints in varying degrees of flexion can bring about a dramatic improvement in hygiene and function.

Finally there is still a spirit abroad that 'hand surgery never did any good for the rheumatoid hand anyway'. This is not true and the horrifying number of hands devastated by neglect and cortisone injections bears testimony to this. There can be no doubt that the physician needs to choose his surgeon with caution, just as that surgeon must be critical and circumspect in his patient selection.

DUPUYTRENS AND OTHER CONTRACTURES

Contractures in the hand are usually due to one of the following:

1. Secondary to trauma, usually of joint or flexor tendon.
2. Secondary to ineffective treatment for same.
3. Secondary to muscle and/or nerve ischaemia, classically Volkmann's ischaemic contracture.

4. Secondary to a neurological lesion, of which cerebral palsy, a CVA, or an ulnar nerve affliction are probably the most important.

5. Rheumatoid, psoriasis, and other arthritides.

6. Dupuytren's disease.

Items 1, 2, and 5 have been dealt with. Prompt diagnosis, repair, or decompression of ulnar nerve lesions have been described. The place of surgery in relieving *gross deformity in spasticity* is controversial. Its benefits are largely those of facilitating hygiene and improving the uncomfortable and grotesque. It has a limited part to play in improving function. Passive stretching with a physiotherapist, plus the possible use of splints may help some of these patients.

Volkmann's ischaemic contracture

Tightness in a muscle group, whether it is due to an excessively tight plaster or bandage, or whether it is due to bleeding into that group, perhaps from a fracture, or even due to a bleeding disorder, can cause a significant rise in compartment pressure so as to shut down the small venules and arterioles. Pulses such as the radial or ulnar, being at higher pressures, may persist, leading treacherously to a false sense of security. Shutting down the small vessel supply to muscle and nerve causes them to become ischaemic and also to swell further, thus worsening the situation. If it remains unrelieved for eight, certainly 12 hours then a permanent Volkmann's ischaemic contracture will be established, which may render the hand a little less useful than that supplied by a good artificial limb centre. THE TREATMENT IS RECOGNITION AND PREVENTION:

Thus patients who have had heavy bruising, fractures, or who have a bleeding disorder and who present with swelling in the forearm, with Pain, Paraesthesia, Partial Paralysis on flexion, Pain on Passive extension of fingers, Poikolothermia (cold) or Pulselessness should be referred as acute emergencies for surgical decompression, which is achieved by dividing the deep fascia (*fasciotomy*). There is probably no more crucial piece of preventive medicine in general orthopaedic practice.

Dupuytren's disease

Dupuytren's contracture is a common disorder, whose features do not require repetition here. In my experience the tragedy is that of the patient told five years previously by his general practitioner that 'it was not yet ready for an operation'. He is likely to be told now that he is 'too far advanced for surgery'. The timing of surgery is all important and in turn the referral of patients to the surgeon by the practitioner is crucial. The poor reputation which fasciectomy undoubtedly enjoys amongst many general practitioners (and some surgeons) is due substantially to surgical help being sought too late. The following therefore are some guide lines, which may help in sorting out which sufferers from a very common condition are likely to benefit from an orthopaedic (or plastic surgery) referral.

1. Many of these patients may be alcoholic, but one must not regard all as so being. Some are epileptic and others diabetic.

2. The younger the patient at onset the more aggressive is the progress of the condition likely to be and thus early referral and surgery are advisable.

3. The first web space can be affected and the span between thumb and index accordingly reduced, causing loss of function.

4. As emphasized earlier the MP joints are safe in flexion, the PIP joints safe in extension. Put another way the MP joints stiffen in extension, the PIP joints in flexion. Accordingly an early MP joint contracture, particularly in an older patient may be watched for progression, before a decision regarding surgery is made, whereas early PIP joint contracture in all but those over 65 or 70 years of age is a prompt indication to consider surgical correction.

5. There is in my opinion no such thing, in either a new patient or a recurrent case, as a Dupuytren's which has progressed too far to consider surgical treatment, although such surgical vigour must be tempered by an appraisal of the patient's physical and psychological fitness for the procedure. MP joint contractures can always be improved and often corrected. Severe PIP joint contractures, even recurrent ones, can at least be improved by arthrodesis. There are rare occasions upon which a finger amputation is useful, to expedite a rapid return to function of a severely deformed hand.

The results of treating a Dupuytren's surgically are good in cases where both patient and surgeon are prepared to make a consistent effort and where there is recourse to good post-operative physiotherapy.

LUMPS, BUMPS, AND TUMOURS

Fortunately malignancy in the hand is very uncommon and most frequent is that of the skin. Lumps, however, are very common and most of them are *ganglia*. These are painful as they first develop and often worry the patient. Reassurance that pain will go and that it is a benign lump of 'gristle' is often sufficient. Many young ladies and some men who attribute certain social differences to the lump on their hand, become less enthusiastic about their capricious prospects following a possible operation, upon learning that they exchange a lump for a scar.

One word of caution, however, is that the rare tumour *synovioma* is often misdiagnosed initially and ganglion is a common misdiagnosis in this unusual circumstance. Lastly it is always as well to think of rheumatoid disease when confronted with almost any lump about the hand or wrist.

CONGENITAL CONDITIONS

This is a vast, complex, and intriguing subject, beyond the scope of this book. Most of these deformities are extremely rare although the incidence of congenital upper limb deformities may be as common as one in 250 of all live births. Generally speaking these conditions can be classified as follows:

1. Too many parts (e.g. fingers).
2. Too few, or absent parts (e.g. fingers or phalanges).
3. Failure of parts to separate into compartments (e.g. syndactyly).
4. Excessive separation of parts (e.g. cleft hand).
5. Parts too large or too small.
6. As a component of an overall skeletal disorder (e.g. Marfan's syndrome or Achondroplasia).

Treatment of these conditions is based upon the following principles:

1. If it is likely to require surgery this is best done with the patient as young as possible, and usually by the age of three.
2. Very early in life splints may be important to prevent secondary deformity developing.
3. Surgery is offered primarily to improve function and secondarily to improve cosmesis, ideally both.
4. A congenital hand case should always be screened for other deformity, both orthopaedic and otherwise.
5. These patients should be cared for by someone with interest and experience in the field.

BRACHIALGIA AND NEUROLOGICAL SYMPTOMS

Causes of pain and sensory disturbance in the upper limb, arising proximally have been outlined in the previous chapter. However, it is very important to remember that central nervous system disease may first be manifest in the upper limbs. Thus confronted by a patient with pain or paraesthesiae in the hand, the following is a useful procedure to follow:

1. Are there signs of upper motor neuron involvement? If there are examine and investigate from the neck proximally.
2. If the problem is lower motor neuron then one must consider the neck, the brachial plexus, and thoracic oulet before considering the individual nerves in turn.

Thus examination and radiographs of neck, including root of neck and chest are vitally important. Reference to charts of dermatones or myotomes will help one to decide whether it is a root problem or a peripheral nerve problem. I find it helpful to have these charts actually pinned up in the examining room.

3. Vascular causes and problems in the shoulder joint must always be excluded at an early stage.
4. If the problem appears to be one distal to the plexus and thus a peripheral nerve lesion, it is useful to have an 'acid test' for each nerve in turn. I recommend the following:

Nerve	Motor	Sensory
1. Axillary (circumflex)	Abduction of shoulder against resistance. Feel for deltoid.	Area of sensation in 'vaccination' mark' area
2. Radial nerve	Dorsiflex wrist and extend tip of thumb against resistance.	Area of sensation around anatomical snuff box
3. Musculocutaneous	Flex elbow against resistance – feel biceps	Sensation of lateral aspect of forearm
4. Median	Oppose thumb to little finger. Thumb must be rotated so that nail faces examiner. Compare with other side. Testing short abductor is most reliable test	Sensation over radial 3 digits and lateral ⅔ of hand
5. Ulnar	Abduct index finger against resistance. Same for little finger	Sensation over ulnar 1½ digits or 2 digits and medial ⅓ of hand

Median nerve compression at the wrist in the carpal tunnel and ulnar nerve at the elbow behind the medial epicondyle are the two commonest nerve compression syndromes. They should be relieved surgically, promptly. If there is doubt regarding their diagnosis then this can be substantiated by means of a nerve conduction study.

SUMMARY

This is a very brief review of some of the commoner conditions in the hand causing locomotor disability. In so short a text there are inevitably many omissions. It is the general principles, however, which matter and they may be applied to almost any problem which one is likely to encounter. The theme however is one of early recognition and prevention of disability, for it is here that the general practitioner has so good an opportunity to help his or her patients.

REFERENCE

Lister, G. (1977). *The hand: diagnosis and indications.* Churchill Livingstone, Edinburgh.

Section III
Disturbances of gait

9 Back pain and sciatica

Malcolm I. V. Jayson

INTRODUCTION

The size of the back pain problem is daunting. Every general practitioner sees large numbers of patients with this as their principal complaint. It is responsible for about 6.5 per cent of all general practice consultations. Overall about two million adults in Britain consult their general practitioner each year for this reason. Nearly one in fifty of the population lose time from work with a total of over nineteen million days lost per annum. It is fashionable to claim that much back pain is a form of malingering and represents a deliberate attempt to obtain some benefit such as time off work or compensation. Individual cases undoubtedly occur but nevertheless the magnitude of this problem is in considerable doubt. The figures for the incidence of back pain and time lost from work are not very different in other countries where welfare benefits are not nearly as generous as in Britain.

Back pain is a symptom with a large number of different causes and with varying forms of management. A discussion of the treatment of back pain is as meaningless as a discussion of the treatment of abdominal pain unless put in the context of the underlying problem. It is always important to consider the underlying diagnosis. In many cases it is not possible to arrive at the definitive cause but nevertheless it is possible to make sure that the problem is mechanical in origin and is not due to inflammatory, neoplastic or other disease.

CLASSIFICATION OF THE CAUSES OF BACK PAIN

Table 9.1 shows the principal causes for back pain. The structural or biomechanical group include not only such well-documented problems as a prolapsed intervertebral disc and spondylolisthesis, but also many other disorders in which the diagnosis is made with varying degrees of certainty. There is a large group of patients, perhaps the majority, with back pain in whom the history and physical findings and other tests point to a mechanical cause but without precise identification of the actual problem. Rather than use terms such as lumbosacral strain or sacro-iliac strain which have no meaning in pathological terms, I prefer to call these 'non-specific back pain' which simply means back pain of uncertain cause but thought to be of mechanical origin.

Inflammatory causes include not only septic arthritis due to bacterial infections such as a staphylococcal abscess and tuberculosis of the spine but also the non-infective chronic inflammatory diseases such as ankylosing spondylitis and

Table 9.1. *Causes of back pain*

Structural	Prolapsed intervertebral disc
	Spondylosis and apophyseal osteoarthrosis
	Spondylolisthesis
	Spinal stenosis
	Other congenital anomalies
	Fractures
	Non-specific
Inflammatory	Ankylosing spondylitis and related inflammatory spondylo-arthropathies
	Rheumatoid arthritis
	Infection
Neoplastic	Primary tumors
	Metasteses, reticuloses, and myelomatosis
Metabolic	Osteoporosis
	Osteomalacia
	Paget's disease
Referred pain	

occasionally rheumatoid arthritis. Neoplastic disease is usually due to secondary metastatic deposits and the commonest sources are carcinomas of the breast, bronchus, thyroid, kidney and prostate. Reticuloses such as Hodgkin's disease occasionally present with spinal deposits and back pain. Metabolic bone disease of various sorts can produce deformity and acute episodes of pain due to fractures. Back pain can arise not only due to spinal disease but also as a referred pain from abdominal or pelvic diseases and felt in the spine.

In this chapter I have concentrated on the common mechanical problems with a short discussion about ankylosing spondylitis and only a brief description of the other conditions indicating how they can be recognized.

DIAGNOSIS AND ASSESSMENT OF THE BACK PAIN PATIENT

First and foremost in making the diagnosis is the clinical skill of the physician. A careful clinical history will usually indicate the nature of the problem and this is aided by a full physical examination. Both should be directed not only at the back problem but also at the patient's general health and other body systems. The temptation is to concentrate on the back alone but by doing this, sooner or later important diagnostic clues about other conditions will be missed. Investigations play a relatively limited role. Most X-ray requests are unnecessary as they rarely influence the management of mechanical back problems and they should only be ordered for selected cases as indicated later.

THE CLINICAL HISTORY

The age of the subject is of some help. Prolapse of an intervertebral disc most frequently occurs in the 30–50 year age group although it often occurs outside these extremes. Ankylosing spondylitis is more common in younger people. Although structural causes of pain causing loss of work are more common in males this probably reflects the predominantly male working population and also that men on the whole undertake much heavier work than women. However, the totality of back pain if anything is greater in females than males. Males are more frequently affected by ankylosing spondylitis although we now appreciate that this disease occurs far more commonly in women than previously thought. Indeed a recent survey of the milder forms of the disease suggested that it occurs in equal incidence in the two sexes.

Table 9.2. *Principal changes used for identifying the sites of lumbar nerve-root lesions (NB This does not list the total distribution of each root)*

Root	Paraesthesiae and sensory change	Muscle weakness	Tendon reflex change
L_2	Upper thigh, anterior medial and lateral	Flexion and adduction of hip	None
L_3	Anterior surface of lower thigh. Anterior and medial surfaces of knee	Adduction of hip Extension of knee	Knee jerk possibly decreased
L_4	Anteromedial surface of leg	Extension of knee Dorsiflexion and inversion of foot	Knee jerk decreased
L_5	Anterolateral surface of e.g. dorsum and medial surface of foot especially dorsal surface of hallux	Extension and abduction of hip. Flexion of knee. Dorsiflexion of foot and toes, especially hallux	None
S_1	Lateral border and sole of foot, back of heel, and lower calf	Flexion of knee. Plantar flexion and eversion of foot	Ankle jerk decreased

STRUCTURAL CAUSES OF BACK PAIN

The mode of onset of pain is helpful. The structural group of conditions produce pain that usually can be related to a mechanical event and sometimes to a particular incident. The patient bends, twists or lifts and develops acute pain and finds that he is unable to straighten up. When there is a prolapsed intervertebral disc this pain is felt initially in the lumbar region and then may radiate down into the lower limb. There may not only be pain but also numbness and tingling and the distribution of symptoms will be a guide to the particular nerve root that is affected (Table 9.2). With these mechanical problems the patient will often suffer acute episodes of pain which may last a few hours or even a few days but then gradually wear off until the next bout occurs. The patient may identify specific movements and in particular bending and lifting that will exacerbate the pain. There may be a problem with chairs and in particular prolonged sitting in a poor posture may aggravate the problem. The pain can be extremely severe and some patients find themselves totally incapacitated.

The history of acute pain in the back followed by sciatic pain in the lower limb is very straightforward in indicating the diagnosis of a prolapsed intervertebral disc but the other different types of structural causes may be more difficult to identify. In one important group, however, the patient describes back pain and sciatica which develop only after walking and are relieved by rest. The sciatic pain felt in the calves is described as a woolly, numb, and unpleasant tingling sensation and may easily be confused with intermittent claudication due to arterial insufficiency. However, the peripheral pulses are usually present excluding vascular claudication. Another characteristic of the symptoms is that they are relieved by bending forwards and aggravated by standing upright. This is a mechanical problem in which there is narrowing of the vertebral canal so that the spinal nerve roots are tightly packed and any intrusion into the canal readily damages them. This is known as spinal stenosis and it is important to recognize this condition as it lends itself to surgical correction.

NON-MECHANICAL PROBLEMS

The clinical history will often indicate whether the back pain is due to some non-mechanical problem. The patient's general health is important. Loss of weight, feeling off colour, poor appetite as well as more specific features such as a cough, difficulty in passing urine, or a lump in the breast can be directly relevant.

ANKYLOSING SPONDYLITIS

In ankylosing spondylitis the symptoms usually start insidiously and gradually progress. The commonest complaint is of pain and stiffness which are felt across the low back and sometimes in the buttocks and backs of the thighs. In contrast with the structural problems, the pain is exacerbated by rest and relieved by exercise. These patients will be at their worst when they awake in the morning and may have to perform some physical exercises in order to relieve their symptoms. The severity of the stiffness and pain may wake them earlier and earlier. They frequently describe tossing and turning in bed and are forced to get up and move around in order to obtain relief. In the early stages of the disease the symptoms usually clear by the time they are seen in the surgery. However, towards the end of the day when they are tired there may be a mild relapse of pain. These symptoms are not infrequent and ankylosing spondylitis is one of the commonest causes of back pain in groups of young adults such as University students and armed forces recruits.

OTHER INFLAMMATORY CAUSES – RHEUMATOID ARTHRITIS AND INFECTION

Rheumatoid arthritis can involve the lumbar spine. However, the arthritis always involves the peripheral joints where the diagnosis is obvious. Septic arthritis fortunately is rare but can produce devastating damage. The patient is ill with progressive and severe pain in the back perhaps with rigors. If anything there has been a recent increase in the incidence of tuberculosis of the spine. This is more frequent in immigrant populations from the Indian subcontinent.

TUMOURS AND RETICULOSES

Neoplasms and reticuloses in the spine produce a picture of back pain of gradual onset but slowly and remorselessly progressive without periods of remission. Increasing back pain may be followed by progressive neurological involvement due to damage to the nerve roots in the vertebral canal. Pain and paraesthesiae develop in the lower limbs and may be followed by muscle weakness. Involvement of the cauda equina may interfere with sphincter function with loss of sensation and loss of control of micturition and defecation. Patients with these complaints must be referred immediately for a specialist opinion as emergency decompression may be required in order to prevent permanent neurological damage.

OSTEOPOROSIS AND OSTEOMALACIA

Osteoporosis is weakening of bone due to deficiency of the matrix. It is most common in elderly females but can also occur in other subjects and particularly in patients with other diseases treated with corticosteroids. Weakness of bones leads to fractures producing acute episodes of back pain which will usually settle over a few days. However, the progressive failure of individual vertebrae leads to the characteristic kyphosis and loss of height so commonly seen in older women. Osteomalacia is due to deficiency of calcium and vitamin D. It may be due to an inadequate diet and is not uncommon in elderly people living on tea and toast who rarely go outdoors. We see it today in immigrant families and also in those with a malabsorption syndrome such as coeliac disease. This latter group may suffer from diarrhoea with characteristic pale frothy stools.

REFERRED PAIN

Pain may be felt in the back but arise due to abdominal or pelvic disease. A posterior peptic ulcer, aneurysm of the aorta, and gynaecological diseases of various sorts can be responsible. There will be features of the abdominal or pelvic disorder indicating the appropriate diagnosis. With most gynaecological pains not only will the patient describe heavy periods and dysmenorrhoea, but also she will relate the severity of the back pain to the time of the periods. Back pain is a common accompaniment of pregnancy. It is due to a combination of the weight of the gravid uterus altering the centre of gravity of the body and throwing strains on the spine, and the laxity of the ligaments of the spine and sacro-iliac joints allowing them to be stretched more readily.

THE PHYSICAL EXAMINATION

The need for a full physical examination must again be stressed. It is important in all patients presenting with back pain for the first time and in patients with recurrent mechanical problems in whom the character of the symptoms change. This can indicate a new reason for the back pain.

The general examination should include the abdomen and if there are any grounds for suspicion a rectal or vaginal examination. When the pain is referred from abdominal or pelvic disorders pressure over the lesion will reproduce the symptoms.

The back should be inspected from behind with the subject standing up. Careful inspection will show the alignment of the vertebrae and any kyphosis or scoliosis will be noted. In patients with a prolapsed intervertebral disc the paravertebral muscles may be in spasm twisting the spine into a 'sciatic scoliosis'. This is a reflex attempt by the body to splint the spine and to protect the damaged nerve roots against pressure from the vertebrae. Scoliosis may be difficult to recognize when standing up but becomes obvious when the subject bends forward. Palpation of the paraspinal muscles with the patient lying prone will demonstrate muscle spasm.

POSTURE

The posture of the lower limbs and the leg length should be checked. Real or apparent shortening of one of the lower limbs may produce a tilt in the pelvis and a scoliosis and be a potent cause of premature degenerative changes in the back. By carefully palpating the iliac spines it is possible to determine whether they are level and the lower limbs are of equal length.

The normal person stands upright with a lumbar lordosis and a slight forward curvature of the dorsal pine. In early ankylosing spondylitis the posture is normal but in more advanced disease a characteristic change is flattening of the lumbar spine with a loss of the normal lordosis and the gradual development of a smooth dorsal kyphosis. A sharp angular kyphosis on the other hand indicates localized disease such as vertebral collapse.

MOVEMENTS

The movements of the spine should be carefully examined. These include flexion, extension, lateral flexion, and lateral rotation about a vertical axis in both directions. Forward flexion is checked by asking the subject to bend forward and touch his toes. This should be done carefully for fear of exacerbating the symptoms. Forward flexion can be acomplished by bending the hips without any lumbar spine movement at all and some subjects are able to touch their toes in this way. Careful observation of the lumbar spine is required to determine exactly how much it is moving and which movements produce pain. It is possible to see the normal change from lordosis to a slight lumbar kyphosis during forward flexion. This can be checked by placing the fingers of one hand lightly over the spines of the lumbar vertebrae and then observing the spreading apart of the fingers as the subject flexes forwards. Extension of the back is observed in a similar way. Lateral flexion in each direction is examined by asking the subject to bend sideways with the hands resting on the thighs and run the fingertips down the side of each lower limb respectively. Rotation is observed by standing behind the patient, fixing the pelvis with the hands on each side, and asking the patient to twist round to look over his shoulder. By comparing lateral flexion and lateral rotation in the two directions it is possible to detect any abnormalities.

In patients with mechanical problems it is characteristic for flexion and extension to be limited and perhaps lateral flexion and lateral rotation in one direction only. However, there is not uniform limitation of movements in all directions. This is in contrast to the ankylosing spondylitic in whom all three types of movements are involved to a similar extent. There is one type of mechanical back problem in which all movements are limited. I call this condition a 'stuck back' and it is readily confused with ankylosing spondylitis. It predominantly occurs in patients with mechanical back problems who have worn a corset for an excessive length of time. They then develop severe back pain and gross limitation of movement in all directions. I think that this is possibly due to adhesions occurring in the spine. At the other extreme we sometimes see patients with back pain whose backs appear excessively mobile. This is demonstrated by asking them to bend forward and they are able to go so far as to place their palms flat on the ground. They often show features of hypermobility in other joints and the ligaments surrounding them. Recognition of this condition is important as it is all too easy to label the patient as hysterical as they complain of severe back pain with so obviously free back movements.

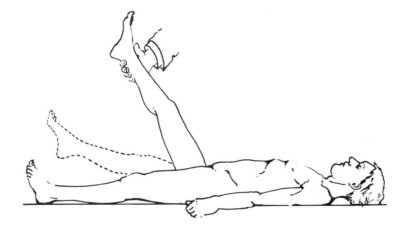

Fig. 9.1. The straight leg raising test.

PALPATION

Careful palpation of the back is important. This should be done with the patient relaxed and lying prone and not sitting or standing upright. The contraction of the paraspinal muscles to stabilize the vertebral column when upright prevents proper palpation of the structures in the back. It is important to seek areas of tenderness as they may indicate the source of the underlying pathology. It is often not a very good guide because of the complex anastamotic network of the nerve supply of the back but nevertheless it can be of help. In particular extreme tenderness in a specific area may indicate the site of a fracture, neoplasm, or an infected abscess. In mechanical problems there is often tenderness over the source of the problem. In early ankylosing spondylitis the tenderness may be localized over the sacro-iliac joints. This may be demonstrated by direct palpation, by pressing on the sacrum or by lateral pelvic compression.

NEUROLOGICAL EXAMINATION

A careful neurological examination should be made with special reference to the lower limbs. The straight leg raising test is an important index of nerve root and dural compression (Fig. 9.1). The patient lies supine with both lower limbs extended and the examiner elevates each in turn by raising the heel. This movement may be limited by reproduction of the back and sciatic pain or by tightness of the hamstring muscles. The normal person can achieve straight leg raising of about 80° or more. Patients with a prolapsed intervertebral disc and sciatica may only tolerate 10°. As the patient improves, so straight leg raising gradually increases. This is a useful way for following the progress of the individual patient. The degree of straight leg raising is checked by elevating the extended lower limb to the maximum allowable and then lowering it a few degrees so relieving the symptoms. In this position dorsiflexion of the foot will reproduce the pain.

Muscle power should be examined in the lower limbs and the distribution of weakness may suggest which nerve root is involved (Table 9.2). Of considerable help is weakness of dorsiflexion of the great toe which indicates an L5 lesion and of plantar flexion of the foot suggesting an S1 lesion. The latter may subsequently be checked by the patient's inability to stand on his toes on that foot. The knee and ankle jerks depend on the integrity of the L3/4 and S1 nerve roots respectively but there is no reflex which depends on the L5 nerve root alone. Sensory examination should include testing for light touch and pin prick and may show reduced sensation over the anteromedial aspect of the leg (L4), the anterolateral aspect of the leg and the dorsum of the foot (L5), and the lateral border and plantar surface of the foot and back of the calf (S1).

The patient is also asked to lie supine. In this position flexion of the knee stretches the femoral nerve and production of pain may indicate an L3/4 lesion. Sensation in the back of the lower limb is easily checked in this position.

If there is any doubt about sphincter disturbance, sensation should be checked over the saddle area for cauda equina damage.

INVESTIGATIONS

The clinical acumen of the physician is all important and the various investigations play only a limited role.

When the patient has a long history of recurrent back pain of mechanical type, it is not necessary to undertake laboratory or radiological tests for further similar recurrences. However, investigation is indicated when back pain develops for the first time, if the character of the problem changes, if the symptoms gradually develop and progress rather than occur as acute episodes often related to mechanical events, and if the patient is unwell or shows other features which casts doubt on the basis of the problem.

BLOOD TESTS

Most back problems are due to mechanical causes. The various blood tests are all normal and they merely serve to reassure the physician that nothing else is going on.

Elevation of the erythrocyte sedimentation rate (ESR) or the plasma viscosity (PV) may indicate some inflammatory disorder or a neoplasm. A very high ESR or PV may suggest abnormal plasma proteins as in multiple myeloma. Patients with these disorders are often anaemic. The white cell count is usually elevated when there is infection and the differential count may indicate leukaemia. A request for a biochemical test usually results in a battery of assays performed in an automated system. These should be examined for abnormalities of the serum calcium, phosphate, and alkaline phosphatase which suggest bone disease such as osteomalacia. In older male subjects elevation of the serum acid phosphatase may indicate carcinoma of the prostate. Plasma protein electrophoresis may show abnormal bands which are the paraproteins of myelomatosis.

X-RAYS

The need for radiographs of the back is extremely limited. When the pain arises due to mechanical problems, the radiographs may or may not show degenerative changes but the significance of these findings is in considerable doubt. The incidence of back pain is almost as high in those subjects without any radiological evidence of lumbar spondylosis as in those with the grossest changes. This means that the finding of radiological evidence of disc degeneration or apophyseal osteoarthritis does not mean that these are the cause of the pain and their recognition plays little part in determining subsequent management.

Similarly, developmental anomalies such as lumbarization of the first sacral vertebra producing six lumbar vertebrae, and sacralization of L5 leaving only four lumbar vertebrae are common and in general have no clinical significance.

Detailed surveys show that one can safely reserve radiographs for the patients in whom there are good grounds of suspicion from the history, physical findings, and blood tests to suggest that there is some cause other than a simple mechanical problem. X-rays of the lumbar spine expose the patient to a very high dose of radiation. In particular the lateral view of the L5/S1 disc space requires the X-rays to penetrate both sides of the pelvis and the mass of soft tissue as well as the bones of the spine itself and exposes the subject and particularly the gonads to a very high dose of radiation. The general practitioner should resist ordering lumbar spine radiographs unless he feels that there is some special indication. One must sympathize with his problem that many

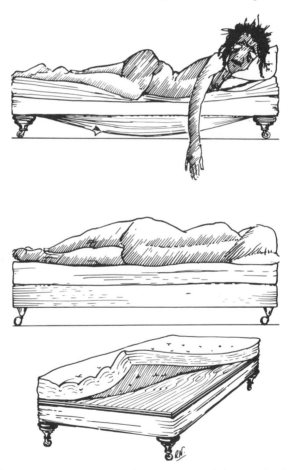

Fig. 9.2. A poor bed that sags readily, a firm bed, and the use of a board under the mattress.

patients feel they are not being treated properly without having an X-ray and so place unfair pressures on him.

Of course there are times when radiographs are important. In ankylosing spondylitis radiological changes are seen in the sacro-iliac joints and elsewhere in the spine. However, in early disease the sacro-iliac joints may appear normal and the typical features only develop at a later stage. If there is a diagnostic problem the consultant may order a bone scan as this is helpful in assessing these more difficult cases.

TISSUE TYPING

Ankylosing spondylitis and related inflammatory spondylo-arthropathies such as occur in psoriatic arthritis, Reiter's disease, ulcerative colitis, and Crohn's disease are all associated with an abnormal tissue type. Human white blood cells can be classified in similar ways as human red blood cells. The human lymphocyte antigens (HLA) were first recognized because of the importance of correct matching for tissue transplantation purposes. One type known as HLA B27 occurs in about 95 per cent of ankylosing spondylitics but only 5 to 8 per cent of the normal population. About 20 per cent of people born with this particular tissue type may develop at least minor forms of ankylosing spondylitis. This means that 80 per cent will not develop this problem so the test is not particularly useful for making the diagnosis of ankylosing spondylitis. It is more helpful in excluding the diagnosis in patients in whom there is some doubt, for a negative result makes ankylosing spondylitis extremely unlikely although not impossible. As the test involves complex and time-consuming laboratory procedures, at the present time it is not recommended or indeed usually not even available in general practice.

MANAGEMENT OF BACK PAIN AND SCIATICA

The management of back pain depends upon the underlying diagnosis. The various types of mechanical problems are all grouped together as the treatment is related to the severity of the symptoms rather than specific structural problem.

BACK PAIN OF MECHANICAL CAUSE

It is characteristic of this type of problem that the individual bouts of severe pain are of relatively limited duration and the patient may make a full or partial recovery until the next episode occurs. Each individual episode may last only a few days or perhaps a few weeks. Treatment therefore requires relief of the acute episode, control of chronic symptoms, and advice and help in an effort to prevent further episodes.

Fig. 9.3. Good and bad postures when standing.

Rest

The most important aspect of treatment of the patient with acute low back pain with or without sciatica is bed rest. I have frequently seen patients who try to struggle on as normal, paying only lip service to bed rest and whose pain persists for months without relief or comes and goes with periods when things seem to be getting better followed by sudden and depressing relapses. When prescribed a period of complete bed rest these symptoms usually will clear within a few days or a couple of weeks. It is in this context that referral to hospital does have real advantages. Although the general practitioner may advise complete bed rest at home, there are so many temptations to get up and around that it may be difficult to achieve. The eagle eye of the hospital ward sister can work wonders.

Complete bed rest means lying flat in bed with only one pillow on a properly supported mattress (Fig. 9.2). Soft beds sag readily and can exacerbate back problems. Although orthopaedic beds seem ideal they are extremely expensive and provide little advantage over a board placed beneath the mattress. This should be strong enouch so as not to bend readily and should run the full length of the bed and be at least as wide as the patient. Another alternative is to place the mattress on the floor although this can be a bit draughty. During the period of complete bed rest I allow my patients up for toilet and washing. This places much less stress on the spine than trying to perch precariously on a bed pan whilst lying flat or to wash without soaking the sheets.

Lumbar corsets

Once the symptoms have remitted significantly, the patient is allowed to remobilize but considerable care must be taken to avoid stressing the back. He should lever himself up carefully using his arms and avoid flexing forwards and doing any heavy lifting. In this acute stage, a lumbosacral corset is helpful. It probably works both by splinting the spine and by increasing the intra-abdominal pressure so relieving the back of a proportion of the total load. The corset is helpful in the short term but its use should not be prolonged as it can lead to permanent stiffness of the back and further back pain.

Analgesics

Pain-relieving tablets are usually required. There is commonly an inflammatory element associated with the structural problems due to the mechanical damage to the tissues so that for many people the non-steroidal anti-inflammatory drugs possess advantages over pure analgesics. Aspirin, the propionic acid derivatives such as ibuprofen, naproxen, ketoprofen and many others seem most suitable. The more powerful anti-inflammatory drugs, indomethacin and phenylbutazone, are more effective but do possess greater risks of side-effects. Their use should be avoided except under various special circumstances. In other patients pure analgesics may be helpful. Paracetamol is a relatively mild analgesic. Dihydrocodeine tartrate and dextropropoxyphene are very effective drugs for patients with more severe pain but carry a very slight risk of habituation. They also tend to constipate. Consequential straining can exacerbate the back problem. Narcotic analgesics are very rarely required and they should only be given to patients with acute severe pain and for a very limited period of time.

Ergonomic advice

Detailed ergonomic advice should be given to patients as they return to their normal occupation and also to those whose back symptoms are never severe enough to require a period of complete rest. The patients should be encouraged to stand upright with the normal hollow in the small of the back and avoiding a forward slouch (Fig. 9.3). High heels tip the spine forwards and can exacerbate back problems. The subject should avoid lifting anything that is too heavy. Objects should be held close to the body (Fig. 9.4) and not at arm's length and should not be lifted above the head as this places enormous strains on the spine.

The usual way to lift from the floor is to flex the spine forwards (straight knees/bent back). In this posture the paraspinal muscles are relaxed and all the stress is carried by the spinal ligaments which are stretched and readily damaged. Many patients are taught to lift by the straight back/bent knees method in which they crouch down with their feet together. This position is also

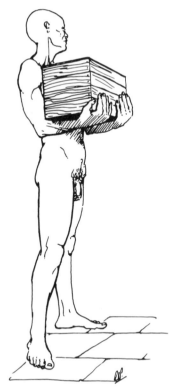

Fig. 9.4. Carrying a heavy object.

incorrect. It is very unstable as the patient is trying to balance on the forefeet and the actual lift from the floor is achieved by reaching forwards. Biomechanical analysis shows that this manoeuver is extremely unsound and likely to exacerbate back problems. The correct method is known as the kinetic method of lifting. The subject stands with one foot behind and the other beside the object and pointing in the direction that the subject wishes to move. He squats down with the object between his knees and with the back straight although it may be inclined forwards (Fig. 9.5). The object is firmly grasped with both hands and then lifted using the powerful hip and thigh muscles to straighten up. Setting down is achieved in a similar fashion. Even when the immediate back problem has passed the patients, and indeed all of us, should use this technique.

The back sufferer should sit in an upright chair with adequate lumbar support. Better quality chairs will provide this support and in some secretarial chairs and car seats the amount is adjustable (Fig. 9.6). A small pillow in the small of the back is a useful compromise and adequate for most people.

Physiotherapy

Many general practitioners have access to physiotherapists either directly attached to their own surgeries or via open access to a hospital remedial therapy department. There is a great tendency to order physiotherapy indiscriminatly without knowing the objectives of treatment, what is being done and the values of the different forms of therapy. As much thought should be given to the prescription of physiotherapy treatment as is given to the choice of drug therapy. Physiotherapy should not be used merely as a placebo for the cost of the services of a physiotherapist far exceeds that of most drugs and they resent rightly the consequent loss of esteem that one normally accords to professional colleagues.

Perhaps the most important role the physiotherapist can play lies in reinforcing the postural and ergonomic advice described earlier. The physio-therapist will have time to explain how the back works and what may go wrong and by demonstration and assistance will teach the patient the right and wrong ways to perform various tasks. In Sweden this educational function is carried to

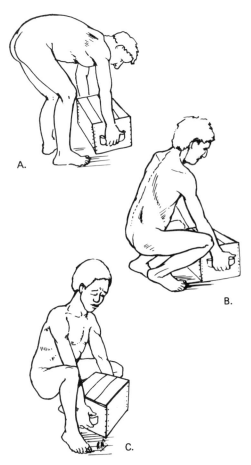

Fig. 9.5. Lifting by the straight knees/bent back method, the straight back/bent knees method, and the correct kinetic technique.

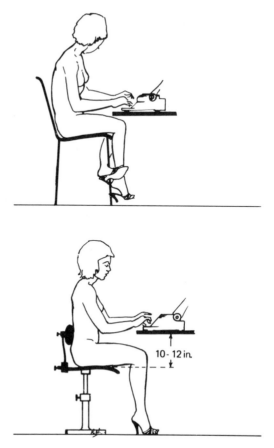

10- 12 in.

Fig. 9.6. Poor and good sitting postures for a secretary.

great lengths in their 'Back Schools' and at the end of a course of lessons the patients take an examination to make sure that they have understood all that has gone on and to act as a revision exercise. The Swedish workers claim that they get as good results from patients attending their Back School as from more conventional physiotherapy.

Exercises

Exercise programmes are often prescribed. Mobilizing exercises are prescribed on the hypothesis that forcibly improving the range of movement of a stiff spine leads to improvement. Although there are certain exceptions, in general this form of treatment seems unwise. If movements of the back are limited by pain, it is likely that the spine and its ligaments are damaged so that forcibly undertaking these manoeuvres can only increase the problem. I frequently hear of patients made worse following this form of treatment. For this reason I feel that mobilizing exercises, flexion exercises, and extension exercises should in general be avoided. There is logic, however, in the use of isometric exercises in which the strengths of the paraspinal and abdominal muscles are built up. The paraspinal muscles help to stabilize the spine and the abdominal muscles will increase the intra-abdominal pressure so relieving the spine of a proportion of its load. The paraspinal muscles are strengthened by lying prone and raising each lower limb in turn by a few degrees or by lifting the head and shoulders off the ground (Fig. 9.7a). The abdominal muscles are strengthened by lying supine, and raising the extended lower limbs in the air and then lowering them gently to the ground or lifting the head and shoulders off the floor (Fig. 9.7b).

Traction

Traction of the lumbar spine is sometimes used. For patients who are admitted to hospital continuous traction whilst lying in bed seems useful although it is not clear if it works simply by being an extremely effective method of confining the patient to bed. Intermittent traction is frequently used in physiotherapy departments. A harness is fixed round the pelvis and another around the upper trunk and the lumbar spine is gradually stretched. Although this technique relieves back pain during treatment, it is doubtful whether it makes any long-term difference.

Minor procedures

There are hosts of minor procedures that are often requested but seem of little value. These include various methods of applying heat ranging from simple application of hot water bottles to more sophisticated forms of heat radiation and short-wave diathermy. Conversely cold is sometimes used in the form of ice packs. Massage in one form or another is frequently requested. These forms of treatment are comforting at the time of administration but do not make any difference to the underlying problem. They have a limited role as preliminary pain relieving procedures before other forms of therapy that have more permanent value but otherwise should not be prescribed.

Manipulation

Manipulation and mobilization to the lumbar spine refer to treatment of the back by manual pressure. There is a wide variety of different methods and they are practiced by physiotherapists, specialists in manual medicine, orthopaedic surgeons, osteopaths, and chiropractors. Anecdotal cases of relief of back pain by these methods are common yet there is no agreement on the values of these forms of treatment and the relative merits of the various types of practitioner. The placebo element in these forms of treatment is high and this combined with the natural process of recovery of individual attacks of back pain casts considerable doubt on many of the claims for manipulation. Careful controlled studies have been carried out on the types of manipulation commonly practiced in hospital physiotherapy departments. They demonstrate a high incidence of resolution of symptoms irrespective of whether patients

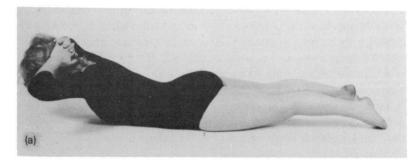

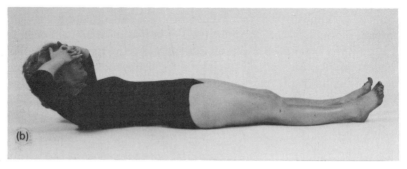

Fig. 9.7. Isometric exercises for the paraspinal (a) and abdominal (b) muscles.

receive manipulation or some placebo form of treatment. There is a slight advantage to those manipulated but in the long run there is no difference. The forms of manipulation practiced by osteopaths and chiropractors have not been subjected to rigorous testing and we still do not know whether the forms of treatment that they offer are of any value. From time to time one meets patients who are made worse by manipulation. In particular, there is the hazard of the patient who is manipulated for back pain due to some other cause when real damage may be done.

Referral to specialist clinics

Referal to a rheumatologist or orthopaedic surgeon may be for a variety of reasons. Doubt about the diagnosis and particularly whether there is some cause for the pain other than a simple structural problem may require a specialist opinion and investigations. In particular pain that is gradually getting worse with the progressive development of neurological signs requires urgent referral and the appearance of bladder or bowel disturbances is an emergency as it may require immediate surgical decompression. Patients may be referred with a view to access to physiotherapy departments or supply of a surgical corset, for consideration for admission or more detailed investigation.

Additional forms of treatment available include local injections into the back. These may be directly into painful and tender nodules. The cause of these localized painful areas is not known but many patients are relieved by injection of a small dose of long-acting steroid and local anaesthetic. It is not at all clear whether it is the steroid, the anaesthetic or even the injection itself which provides the relief. Epidural injections are also useful for some back problems. Usually these are given by anaesthetists by the lumbar or sacral routes into the epidural space and for some patients can provide dramatic relief of symptoms although this is often temporary.

Surgery

Surgery is required only very rarely. It has been calculated that out of every 10 000 people who consult their general practitioner most get better and the practitioner will only refer about 2000 to hospital rheumatology or orthopaedic clinics. Long waiting lists occur in many departments but an interesting benefit from this is that many back sufferers recover before seeing the specialist. One estimate is that two-thirds get better in this way leaving about 700 who are actually seen in the clinic. About 1 in 10 of these or 70 will require admission to hospital. Most will recover with conservative treatment or are unsuitable for surgery and only 1 in 10 of those admitted or seven patients will need an operation. So, we are left with a figure of about seven operations for every 10 000 patients seen by their general practitioner because of back pain.

The surgical procedures depend upon the precise anatomical problem but include laminectomy, removal of the protruding disc and sometimes fusion of the vertebrae. The indications for operation are as follows:

1. Significant and unremitting symptoms despite an adequate period of proper conservative medical treatment including a period of supervised complete bed rest.

2. Persisting or advancing neurological signs. Of particular importance are cauda equina lesions with bladder or bowel involvement which may require emergency surgical decompression.

3. In lesser cases surgery may be considered for recurrent attacks of severe pain and persistent disability such as inability to perform normal work.

In general, surgery is much better for the sciatic pain than for back ache alone. As a rough guide, out of every 10 patients eight will be completely relieved of the lower limb pain but only five or six of back pain. Although these figures do not seem quite as good as one might like it should be remembered that many of these patients have some of the worst back problems and other types of treatment have failed.

Acupuncture and transcutaneous nerve stimulation

Acupuncture is a favourite topic for discussion by back sufferers. After initial disbelief by orthodox medical opinion, there is now considerable interest in understanding how the perception of pain is modified by the release of endorphins and enkephalins in the brain stem and how the formation of these substances may be stimulated by acupuncture. A new development of the method employs transcutaneous electrical nerve stimulation (TNS). A small box with surface electrodes is placed over the skin of the back and repetitive low voltage electrical stimuli are used instead of the acupuncture needle. Although some patients claim relief, controlled surveys so far have not demonstrated any clear value for acupuncture or TNS.

OTHER CAUSES OF BACK PAIN

Recognition of the underlying diagnosis is essential in patients with inflammatory, neoplastic, metabolic, or referred causes of back pain. Each of these will require the appropriate specific treatment.

In ankylosing spondylitis the principal emphasis is on attention to posture and maintaining a full range of movements of the whole spine, the limb joints particularly including the shoulders and hips, and maintaining a full chest expansion. The patient must be taught these exercises and encouraged to perform them every day. Sports should be encouraged but contact sports in which trauma may occur should be avoided if possible. Swimming is a particularly good exercise because of the range of movements required in the spine and limb girdles. Anti-inflammatory drugs are particularly effective in relieving the symptoms. However, they should be used to allow the patient to perform exercises more thoroughly and not simply to let him slouch and stiffen in comfort.

Details of the management of the other less common causes of back pain will be found in appropriate textbooks.

CONCLUSION

The responsibility of the general practitioner lies in differentiating the common structural problems from other disorders which require specific treatment. A careful history and physical examination with selected investigations will enable him to reach a working diagnosis. Practical advice and accurate prescription of drugs and various forms of physical therapy will cope with the majority of back problems.

FURTHER READING

Evans, D. P. (1982). *Backache: its evolution and conservative treatment.* MTP Press, Lancaster.
Farfan, H. F. (1973). *Mechanical disorders of the low back.* Lea and Febiger, Philadelphia.

Grieve, G. P. (1981). *Common vertebral joint problems.* Churchill Livingstone, Edinburgh.

Jayson, M. I. V. (ed.) (1980). *The lumbar spine and back pain,* 2nd edn. Pitman Medical, London.

—— (1981). *Back pain: the facts.* Oxford University Press.

10 The painful hip and knee

C. S. B. Galasko

Pain in the hip and knee are common complaints at all ages but the causes differ in different age groups. It usually is possible to make the diagnosis from the history and clinical examination although X-rays and other investigations are often essential to confirm the diagnosis.

SYMPTOMS

Lesions affecting the hip joint usually produce pain in the groin, along the anteromedial aspect of the thigh, and sometimes referred to the knee. Occasionally the latter may be the only symptom. Lesions affecting the knee joint usually produce pain in the knee or radiating down the front of the lower leg. Pain in the buttock usually is referred from the back.

Lesions affecting either joint may produce stiffness which often affects the patient's functional ability. Patients may have difficulty or find it impossible to cut their toe nails; put on their socks, stockings, or tights; walk up and down stairs; climb onto a bus or train; sit on a low toilet seat or climb in and out of a bath tub. When examining the patient the range of movement, and the amount of fixed deformity should be recorded.

ARTHRITIS

It is beyond the scope of this chapter to discuss the differential diagnosis of arthritis of the hip or knee joint. However, it is essential to diagnose the underlying condition, which may require specific therapy, e.g. septic arthritis, tuberculous arthritis, gout, etc.

In many instances the treatment depends on the functional disability. For example total hip arthroplasty would be contraindicated in a patient with arthritis of the hip joint whose only complaint is pain after playing 36 holes of golf. On the other hand it may be urgently required in a patient with pain at rest and gross functional disability.

The treatment depends on the underlying cause and on the symptoms. From the surgical point of view there are two forms of arthritis. (i) Arthritis isolated to one or only a few joints and without a systemic illness, e.g. osteoarthritis. If the symptoms are cured with treatment the patient is 'cured'. (ii) Polyarthritis associated with a systemic disease, e.g. rheumatoid arthritis. The arthritis is only part of a generalized illness, and management of the arthritis only forms part of the total treatment of the patient.

The development of pain in the knee or hip joint may be the first presentation of a polyarthropathy and therefore, it is important to examine the entire locomotor system.

X-rays form part of the examination. They confirm the clinical diagnosis, may indicate the type of arthritis and demonstrate the amount of joint destruction. Weight bearing films give a more accurate assessment of the amount of joint destruction and deformity.

OSTEOARTHRITIS

Osteoarthritis is probably the commonest cause of a painful hip or knee joint in the adult. There are three broad causes, although in some patients it is not possible to categorize the underlying lesion.

1. Normal articular cartilage but abnormal load, e.g. following slipped upper femoral epiphysis.

2. Normal load but abnormal articular cartilage, e.g. following septic arthritis. Abnormalities of the articular cartilage may be responsible for many so-called 'primary' cases.

3. Normal articular cartilage and normal load but defective subchondral bone, e.g. avascular necrosis following irradiation, renal transplantation, etc.

The primary pathology is in the articular cartilage which gradually is destroyed, the underlying bone becoming pitted and eburnated. New bone formation at the joint margin leads to the development of osteophytes.

The incidence increases with increasing age. 'Primary' osteoarthritis tends to occur in the older patient whereas secondary osteoarthritis often occurs in the third, fourth, or fifth decades. The patients usually complain of pain which initially is aggravated by weight bearing and relieved by rest. The pain gradually increases in severity, becomes more constant and may even wake the patient at night. There is progressive restriction of movement which often is associated with fixed deformity and may produce progressive functional disability. In the hip joint the common deformity is flexion and adduction whereas in the knee joint it is usually a flexion deformity. Secondary muscle weakness and a limp usually develop.

The radiographic features consist of narrowing of the joint space, osteophyte formation, sclerosis of the joint margin, and the development of cysts.

Treatment

The treatment depends on the severity of the symptoms. Conservative measures are often of help in the mild to moderate case. The load acting on the joint can be reduced by weight reduction in the obese patient, the use of a walking stick and modification of the patient's mode of life. Muscle strengthening exercises are particularly helpful. Local heat and short-wave diathermy are of little long-term benefit but often give temporary relief of symptoms and allow the patient to carry out the exercises. Mild analgesics and non-steroidal anti-inflammatory agents often are of benefit.

Surgery is indicated if the symptoms are severe, i.e. They seriously interfere with the patient's ability to walk or work, interfere with sleep, are associated with severe functional disability, and have not responded to conservative treatment.

Surgical treatment

A variety of surgical procedures are used in the management of patients with severe osteoarthritis.

1. Neurectomy, cyst obliteration, forage (muscle dividing procedures), capsulectomy. Although these operations are still occasionally carried out it is the author's opinion that they are no longer indicated.

2. Osteotomy is indicated in the younger patient with moderate disease. Upper femoral osteotomy is used for osteoarthritis of hip joint providing there is at least 90° of flexion and a reasonable joint margin. In the knee it is particularly useful when the osteoarthritis is associated with a varus deformity.

There are several hypotheses as to why osteotomy works. It may re-align the joint surfaces, redistributing the forces across the joint; the remodelling of bone trabeculae necessitated by the osteotomy may stimulate repair processes in the damaged articular cartilage; osteotomy also relieves the intraosseous venous hypertension that frequently is present in osteoarthritis.

With careful selection of patients osteotomy is associated with relief of symptoms in the vast majority of patients and often lasting for many years. In a proportion of patients the improvement in symptoms is associated with widening of the joint space, which may represent cartilage regeneration.

3. Arthroplasty (i.e. the construction of a new moveable joint).

(ii) Excision arthroplasty, i.e. excision of one or both sides of a joint. Excision of the head and neck of the femur (Girdlestone operation) has been used as the initial form of surgical management of osteoarthritis but today is only indicated as a salvage procedure after other methods of treatment, particularly total replacement arthroplasty have failed.

(ii) Interposition arthroplasty. After reshaping the two joint surfaces some material is interposed between them to prevent fusion. The cup arthroplasty which was used in the hip joint has been superseded by the double-cup arthroplasty. The femoral head is shaped and a metal cup cemented onto it. A high-density polyethylene cup is cemented into the prepared acetabulum. The results are not as good as with total hip replacement arthroplasty, but it is easier to salvage a failed double-cup arthroplasty. It should be regarded as an experimental procedure since its exact place in the treatment of arthritis has not yet been established although this is likely to be in the younger patient.

(iii) Hemiarthroplasty, i.e. replacement of one half of the joint. Hemiarthroplasty of the hip joint is no longer indicated in the treatment of arthritis although it is still used in the treatment of the displaced subcapital fracture in the elderly. In the knee joint some surgeons prefer replacement of one or both tibial plateaus to osteotomy for moderate disease associated with deformity.

(iv) Total joint arthroplasty. Both sides of the joint are replaced. The development of total joint arthroplasty must rank as one of the major medical advances during the past two decades. The two components are made usually from different materials. In the hip joint the femoral head and neck are replaced by a metal component whereas the acetabulum is replaced by a high-density polyethylene cup. In the majority of total knee prostheses the femoral component is metal and the tibial component made from polyethylene. The vast majority of prostheses are cemented to the bone with methylmethacrylate, although some have been designed to avoid cement fixation. Much research is being carried out to develop prostheses that will allow direct ingrowth of bone, will not rely on acrylic cement for fixation and yet will give the same strength as existing prostheses.

4. Arthrodesis. Surgical fusion of the joint is particularly useful in young patients whose other joints are normal, whose job is heavy, who because of their job and age are unlikely candidates for a total joint arthroplasty but who are not suitable for osteotomy because of the gross destruction of the affected joint.

Choice of procedure

Arthrodesis is considered in younger patients under the age of 40 with severe unilateral osteoarthritis and normal joints above and below the affected joint, with gross restriction of movement and marked radiographic changes. It usually is associated with total relief of pain. The main drawback is the difficulty patients experience with sitting in a confined space, so that they may have to obtain aisle seats in the cinema, theatre, bus, etc. to allow them to stretch out the affected limb. There is an increased tendency for the development of degenerative changes in the adjacent joints after 20 years.

Osteotomy is considered in patients under the age of 55 who have at least 90° of flexion and a joint space which is still preserved although it may be narrow. It gives good pain relief which may be lasting in over 70 per cent of patients (Fig. 10.1).

Total joint replacement is the treatment of choice in patients over the age of 60, who have disabling symptoms especially if they have multijoint disease. Good to excellent results are obtained in approximately 90 per cent of total hip and 85 per cent of total knee arthroplasties. However, the procedure carries a significant risk of infection or loosening. The development of a deep infection may require an excision arthroplasty which may leave the patient worse off, and occasionally an amputation. It is for this reason that the procedure should never be undertaken for mild symptoms.

Long-term results in the younger patient are not yet known and, therefore, it is necessary to exercise caution in patients under the age of 55 whose other joints are normal. This is even more important for total knee arthroplasty where the long-term follow-up is much shorter than for total hip prostheses.

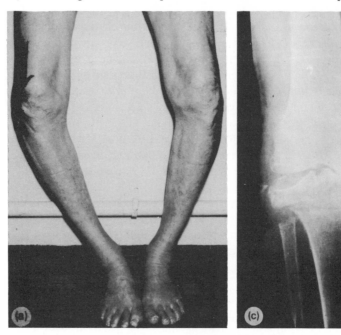

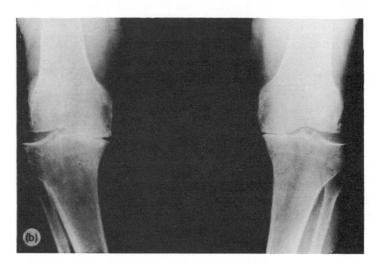

Fig. 10.1. Osteoarthritis of the knee. (a) Clinical photograph. The arthritis is associated with a marked varus deformity. (b) Pre-operative X-ray. The medial compartments are affected to a greater extent than the lateral compartments. (c) Post-operative X-ray. A laterally based weight of bone has been removed from the tibia, correcting the deformity and re-aligning the stresses across the joint. The head of the fibula has been excised.

RHEUMATOID ARTHRITIS

These patients are best managed by a team and in many instances should be seen in a combined clinic. The rheumatologist is responsible for their systemic treatment, the physiotherapist and occupational therapist for rehabilitation and the provision of aids, and the orthopaedic surgeon for the surgery. Rehabilitation is an essential part of therapy whether surgery is carried out or not.

The patient complains of pain and stiffness which often are worst first thing in the morning and initially may improve with activity. Rheumatoid arthritis of the knee joint usually is associated with an effusion and synovitis except in the late case where the disease is 'burnt out'. The joint may be warmer. Movements are impaired in all directions and are painful if forced. Fixed flexion or adduction deformity of the hip and fixed flexion deformity of the knee may occur and muscle wasting is frequent. There usually is a polyarthropathy.

Initially there are no radiographic changes. The earliest radiographic sign is periarticular rarifaction, followed by narrowing of the joint space due to destruction of the articular cartilage, and subsequent destruction of subchondral bone with deformity. Skeletal scintigraphy shows increased uptake and indicates the extent of the polyarthropathy.

The patient needs systemic treatment, the choice of therapy depending on the generalized condition. In many instances the inflammatory symptoms can be managed by simple anti-inflammatory drugs including aspirin. If not second-line drugs such as penicillamine and gold may be indicated. Systemic steroids should rarely if ever be used although intra-articular injections of hydrocortisone may give relief. There is some evidence that the latter may be responsible for progressive bone destruction and their number should be restricted to two or three. In acute inflammatory episodes, bed rest and immobilization with plaster of Paris splints may be extremely helpful. Physiotherapy is important, particularly active exercises to improve muscle function. Short-wave diathermy may be useful as an adjunct to physical exercises.

Surgical procedures

Soft-tissue procedures

Excision of painful subcutaneous nodules and tendon and nerve release may be extremely helpful especially in the upper limb. Excision of a large popliteal cyst may be necessary. Occasionally these cysts rupture into the calf producing signs and symptoms identical to a deep-vein thrombosis with localized pain, tender calf swelling, and a positive Homan's sign. The condition is always associated with an effusion in the knee joint and arthrography will confirm the diagnosis.

Soft tissue release of contractures is often of benefit.

Synovectomy

In rheumatoid arthritis the primary pathology is in the synovium which may become thick and bulky, particularly in the knee joint (Fig. 10.2). A florid

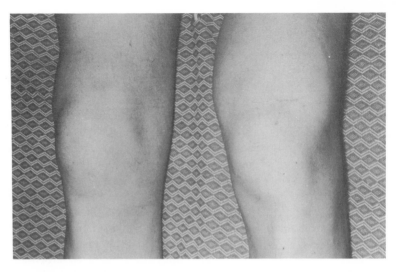

Fig. 10.2. Patient with rheumatoid arthritis associated with boggy synovitis of the left knee, mainly affecting the suprapatellar pouch.

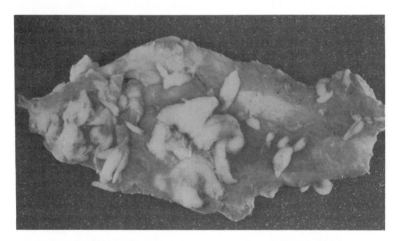

Fig. 10.3. Synovectomy specimen. Note the large fibrinous bodies most of which are still attached to the synovium although some are loose within the joint.

synovitis often occurs early in the disease and is associated with pain and stiffness. It often is associated with large fibrinous loose bodies (Fig. 10.3) which are radiolucent and not apparent on X-ray. At this stage synovectomy is associated with relief of pain and improvement of symptoms in the majority of patients. There is some debate as to whether synovectomy is prophylactic. Despite a successful result, with relief of pain and maintenance of movement, progressive joint destruction may be seen on X-ray, although the progression of the disease may be slower than in the contralateral joint.

Osteotomy

Osteotomy of the upper tibia, or double osteotomy of the lower femur and upper tibia is often indicated in moderate rheumatoid arthritis of the knee associated with some joint destruction and deformity, but maintenance of at least 70–80° of flexion.

Arthroplasty

(i) *Excision arthroplasty.* As in osteoarthritis this is usually a salvage procedure for failed joint replacement.

(ii) *Interposition arthroplasty.* There is no indication for this procedure in rheumatoid arthritis affecting the hip or knee.

(iii) *Hemiarthroplasty.* Some authors prefer this to osteotomy for patients with moderate disease of the knee.

(iv) *Total joint arthroplasty.* This is the commonest surgical procedure carried out in the rheumatoid hip or knee. In general, patients with rheumatoid arthritis require less from a joint prosthesis than patients with osteoarthritis. The former usually have multiple joint disease, the other affected joints acting as a 'brake' and limiting the patient's total activity. For this reason the age restriction does not apply to patients with severe rheumatoid arthritis and multiple joint disease.

Rheumatoid arthritis of the knee often is associated with gross laxity of the joint. This may be due to rupture or stretching of the ligaments and capsule. The choice of prosthesis often is different to that for an osteoarthritic knee, since it may have to provide stability as well as mobility (Fig. 10.4).

Arthrodesis

Arthrodesis is rarely considered in a patient with rheumatoid arthritis since the other joints are frequently affected.

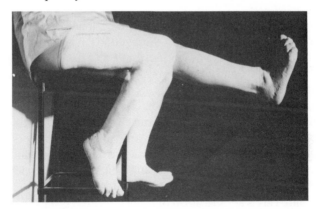

Fig. 10.4. Patient with seropositive rheumatoid arthritis, two years following bilateral total knee replacement arthroplasty. Multiple exposure photograph indicating the range of movement in his right knee.

Choice of procedure

Synovectomy of the knee is indicated for a 'boggy' synovitis that has not responded to conservative measures and is associated with no or minimal radiographic changes. Upper tibial or double osteotomy is often considered for a painful knee associated with narrowing of the joint space, early destruction of bone and the presence of at least 70° of flexion. Where there is gross joint destruction, marked instability, and severe loss of flexion total knee arthroplasty is indicated providing the patient's symptoms warrant major surgery.

Arthrodesis may be considered for a patient with an extremely unstable knee joint that is not controlled by an orthosis and who is unsuited for a total knee replacement because of an excessively high risk of infection (e.g. due to a recent infection) or loosening (e.g. patient with parkinsonism plus rheumatoid arthritis). It is sometimes used as a salvage procedure following failed total knee replacement. Arthrodesis of the hip joint is not indicated in rheumatoid arthritis.

OTHER ARTHRITIDES (SEE CHAPTER 3)

The hip and knee joint can be affected by a variety of other forms of arthritis.

Ankylosing spondylitis affects the hip more commonly than the knees. It is usually associated with marked stiffness and total hip replacement may be necessary.

Septic arthritis (Chapter 13) usually affects the hip in the neonate but most commonly affects the knee. It may complicate rheumatoid arthritis, the symptoms and signs often being dampened. The affected joint is painful, there is an effusion and the joint may be warm. Movements are limited by pain. The diagnosis is made by aspiration and should be suspected in any rheumatoid joint associated with a sudden increase in pain.

Haemophilic arthropathy affects the knee more commonly than the hip joint (Chapter 13).

Gouty arthritis may affect the knee and, occasionally this may be the first presentation. Acute gout has to be distinguished from the other forms of arthritis of sudden onset such as septic arthritis and rheumatic fever. The joint is painful and swollen and often there is a history of previous attacks with symptom-free intervals.

The diagnosis is made by the detection of urate crystals in the synovial fluid aspirate, the presence of tophi in the ears and elsewhere, and a raised plasm uric acid. The acute symptoms are treated with phenylbutazone or other anti-inflammatory agents. Long-term drug therapy is designed to reduce the plasma uric level either by reducing tubular reabsorption of urate (e.g. probenemid) or reducing the formation of uric acid (e.g. allopurinol).

Pseudogout. This is another form of acute arthritis, crystals of calcium pyrophosphate being deposited within the joint. The symptoms are essentially similar to gout and the diagnosis is made by detecting pyrophosphate crystals in the synovial fluid. Calcification of the menisci may be seen on X-ray but this is not always present. Treatment is by rest, aspiration of joint fluid and anti-inflammatory agents.

Neuropathic arthritis (Charcot's arthropathy) is secondary to a neurological disorder affecting the deep pain fibres, e.g. tabes dorsalis, diabetic neuropathy, or a cauda equina lesion. The knee is one of the most frequently affected joints, the patient complaining of swelling, instability, and often of pain. X-rays demonstrate gross disorganization of the joint with severe destruction and considerable bone formation. The joint requires stabilization either with an orthosis or arthrodesis.

Arthritis of rheumatic fever is uncommon but still occurs. The affected joint is swollen and warm with painful restriction of movement. X-rays are normal. The ESR is elevated and there may be a raised white cell count. Frequently the patient presents with a flitting arthritis. As the symptoms settle in one joint they develop in another. The joint involvement responds to salicylates, the patient is placed on penicillin to prevent further attacks and the other manifestations of the rheumatic fever are treated as necessary.

INTRA-ARTICULAR LESIONS AFFECTING THE HIP JOINT

CONGENITAL DISLOCATION

See Chapter 13.

TRANSIENT SYNOVITIS

This is a condition of unknown aetiology that affects young children. It is rarely seen after the age of 10 and occurs more commonly in boys. The child develops a limp and complains of pain in the region of the hip joint. Movements of the hips are limited, particularly flexion and internal rotation. X-rays are normal.

The signs and symptoms are identical to early Perthes' disease. The scintigram is normal in transient synovitis, whereas areas of increased uptake may be noted in early Perthes' disease although the X-rays may still be normal.

Treatment

The patient is treated with bed rest and bilateral skin traction until the signs and symptoms settle. This usually takes 10–20 days. The diagnosis is made in retrospect after the condition has totally subsided. The differential diagnosis includes Perthes' disease, early septic or tuberculous arthritis, and bone infarct (usually secondary to sickle-cell anaemia). Because the signs and symptoms could be due to these more serious conditions, the child should be admitted to hospital, X-rays of the hip obtained, and blood taken for erythrocyte sedimentation rate, white cell count, and haemoglobin type.

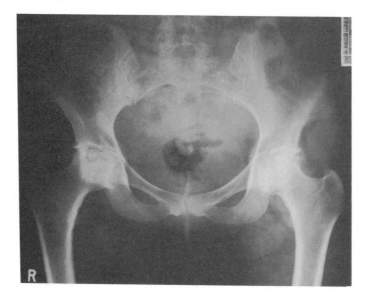

Fig. 10.5. Avascular necrosis. Collapse of both femoral heads with fragmentation in a patient with Henoch–Schonlein purpura who had been treated with high-dosage steroids.

Occasionally the condition is recurrent, the child presenting with a second episode one to two years later. Very rarely X-rays taken six months later show early Perthes' disease.

PERTHES DISEASE

See Chapter 13.

AVASCULAR NECROSIS (OSTEONECROSIS) (FIG. 10.5)

The femoral head is the commonest site of involvement. However, many other bones can be involved including the distal femoral condyle, humeral head, dome of the talus, scaphoid, etc. The blood supply can be interrupted for a variety of reasons although in some instances the pathogenesis is not understood.

(i) fracture;

(ii) dislocation;

(iii) osteoarthritis: this may be related to the use of analgesic agents;

(iv) Caisson's disease which occurs most commonly in deep sea divers but can occur in tunnellers and other people working at deep levels without adequate decompression when they ascend;

(v) chronic alcoholism;

(vi) sickle-cell anaemia and other forms of haemoglobinopathy associated with bone infarcts;

(vii) renal transplantation: there is some relationship between the incidence and the amount of steroids used to suppress rejection;

(viii) other uses of steroids, e.g. systemic lupus erythematosus;

(ix) Gaucher's disease;

(x) idiopathic

The patient may complain of pain before there are any radiographic signs. Serial X-rays show an increase in sclerosis affecting the anterosuperior aspect of the femoral head, which is followed by collapse of the segment. Histological examination shows necrosis of the subchondral bone with intact overlying articular cartilage, the dead bone being surrounded by a zone of hypervascularity.

Early cases may respond to rest. Once there are radiographic signs a varus/rotation osteotomy of the proximal femur may prevent collapse of the femoral head but once deformity has occurred total hip replacement is indicated.

SLIPPED UPPER FEMORAL EPIPHYSIS

See Chapter 13.

TRANSIENT OSTEOPOROSIS

This is a condition of unknown aetiology. It most commonly affects the femoral head but the patella is another common site. The patient complains of pain; serial X-rays showing progressive osteoporosis of the affected bone.

The disease is self-limiting and after weeks to months the pain subsides, the radiographic appearance returning to normal. Skeletal scintigraphy is more sensitive than radiography. In the early stages there is an increased uptake of isotope, before the radiographic appearances develop. This increased uptake returns to normal before reconsolidation of the bone is seen on X-ray.

Often as the disease is subsiding in one area the patient starts to develop symptoms at another site.

Treatment is by resting the affected part until the bone has reconsolidated.

ARTHRITIDES

See Chapter 3.

INTRA-ARTICULAR DISORDERS OF THE KNEE JOINT

ARTHRITIS (SEE ABOVE)

The knee is the commonest joint affected by septic arthritis, osteoarthritis, and haemophilic arthropathy. It is commonly affected in rheumatoid arthritis and one of the joints most commonly affected by neuropathic arthritis.

PATELLOFEMORAL PAIN

Chondromalacia patellae

The patellar articular cartilage is fibrillated, the condition affecting adolescents and young adults, particularly females. The patient complains of an aching pain behind the patella, which is worsened by climbing or descending stairs. Kneeling is uncomfortable. Clinical examination usually reveals fine patellofemoral crepitus with retropatellar tenderness. The symptoms are often improved by curtailing sporting activities, particularly those associated with extending the flexed knee against resistance, and by regular quadriceps exercises with the knee in extension. A variety of surgical procedures have been described but the results are uncertain. These include excision of the damaged articular cartilage (if it is less than 1 cm in diameter) with drilling of the underlying subchondral bone, drilling of the prominent ridge in the front of the medial femoral condyle, patellectomy and tibial tubercle osteotomy.

Patellofemoral malalignment

Identical symptoms can be produced by malalignment between the patella and the femur. X-rays of the patellofemoral joint (skyline patellar views) taken in 20°, 30°, and 60° of flexion may demonstrate the malalignment, the patella lying laterally. Treatment is surgical either by releasing the lateral expansion of the quadriceps aponeurosis, medial transfer of the insertion of the patella tendon, or a combination of both.

Patellofemoral osteoarthritis

This occurs in older patients and is predisposed to by chondromalacia patellae.

Recurrent subluxation of the patella

This usually affects adolescents, and is commoner in girls. There is recurrent subluxation or frank dislocation of the patella, the patella always subluxing laterally. The condition is predisposed to by a high-riding or small patella, a small lateral femoral condyle, ligamentous laxity, or genu valgum. The dislocation usually occurs when the patient flexes the knee and often reduces spontaneously. In the dislocated stage the patella will be felt laterally. Immediately after the dislocation the knee is swollen and an effusion may be present. Treatment is by realigning the quadriceps mechanism by transfering the tibial tendon medially, by releasing the lateral quadriceps expansion and tightening the medial quadriceps expansion, or by a combination. Patellectomy should not be carried out without realigning the quadriceps mechanism; patellectomy alone may lead to recurrent dislocation of the quadriceps tendon. The condition predisposes to osteoarthritis.

Habitual dislocation is a variant in which the patella dislocates laterally every time the knee is flexed, whereas in recurrent dislocation the knee may seem normal for weeks or months between dislocations. Habitual dislocation usually presents in childhood and is due to a fibrous contracture of the vastus lateralis or fibrous tethering of the vastus lateralis to the iliotibial tract, the shortened vastus lateralis pulling the patella laterally every time the knee is flexed. The patella may eventually become permanently dislocated. Treatment is by division of the tight muscle or fibrous band.

MENISCAL LESIONS

These occur commonly in males between 17 and 45 years. The tear usually is caused by a twisting injury when the knee is flexed, and occurs most commonly as a result of a footballing injury. It also occurs in miners and other who work in a squatting position. The medial meniscus is torn more frequently than the lateral meniscus, the tear usually being longitudinal. If it extends the entire length of the meniscus a buckle-handle tear is produced, 'the handle' or the central fragment being displaced towards the middle of the joint (Fig. 10.6). Anterior or posterior horn tears are produced when the tear only involves the anterior or posterior portion of the meniscus. Another variety is the horizontal cleavage tear which occurs more commonly in the older patient and is associated with degeneration of the meniscus.

The menisci are avascular. Tears are associated with an effusion which may take up to 24 hours to develop but not an haemarthrosis. Torn menisci do not heal spontaneously. Occasionally the meniscus may be detached from the synovium. Under these circumstances healing may occur if the meniscus is sutured back.

The patient usually gives a history of pain and swelling of the knee following a twisting injury, the symptoms having settled over 10–14 days. After an interval of weeks or months he develops recurrent episodes of pain, swelling,

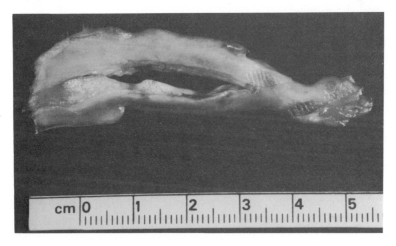

Fig. 10.6. Bucket-handle tear of a meniscus. The longitudinal tear extends the entire length of the meniscus.

instability during a twisting movement, and locking. By 'locking' is meant the inability to extend the knee fully. The history of the manner in which patient unlocks the knee is important. He usually has to flex the knee even more or rotate it before he feels something 'give' and is then able to extend the knee fully. This ability to extend the knee fully after unlocking it differentiates true locking from an effusion or haemarthrosis which also may prevent full extension.

The clinical findings depend on the stage of the condition. Initially the knee is swollen and tender along the medial or lateral joint margin, depending on the site of the tear. Limitation of the last few degrees of extension with sharp pain if passive extension is forced is not always present but is virtually diagnostic. The McMurray test may be positive and the quadriceps are often wasted.

In the 'silent' stage between attacks there often are no physical findings other than wasting of the quadriceps and occasionally a positive McMurray test, the diagnosis being made from the history.

Quadriceps exercises are important to restore the muscle power. If the diagnosis is suspected arthroscopy is indicated. Arthrography may be useful. Once the diagnosis is confirmed the meniscus should be excised; in a bucket handle tear only the 'handle' need be removed. In select cases the meniscus can be removed by arthroscopy. A torn meniscus predisposes to osteoarthritis but so does meniscectomy. The meniscus plays an important role in spreading the load across the knee joint. Conservative meniscectomies, preserving the normal portion of the meniscus are now being carried out more frequently.

Horizontal cleavage tear of the degenerative medial meniscus usually occurs after the age of 50. The patient complains of pain on the medial aspect of the knee and which usually settles with quadriceps exercises and supportive bandaging. Meniscectomy is advised if the symptoms fail to subside with conservative treatment.

Cysts of the meniscus (usually the lateral) and a discoid lateral meniscus may produce recurrent discomfort in the knee. Usually, the cyst is more obvious when the knee is slightly flexed and is tender. A discoid meniscus may produce a loud cracking noise on movement of the knee. If troublesome a cystic or discoid meniscus should be excised.

OSTEOCHONDRITIS DISSECANS

This is a localized disorder of a convex joint surface characterized by avascular necrosis of a segment of subchondral bone which may gradually separate together with the overlying articular cartilage to form a loose body, the resulting cavity being fitted with fibrous tissue. The medial femoral condyle is the commonest site although the lesion does occur in the hip and other joints. The aetiology is not known and suggested causes include thrombosis of an end artery or injury. The condition may be familial. The affected segment varies in size and in the knee may be up to 3 cm in diameter.

The condition occurs in adolescents and early adulthood. Initially the patient complains of aching after use and recurrent swelling. Once the fragment has separated recurrent acute locking of the joint occurs, associated with pain and followed by an effusion. The diagnosis is confirmed on radiographs.

Treatment

In the developing stage the treatment is conservative. The knee is supported in a crepe bandage or plaster cast and the patient carries out quadriceps exercises with the knee in extension. If the lesion is 'ripe', i.e. there is a clear line of demarcation between the separating fragment and the remainder of the medial femoral condyle, it should be removed if small. If large it should be fixed with a pin, after its bed has been freshened. Once the fragment has separated and become a loose body it should be excised, even if large.

Osteochondritis dissecans predisposes to the development of osteoarthritis in later life.

LOOSE BODIES

The knee is the commonest joint to be affected. The main causes are:

(i) *Osteochondritis dissecans.*

(ii) *Osteoarthritis.* Although detached marginal osteophytes appear loose on X-ray they often retain a synovial attachment, but may separate to form loose bodies.

(iii) An *osteochondral fracture.*

(iv) *Osteochondromatosis.* This is an uncommon synovial disorder of unknown aetiology, characterized by the formation of multiple small villous processes which become pedunculated. Eventually their bulbous ends become cartilaginous, and may calcify or ossify and may become detached.

Irrespective of the cause the patient develops recurrent attacks of acute locking when the loose body is trapped between the articular surfaces. Each episode is associated with severe pain, the knee swelling over the ensuing 12 to 24 hours. Occasionally a loose body may be palpable. It will be seen on X-ray but the fabella (the sesamoid bone in the lateral head of the gastrocnemius) should not be confused with a loose body. Loose bodies should be removed if they produce symptoms.

EXTRA-ARTICULAR DISORDERS INTRINSIC TO THE HIP REGION

COXA VARA

See Chapter 13.

TROCHANTERIC BURSITIS

The bursa overlying the greater trochanter may become inflamed or infected, the patient complaining of pain over the lateral aspect of the thigh. Examination reveals a localized area of tenderness. In the presence of infection the overlying skin is usually red and warm. X-rays are normal. Inflammatory bursitis is treated by hydrocortisone and local anaesthetic injections or ultrasound; infection by incision and drainage.

SNAPPING HIP

This is a harmless condition in which a distinct snap is felt and often heard when the patient actively flexes the hip. Often it is not reproduced by a passive movement with the muscles relaxed. It usually is due to slipping of the aponeurosis of the gluteus maximus over the greater trochanter. Treatment is very rarely indicated. If required the affected part of the aponeurosis is divided.

GROIN STRAIN

This is a tendonitis of the origin of the adductor tendons. It usually responds to local injection of hydrocortisone and local anaesthetic or ultrasound. Occasionally surgical release of the fibrotic tendon origin may be required. It should not be confused with a *femoral hernia* which lies close to the adductor tendon and which can also produce pain in the region of the hip joint.

HAMSTRING TENDONITIS

This usually affects the distal insertion rather than the pelvic origin but occasionally the latter may be painful. Tight hamstrings can produce pain in the region of the buttock, or posterior thigh.

EXTRA-ARTICULAR LESIONS INTRINSIC TO THE REGION OF THE KNEE JOINT

RUPTURE OF THE QUADRICEPS APPARATUS

This is due to an unexpected flexion force, the rupture occurring at the point of the attachment of the quadriceps tendon to the upper pole of the patella, through the patella and surrounding quadriceps expansion or at the attachment of the patella tendon to the tibial tubercle. The proximal rupture usually occurs in the elderly in whom the quadriceps tendon is often degenerate, whereas the distal rupture often occurs in children or young adults. The patient complains of pain and is unable to straight leg raise.

Treatment

The patellar mechanism should be reconstituted surgically. A ruptured quadriceps tendon or patellar tendon should be sutured. If possible a fractured patella should be fixed but if too comminuted may be excised and the tendon reconstructed.

OSGOOD–SCHLATTER'S DISEASE

This occurs in adolescents and boys are affected more commonly. It is an apophysitis of the developing tibial tubercle due to the pull of the patellar tendon. The patient complains of a painful lump in front of the knee, the pain being aggravated by activity. The tibial tubercle is enlarged (Fig. 10.7a) and tender and X-rays frequently show fragmentation of the tubercle (Fig. 10.7b).

The disease is self-limiting. The patient is advised to avoid strenuous exercises whilst he has the symptoms. If very acute the knee may be rested in a plaster cylinder for four to six weeks. Very occasionally a loose fragment remains which may be tender. This should be removed after growth has ceased.

BURSITIS

Either the prepatellar bursa (which is anterior to the distal patella and proximal patellar tendon) or infrapatellar bursa (which lies in front of the tibial tubercle) may be affected. The bursitis may be due to repeated friction, e.g. produced by frequent kneeling (housemaid's or clergyman's knee) or the bursa may become infected. The former presents with a fluctuant swelling which may be tender, the latter with an acutely tender painful swelling, the overlying skin is

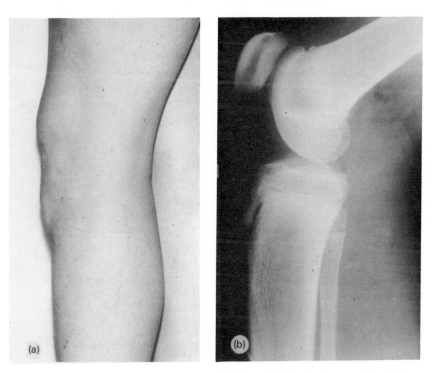

Fig. 10.7. Patient with Osgood–Schlatter's disease. (a) Clinical photograph indicating the enlarged tibial tubercle. (b) X-ray showing fragmentation of the tibial apophysis.

warm and reddened and the patient is pyrexial. An irritative bursitis is treated by excision, a suppurative bursitis by incision and drainage and antibiotics.

POPLITEAL CYST

These do not usually produce pain. The semimembranosus bursa, which lies between the medial head of gastrocnemius and semimembranosus may become distended with fluid, the lesion presenting with a soft cystic swelling behind the knee. This often occurs in children and frequently the cyst disappears. However, if the swelling becomes uncomfortable the bursa should be excised.

A Baker's cyst is a herniation of the synovial cavity of the knee into the popliteal fossa. It is always secondary to a condition of the knee associated with a persistent synovial effusion. It may be uncomfortable if large. Treatment is aimed at the primary condition in the knee. If the cyst is extensive and uncomfortable it may require excision.

Occasionally a popliteal cyst may rupture into the calf producing signs and symptoms similar to deep-vein thrombosis. The diagnosis is made by arthrography. The dye is seen to pass from the knee joint into the cyst and then into the calf.

LIGAMENTOUS LESIONS

The ligaments of the knee are injured more frequently than in any other joint. The patient complains of instability, often associated with pain. Clinical examination will reveal which ligaments are affected. The cruciate ligaments are examined with the knee flexed. The patient must be completely relaxed. The foot is held firmly on the couch, the patient's proximal tibia is held by the examiner's hands with the thumbs on its anteromedial and anterolateral surfaces and the upper end of the tibia is alternatively pulled forwards and pushed backwards. Normally there is an anteroposterior glide of up to half a centimetre. Excessive anterior glide (anterior drawer sign) indicates damage to the anterior cruciate ligament, excessive posterior glide (posterior drawer sign) indicates damage to the posterior cruciate ligament.

The medial and lateral ligaments are examined with the knee in 15–20° of flexion. The medial ligament is examined by putting a valgus strain on the knee, the lateral ligament by putting a varus strain on the joint. The degree of stability should be compared with that on the normal side.

If there has been a recent tear it may not be possible to examine the ligaments adequately because of associated muscle spasm and pain. If there is any doubt about the integrity of the ligaments, the knee should be examined under a general anaesthetic. Acute tears should be repaired as soon as possible, otherwise they often heal with lengthening. Chronic tears and instability may require ligamentous reconstruction, depending on the symptoms. Recurrent instability will predispose the patient to the development of osteoarthritis.

Anteromedial instability is usually due to a combination of anterior cruciate and medial collateral ligament damage, whereas anterolateral instability is frequently due to a combination of anterior cruciate and lateral collateral instability.

PELLEGRINI–STIEDA'S DISEASE

This is due to partial avulsion of the origin of the medial collateral ligament from the medial condyle of the femur with ossification of the haematoma at the site of injury. The patient complains of persistent discomfort, the area is tender and may feel thickened. X-rays show a thin plaque of new bone at the origin of the medial ligament. Treatment is by active mobilizing and muscle strengthening exercises. Ultrasound may help.

HAMSTRING TENDONITIS

The commonest site is the knee joint. The patient complains of pain, precipitated by activity. Clinically the tendon sheath is tender and a nodular thickening may be palpable. The symptoms may respond to conservative treatment including rest and ultrasound but if this fails surgical division of the tendon sheath may be necessary. Hydrocortisone injections must be given with great care. It should never be injected into the tendon as this may cause rupture, but only into the sheath.

Tight hamstrings and hamstring strain usually respond to physiotherapy.

EXTRA-ARTICULAR LESIONS NOT LOCALIZED TO HIP OR KNEE

There are a number of conditions that can occur anywhere but if they occur in the region of the hip or knee will produce pain at these sites.

FRACTURES

The femoral neck is a common site of stress fracture. This often occurs in army recruits and other individuals suddenly faced with prolonged activity. The proximal fibula is a less common site.

The most important fracture in the elderly is a fracture of the proximal femur. It often is secondary to osteoporosis, usually occurs after a fall but occasionally the weakened bone fractures, the patient falling as a result of the fracture. The patient presents with pain, shortening and external rotation of the limb. Treatment depends on the site of the fracture and the general condition of the patient. The patient should be mobilized as soon as possible, since prolonged bed rest is associated with the development of pressure sores, urinary tract infection, respiratory infection, incontinence, mental confusion, and pulmonary embolism. Providing the patient is sufficiently fit subcapital and transcervical fractures associated with gross displacement are usually treated by hemiarthroplasty because of the risk of non-union and avascular necrosis, other proximal femoral fractures being internally fixed.

ACUTE OSTEOMYELITIS (SEE ALSO CHAPTER 13)

If it occurs in the proximal femur it may be difficult to differentiate from septic arthritis of the hip joint. Osteomyelitis of the distal femur or proximal tibia is more readily differentiated from septic arthritis of the knee joint.

TUMOURS OF BONE

These can affect the hip region but the commonest site for many primary bone tumours is around the knee. Giant-cell tumour and osteosarcoma occur most frequently in the region of the knee joint, but Ewing's sarcoma affects the diaphysis of a long bone rather than the metaphysis. The proximal femur and pelvis are amongst the commonest sites for skeletal metastases and chondrosarcoma.

Most tumours present with localized pain, but occasionally a pathological fracture may be the presenting feature. The patient rarely presents with a swelling.

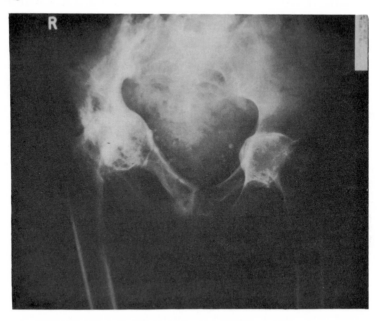

Fig. 10.8. Patient with Paget's disease affecting the pelvis. There is associated osteo-arthritis of the hip joint. In this patient the pain was due to the osteoarthritis and was relieved by total hip arthroplasty.

OSTEOID OSTEOMA

This is a benign lesion of uncertain origin. The characteristic feature is the development of a nidus of osteoid tissue, seldom more than half a centimetre in diameter, and surrounded by an extensive zone of dense sclerotic bone if it develops in the cortex, but little reactive bone if it develops in the medulla. The patient complains of a poorly localized pain, worse at night and usually responding to aspirin. There may be localized tenderness. The lesion is often obvious on straight X-ray but tomograms may be required. Skeletal scintigraphy is more sensitive, the lesion appearing as a localized area of increased uptake.

FIBROUS DYSPLASIA

One of the commonest sites of fibrous dysplasia is the proximal femur. The bone is replaced by an expanding fibrous lesion and pathological fractures may develop. The aetiology is unknown, the condition may be painful and the diagnosis is made on X-ray, the affected area of bone having an homogenous, ground-glass appearance.

PAGET'S DISEASE

The pelvis is one of the commonest sites and Paget's disease is associated with an increased incidence of osteoarthritis of the hip. It is important to determine whether the pain is due to the Paget's disease or the associated osteoarthritis. If the former, the patient requires treatment with calcitonin, diphosphonate, or mithramycin. If the latter, the osteoarthritis must be treated and often a total hip arthroplasty is indicated. (Fig. 10.8).

GAUCHER'S DISEASE

This also commonly affects the proximal femur. It is a rare lipoid storage disease of autosomal recessive inheritance. The spleen and liver are enlarged and skeletal involvement takes the form of infiltration and replacement of bone by masses of lipoid-laden reticulo-endothelial cells. Pathological fractures frequently develop and avascular necrosis of the femoral head is common.

CALCIFIC PERICAPSULITIS

This occurs most commonly in the shoulder joint. It is an extremely painful condition of acute onset but responds rapidly to immobilization, hydrocortisone injections, or removal of the calcific material. It can affect the hip joint producing acute localized pain, tenderness, and limitation of movement.

OSTEOMALACIA

Looser's zones or stress fractures occur in osteomalacia and can be painful. They not infrequently affect the proximal femur or pelvic rami.

REFERRED PAIN

TO THE HIP

(i) *Lumbar spine pathology.* Prolapsed intervertebral discs, degenerative disc disease, spondylolisthesis, etc. may produce pain radiating to the buttock. Occasionally buttock pain may occur in the absence of back pain.

(ii) Conditions affecting the *sacro-iliac joint*, such as arthritis or ankylosing spondylitis may produce pain in the buttock.

(iii) *Abdominal lesions.* Appendicitis or pyosalpinx may produce symptoms similar to hip pathology partly due to irritation of the obturator nerve causing pain in the thigh and partly from spasm of the hip muscles that arise within the abdomen or pelvis, particularly the psoas and iliacus. An iliopsoas bleed or abscess will also produce pain referred to the front of the hip joint.

(iv) *Arterial occlusive disease.* Occlusion of the lower abdominal aorta or common iliac arteries may produce ischaemic pain in the buttock or thigh. The pain is brought on by activity and relieved by rest, the femoral pulses are weak or absent and a bruit may be heard.

TO THE KNEE

(i) Referred pain to the knee may be the only symptom arising from disorders of the hip joint. Usually the pain radiates from the groin along the anteromedial aspect of the thigh to the knee.

(ii) Disorders of the spine or pelvis causing pressure on the lumbar nerve roots or sciatic nerve may produce pain referred to the knee or upper calf.

ACKNOWLEDGEMENT

I am grateful to the Department of Medical Illustration, Hope Hospital, for the illustrations.

11   Foot problems

L. A. Smidt

Once babyhood is passed, there is little that is particularly pleasing in the appearance
of most people's feet.   (Wood Jones 1944)

From birth there is a tendency for the soft, developing foot to be clothed in
garments which are too tight and unyielding. The natural consequence of this is
for the toes to be prevented from lying in a straight position and for the normal
anatomy of the foot to become distorted.

There are 26 bones in the foot, which is a highly specialized mechanism
designed for both weight-bearing and propulsion. Relatively the feet are small
structures when their function is considered, and they are liable to stresses of a
huge magnitude. It is not unusual for individual areas of the foot to bear loads
in excess of 690 kPa (100 lbf/in^2) for short periods during the walking cycle.

CHILDREN'S FEET

In children carefully fitted footwear is essential. This includes hosiery as well as
shoes. It is important to ensure that hosiery is of adequate length and that shoes
are neither too short nor too long, and that they are of adequate girth.

Digital deformities are caused by shoes which are too short and equally easily
by shoes which are too long and which do not adequately grip the foot around
the instep, thus allowing it to slide forward and cause impaction of the toes.

Assessing shoes for adequate volume is often difficult. Fig. 11.1 illustrates
how two shapes may have the same perimeter but different volumes.

The deformities caused by bad footwear include hammer or claw toes, and
in addition toenail problems may result from multiple minor traumata.

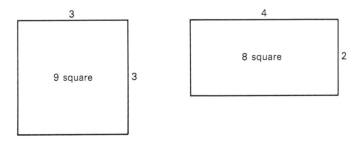

Fig. 11.1. Perimeters are equal but volumes differ.

Most lesser toe deformities will correct themselves in the very young if they are diagnosed early enough and whilst the foot remains supple and adequate footwear is unfailingly worn. In those which are not self-correcting the methods of correction include strapping, exercises, and digital silicone rubber orthoses. It is, however, imperative to ensure that adequate fitting footwear is worn during any attempt to correct lesser toe deformities.

Many feet in the very young appear to be either pronated or supinated and the majority of these correct themselves, again, providing good footwear is worn. These conditions are usually relatively mild in degree and are almost always asymptomatic. It is usually the parents who seek treatment on the basis of appearance. If these anomalies persist beyond about the age of seven years they should be investigated further and assessed biomechanically to ensure that the subtalar joint is in a neutral position. If it is not, then exercises and possibly the provision of a biomechanical orthosis may prove necessary. The object of providing an orthosis for this type of anomaly is to re-align the foot into the neutral subtalar position and thereby reduce stresses both intrinsic and extrinsic to the foot. Torsion anomalies are not within a chiropodists's remit. Causes of pain in the feet of children are few and may be due to postural problems or to osteochondritis of one or more bones. Osteochondritis occurs commonly in the navicular, the metatarsals (commonly the second), or the calcaneum, If present in the navicular it is usually found with an equal sex distribution in children of four to seven years of age. In the metatarsals it is most common in 10- to 15-year-old females. In the calcaneum it is most common in boys nine to 13 years of age with the site of maximum pain being at the epiphysis of the calcaneum at the posterior of the heel. The basic element of treatment for these conditions is to encourage rest and if this is not accepted, then immobilization in plaster of Paris may be indicated.

Metatarsus primus varus and adducted forefoot are relatively common and correct footwear fitting is essential as this type of foot has a predisposition to hallux valgus. Congenital mobile flat foot is uncommon but where it does exist it is usually a strong, well-functioning structure which does not require treatment, and is not prone to the long-arch postural problems of the much weaker high-arched foot.

Rigid flat-foot conditions are often the result of bony anomalies such as 'tarsal bars' and unless they produce symptoms, often do not require treatment. Where symptoms present these patients require an orthopaedic opinion.

Fashion-conscious adolescents are generally unconcerned in regard to the future of their feet and will continue the damaging process started in childhood, by wearing even more unsuitable footwear. Toe deformities become more marked and associated corns and callosities will combine to make for shoe fitting difficulties, and pain in adulthood. It is of paramount importance, if many foot problems are to be avoided in later life, that the feet of teenagers and young adults be as carefully measured and fitted when purchasing shoes, as children's feet.

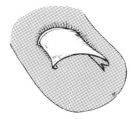

Fig. 11.2. Ingrown nail (onychocryptosis).

Verrucae are common in childhood and adolescence and are usually easily eradicated by the use of strong caustics, cryosurgery, or electrosurgical methods. Few verrucae do not succumb to these methods of therapy, although there is strong epidemiological evidence to suggest that only those verrucae which present painful symptoms are worthy of active treatment. It is suggested that all verrucae have a limited life-span and will resolve spontaneously, given adequate time.

Ingrown toenails (Fig. 11.2) occur most frequently in adolescent males, although no section of the population is immune from this disabling condition. The true ingrown toe nail or onychocryptosis has as an invariable feature a sharp splinter of nail which has penetrated the skin and which may have caused hypergranulation tissue to result from the body's attempts to heal the condition. This condition is commonly complicated by having a secondary bacterial infection. Antibiotics alone will not be very helpful in dealing with onychocryptosis as the offending section of nail needs to be removed before healing is possible. As a high percentage of onychocryptoses recur following simple removal of the splinter, the procedure of choice is to undertake a nail border resection with partial matrix destruction. This is undertaken as an out-patient procedure under local anaesthesia and the portion of matrix is destroyed, most commonly using phenol. A similar procedure is used where the whole nail plate needs to be removed and prevented from regrowing. These operations are commonly carried out by chiropodists and the results have proved to be excellent.

Hyperhidrosis (excessive sweating) is a phenomenon experienced by large numbers of adolescents although predominantly by males. It is often exacerbated by foot strain caused by changing from the sedentary life of school to the more active roles of adulthood. The treatment for such cases is usually to prescribe exercises and/or to supply some form of support to alleviate the foot strain which produces the excess of sweating, and to treat the skin with anhidrotic or astringent preparations. Advice on general hygiene may be necessary. Hyperhidrosis of the feet may also occur as part of the generalized hyperhidrotic state resulting from anxiety which will be unlikely to respond to local treatment.

Fungus infections of the skin and nails may be an accompaniment to hyperhidrotic feet, although it is not uncommon to find particular strains of fungal infection in anhidrotic skins. A definite diagnosis of fungal infection can only be made by sending a skin or nail sample to the laboratory. Some of the more recent topical antifungal drugs are extremely efficient for the skin lesions but where the nails become fungally infected the practitioner must decide whether the symptoms are severe enough to embark on the very lengthy process using oral antifungal drugs which often require to be taken for periods in excess of one year. Evidence so far indicates that topical preparations are not satisfactory.

ADULT FEET

There are a multitude of abnormalities which can affect adult feet. Most of them will be adversely affected by ill-fitting footwear, if not actually caused by this.

Metatarsalgia is the term used to describe pain in the forefoot. It may be the result of degeneration of the normal fibrofatty pad which protects weight bearing under the metatarsophalangeal joints. This is a common phenomenon which occurs with advancing age. Fibrofatty pad degeneration is usually the result of a lack of adequate circulation to the feet, but it may be complicated by mechanical factors. These factors include retraction of the lesser toes which will result in back pressure on to the metatarsal heads thereby forcing them into prominence on the sole of the foot. The normal method of treatment is to aim to replace the natural 'shock-absorber' with an artificial one made from felt or rubber. This can be in the form of adhesive dressings or by the use of orthotic insoles made to fit the patient's shoes. Patients will liken their symptoms to walking on pebbles or marbles or they may describe symptoms of hot burning feet.

Metatarsalgia may have a more specific cause. Morton's neurofibroma is a condition in which an old organized thrombus blocks a digital artery and the associated nerve bundles show perineural fibrosis and nerve fibre degeneration. The most common site of this condition is in the fourth interdigital cleft and a definitive diagnosis is usually possible when there is absence of an ability to differentiate pinprick sensation in the skin in the web between the third and fourth toes. It may occur in other interdigital clefts. It is possible to alleviate symptoms by re-aligning the metatarsal bones by placing an insole with a closed-cell rubber pad beneath one of the metatarsal heads. The only curative measure is the removal of the neurofibroma by surgical means, but unfortunately the resultant scar may be the site of corn and callosity formation.

March fracture is a condition which affects the metatarsals (commonly the second and third) and which derives its name from the army where many such fractures arose after marching. This condition is a fatigue fracture which will not usually be diagnosable from X-ray until healing has commenced and the callus has started to form. Since march fracture is a transverse break there is generally no displacement of the fragments and treatment is strapping on the foot with elastic metatarsal adhesive plaster and rest.

Inflammation of the joint capsule which may arise because of a toe deformity causing back pressure and over-loading on the metatarsophalangeal joint may be another cause of pain in the metatarsal region. The strain on the metatarsophalangeal joint may be alleviated by the provision of chiropodial padding or insoles or the condition may be helped additionally by ultrasound or heat. If the pain still persists, intra-articular steroid injection may prove necessary.

A short first metatarsal bone may cause the second metatarsophalangeal joint to bear undue weight – weight which it was not designed to carry. The consequence of this is often that the head of the second metatarsal becomes thickened and results in osteoarthrosis of this joint. If the stresses involved are very great, the third metatarsal segment may be similarly affected. The treatment of this type of condition is to compensate for the malfunctioning short first metatarsal in the form of an insole which would provide a weight-bearing surface beneath the first metatarsal or spread the load more evenly over the metatarsal area. The sesamoid bones which lie beneath the first metatarsal head may be subjected to undue stress as a result of malfunctioning in the forefoot particularly in patients who are overweight and this may give rise to sesamoiditis. This condition often responds well to a single steroid injection or it may respond to the provision of a weight redistributive orthosis.

Pes cavus is the term used to describe an abnormally high arched foot. Apart from those which are described as being idiopathic the majority of pes cavus feet are associated with some neurological abnormality. They are commonly associated with spina bifida, anterior poliomyelitis, peroneal muscular atrophy, or Friedreich's ataxia. The high-arched foot with its accompanying retraction of the toes gives rise to areas of unusually high pressure underlying the metatarsal heads. This invariably results in callosities over the first and fifth metatarsal heads which are prone to ulcerate where there is an attendant neuropathy and especially if the foot is rigid. The dorsally displaced toes give rise to a shoe-fitting problem combined with the pain which is experienced on the plantar metatarsal areas of these feet unless there is an accompanying neuropathy, which may of course complicate the secondary resultant lesions. The hyperextended toes are likely to be subjected to pressure from the shoes and they too may ulcerate on the dorsal surfaces or at the apices of the toes if they are bearing weight. Special shoes will often be necessary to accommodate these feet. With the provision of shoes which are deep enough to accommodate the pes cavus deformity, it is advisable to provide an insole made to a cast of the patient's foot which will redistribute the weight away from the areas of high load to those where little or no weight is being taken (Fig. 11.3).

The first metatarsal segment of the foot is the site of many painful foot problems. Hallux valgus is the most common and it is said that almost all women in Britain over the age of 40 have a significant degree of hallux valgus. Most cases of hallux valgus are not symptomatic in terms of pain. They do, however, present problems of footwear fitting. It is difficult but not impossible to find footwear to accommodate hallux valgus in normal shoe shops and patients should be advised accordingly. There are a number of secondary effects of hallux valgus which may be troublesome. These include bursitis at the medial aspect of the first matatarsophalangeal joint, subluxed or overriding

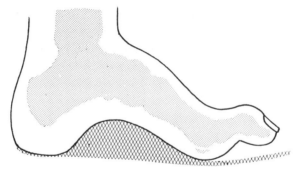

Fig. 11.3. Position of the insole made to a cast of the patient's foot in pes cavus.

second toes, hammer toes, deep painful corns underneath the second metatarsal head as a result of overloading and of back pressure from the second toe. Interdigital corns may arise as a result of the crowding of the foot into the shoe. As the hallux rotates medially in hallux valgus the first toenail is often the site of problems, for example, corns beneath the nail following repeated minor traumas.

THE ARTHROPATHIES

The most common site of osteoarthritis in the body is the first metatarsophalangeal joint where the condition is known as hallux rigidus. It gives rise to pain when the patient needs to extend the hallux in walking. In some cases as the joint becomes progressively rigid over a period of time the pain will decrease. It is during the progression of this condition before it reaches rigidity that most pain is felt by the patient. The available methods of alleviating the symptoms are to stiffen the sole of a patient's shoe, or to provide rocker soles which will prevent the patient having to bend the first metatarsophalangeal joint. Surgical intervention is often resorted to because of the disabling nature of this condition and the methods available are to arthrodese the joint or to provide an arthoplasty.

Often associated with hallux rigidus is hallux flexus (Fig. 11.4). This is a compensatory flexion of the hallux in order to stabilize the foot when the first metatarsal or the first toe is short. By flexing the first toe at the interphalangeal joint the foot is prevented from rolling into a valgus position. It may be necessary in cases of hallux flexus to compensate for the shortness of the first metatarsal segment of the foot by providing appropriate insoles.

Osteoarthritis is the most common of the clinical rheumatic disorders which affect the feet. Though the physical problems produced are considerable, the clinician may be able to alleviate symptoms dramatically by physical measures alone. Osteoarthritis is mechanical or degenerative in origin and is often the result of multiple minor damage to small joints in the foot from ill-fitting footwear or from undertaking particularly physically damaging activities. The second metatarsophalangeal joint may develop osteoarthritis as the result of continuous back pressure on a long or retracted second toe from footwear impaction. It may also be the consequence of an early osteochondritis on the second metatarsal head which will show on X-ray as being flattened. Any metatarsophalangeal joint may be affected by osteoarthritis but the first and second are most commonly so. Heberden's nodes, which are so common in the fingers of those affected by osteoarthritis, may be seen in the toes also, although much less commonly so.

Osteoarthritis may affect any of the tarsal joints as a result of strain. It may be the result of valgus or varus feet which have been inadequately maintained in a functional position. Pes cavus often leads to tarsal osteoarthritis as the result of mechanical strain on the joint facets. Subtalar osteoarthritis is the likely end result following fractures or other damage around the ankle. It may also result from any condition causing undue inversion or eversion of the foot. The midtarsal and subtalar joints may become the site of osteoarthritic changes as a result of over-correction by mechanical devices in earlier life. The mechanical treatment of tarsal osteoarthritis is provided by the use of orthoses made to a cast of the foot which will maintain the foot in its normal anatomical position, thereby preventing the continued strain on the affected joints. It may be advantageous to provide heat treatment or ultrasound in some cases of osteoarthritis.

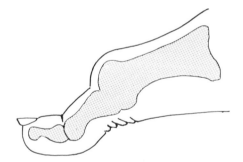

Fig. 11.4. Hallux rigidus with compensatory hallux flexus.

RHEUMATOID ARTHRITIS

Rheumatoid arthritis is a diffuse systemic disease in which chronic inflammation of the synovial joints is the dominant clinical feature. Local lesions resulting from the chronic inflammation are the major cause of physical disability. The normal articular anatomy is destroyed as a result of erosions in the affected joints, and the feet are a common site of erosive articular damage often resulting in gross deformity. It is important to remember that rheumatoid arthritis is a generalized systemic disease and that the management of foot problems cannot be undertaken in isolation of the general management of the disease.

The use of systemic anti-inflammatory drugs may be of great benefit to patients suffering from rheumatoid arthritis and will greatly alleviate painful symptoms. Diuretics too may be useful in helping to reduce the swelling which so often accompanies the effects of the disease. It is this swelling of the articular and periarticular structures which is often first noticed by patients as their shoes become tight.

The deformities which result – ranging from hallux valgus and the secondary digital deformities which often follow this disorder, to total valgus deviation of the feet with their characteristic fibular deviation – can be aided significantly by the physical measures available to and provided by chiropodists. Of course, reconstructive surgery is often desirable, but not always practicable because of circulatory insufficiency which often accompanies rheumatoid arthritis. For convenience the deformities resulting from active rheumatoid disease which affect the feet are best dealt with by dividing them into those in the forefoot and those in the hind foot.

Forefoot problems

The four lesser metatarsophalangeal joints often become subluxated with dorsal displacement of the proximal phalanges. This in turn leads to soft-tissue problems such as corns, callosities, and ulceration on the dorsum of the affected toes which are mainly the result of shoe pressure. The displaced toes cause back pressure on to the metatarsal heads leading to similar problems on the plantar surface. The application of plantar padding in the form of felt plantar metatarsophalangeal pads made from semi-compressed felt will protect the plantar surface of the foot and aid in forcing the toes towards a normal anatomical position. This form of adhesive padding must be regarded as being of a short-term nature as the skin will not react well to constant adhesive dressings, but similar padding may be used as a detachable orthosis covered with soft leather or made on to an insole to be worn in the shoe by the patient. It is now common practice for chiropodists to manufacture silicone rubber digital orthoses for patients with dorsal lesions of the feet. Silicone rubber is mixed with a catalyst and moulded to give an exact fit to the areas where it is required. The improvement in superficial skin lesions and foot function is often remarkable with the use of silicone digital orthoses.

It is extremely common in patients suffering from rheumatoid arthritis to find that the fibrofatty pad which normally lies beneath the metatarsophalangeal joints has become anteriorly displaced and a patient complains, as mentioned earlier in this chapter, of pain similar to the sensation of walking on pebbles. The patient therefore requires some compensatory cushion padding to be provided in combination with other orthoses. Patients suffering from rheumatoid arthritis or other collagen vascular diseases may be receiving systemic drugs such as steroids which may affect the ability of superficial skin lesions to heal normally. It is therefore extremely important that the use of topical medication be confined to bland antiseptics combined with appropriate padding in order to minimise the likelihood of ulceration and to advise the patient regarding scrupulous hygiene. Failure of superficial lesions to heal in such patients may be further complicated by the presence of vasculitis which is a not uncommon secondary manifestation of rheumatoid arthritis.

Hindfoot problems

The tarsal joints particularly the midtarsal and subtalar joints are frequently involved in rheumatoid arthritis. They may be painful and stiff on weight-bearing. The therapeutic objective in physical terms is to provide support for the foot, stabilizing the affected joints in a normal anatomical position. The resultant deformities will thereby be minimized. A cast should be taken of the patient's feet in a neutral subtalar position and an orthotic device in the form of an insole should be manufactured to stabilize the foot in a position of maximum rest.

In the hindfoot too, the fibrofatty protection normally present beneath the calcaneum in order to act as a shock absorber may be reduced or absent as the result of peripheral vascular insufficiency. This may be an area of considerable pain, and can be compensated for by placing a rubber pad in the heel of a patient's shoe. At the junction of the plantar fascia and the medial tubercle of the calcaneum there is frequently an area of increased metabolic activity. This enthesopathic lesion which can be acutely painful needs to be protected and treated. The treatment for pain in the heel of this type may be provided by localized heat or ultrasound, but it may prove necessary to use a corticosteroid injection to relieve the symptoms of this enthesopathy. Similar problems in the heel to those produced by rheumatoid arthritis may result from Reiter's syndrome and psoriatic arthropathy.

Diabetes is notorious because of its secondary manifestations which adversely affect the feet. The most notable of these is peripheral neuropathy. The result of reduced sensation in the feet may lead to patients acquiring small ulcerations of which they are unaware, which may later become infected. These small ulcers can eventually become perforating ulcers which often fail to respond to treatment, since they are complicated by the poor peripheral circulation which often accompanies diabetes mellitus. A further complication of diabetes mellitus is that of affected vision which makes the patient's care of their own feet difficult.

It is therefore necessary for regular chiropody treatment to be provided for diabetic patients, and for regular foot inspection to take place. Diabetic patients should never attempt to cut their own toenails or undertake even simple foot care. They should never use proprietory corn 'cures'. Poor eyesight, arterial insufficiency, peripheral neuropathy, combined with high pressures on areas of the foot may result in gangrene and consequent amputation unless adequate care is provided for the diabetic foot. It is therefore essential that diabetic patients should receive regular chiropody.

Patients suffering from poor peripheral circulation to the feet are subject to chilblains, dry fissuring skin, possible ulceration, and unless suitable care is provided they may too suffer ischaemic skin lesions leading ultimately to gangrene. Patients who present with poor peripheral circulation should be advised to keep the lower limbs warm, to wear adequate spacious footwear which does not cause pressure lesions, and to seek professional help the moment they notice any adverse skin discoloration. Lesions of this type are often the first indication for lumbar sympathectomy. Those with inadequate peripheral circulation may complain of the classic symptoms of intermittent claudication of the legs, but it is sometimes possible for intermittent claudication to affect only the feet, and in this situation it could be all too easy to diagnose and treat quite wrongly as a case of footstrain.

Fig. 11.5. Involuted toe nail which has been corrected by use of a stainless steel spring wire brace.

Toe-nail problems are numerous. Ingrown toe nail, probably the most debilitating of the nail conditions has already been dealt with. Involuted toe nails will cause pain in the sulci of the toe and may be treated by the use of toe-nail braces (Fig. 11.5) made from stainless steel spring wire similar to that used by dentists. The process of correction takes the same time as it takes for a nail to grow from eponychium to the free edge (i.e. approximately a year). This form of treatment requires careful patient selection as co-operation regarding the type of shoes worn is essential. It may prove simpler and more satisfactory to undertake a nail-border resection and partial matrix destruction using phenol or negative galvanism.

Dystrophic nail conditions are often associated with poor peripheral circulation and the only treatment usually required, apart from treatment for circulation, is reassurance. Hypertrophic nail conditions, however, often require more active treatment. Hypertrophy of toe nails occurs frequently as part of the process of aging and their management is often difficult for patients who lack the strength required for pedicure. They also result from damage. These nails may be complicated by the presence of corns and callous in the sulci or under the nails. Regular chiropody will avoid any discomfort from these nails as they can be reduced using high-speed drills and the corns removed at the same time. This treatment needs to be repeated about every six months and the patient advised to use an emery board between visits. Of course, assuming adequate vascularity the nails can be removed and the matrix totally destroyed, thereby avoiding the need for regular treatment.

AKNOWLEDGEMENT

Grateful acknowledgement is made to A. W. Swallow F.Ch.S., S.R.Ch. for the diagrams presented.

REFERENCES

Neale, D. E. (1981). *Common foot disorders: diagnosis and management.* Churchill Livingstone, Edinburgh.
Read, P. J. (1979). *Introduction to therapeutics for chiropodists.* Actinic Press, London.

12 Neurological disorder of gait

A. C. Young

In health the individual's gait often has a characteristic appearance and even sound. As the normal human gait depends on a highly integrated system, it is not surprising that failure in any part of the system produces difficulty in walking at an early age. However, dysfunction in the neurological component of the system is not infrequently overlooked and patients with neurological disorders of gait are commonly referred to rheumatological, orthopaedic, or vascular clinics. Too often a gait disorder is considered only in terms of local disorder of the limbs, namely joint or bone disease. It is very important for the general practitioner to consider a neurological cause in any patient presenting with difficulty in walking. Some knowledge of the probable and commoner neurological disorders is necessary and it must be remembered that some neurological diseases are eminently treatable. Failure to recognize certain conditions may be disastrous and the consequences may be irreversible. Although lesions at many points in the nervous system, from the cerebral cortex to the muscles of the lower limbs produce a gait disorder, a lesion at any one site, irrespective of its nature, will produce a similar pattern of neurological signs. A careful history, by establishing the speed of onset and the evolution of the disorder, may be sufficient to establish the pathological process, which in some circumstances may be confirmed by appropriate investigations. Although the general practitioner may not have the time or experience to perform a detailed neurological examination, he should be familiar with the symptoms and signs associated with various anatomical lesions which produce gait disorders. An anatomical classification (Table 12.1) has the merit of logic and simplicity. Sophisticated investigations may be necessary eventually for the patient with neurological disease but the initial process in diagnosis is a very clinical one,

Table 12.1. *Classification of gait disorders*

Site of lesion	Gait disorder
(A) **Organic**	
Pyramidal pathway	Spastic
Extrapyramidal pathway	Shuffling
Cerebellum	
Posterior spinal columns	Ataxic
Lower motor-neuron pathway	High stepping
(anterior horn cell to peripheral nerve)	
Muscle	Waddling
(B) **Non-organic (hysterical)**	

and the general practitioner, using the anatomical scheme, and willing to listen to and examine his patients, is just as well equipped as his hospital colleague to offer a differential diagnosis, provided of course that the possibility of a neurological disorder has been entertained in the first place. Occasionally it will be obvious as soon as the patient enters the surgery that a neurological disorder is present, but more frequently the practitioner will first have to listen to the patient's description of the problem.

TAKING THE HISTORY

The first essential is to establish whether or not a gait disorder has a neurological basis. Patients differ in their descriptions of symptoms depending on their previous experience, contacts with others, and even their preconceived ideas on diagnosis. It may be suggested that a limp is the result of 'rheumatism' or 'arthritis'. When there is sudden loss of leg movement, a neurological cause is unlikely to be overlooked, but in the elderly a hip fracture may be missed and the loss of mobility attributed to a cerebrovascular episode. A neurological disorder is more likely to be overlooked when the onset is gradual, the disability mild, and the progression slow. The patient may describe heaviness or dragging of the leg. This description can apply to upper motor neuron lesions or extra-pyramidal disease. It is important to discover how movements other than walking are affected and, for example, enquiry must be made about a patient's ability to rise from a chair, get in and out of bed, turn in bed, and climb stairs. Spontaneous jerking of the limbs, especially when the patient is sitting quietly or when in bed, suggest an upper motor neuron lesion. Complaint of the foot or toe catching or the ankle giving way suggests a foot drop but it may result from an upper or lower motor neuron lesion. A paralysed limb is frequently cold, discoloured, and sometimes oedematous, and the patient's problem may be attributed to peripheral vascular disease. The presence of sensory symptoms, such as numbness and tingling, will suggest a neurological lesion, but a word of caution is necessary as many patients equate numbness with 'deadness' by which in turn they mean weakness. Careful and repeated questioning is often necessary to clarify the patient's use of these terms. Loss of temperature sensation is not often volunteered and direct questioning should be used. When proprioception is impaired, patients often give bizarre descriptions which may suggest a neurotic tendency or even hysteria to the uninitiated. Tightness in the limb, swelling, stiffness, and a feeling of wearing stockings, tights or even an extra pair, are all descriptions suggesting damage to proprioceptive pathways. Some patients will describe how they lose control of their legs unless they actually watch their movements and this should always suggest loss of proprio-ception. Vehicle drivers will of course complain that they depress the wrong pedal unless they look at their feet.

Burning pain in the feet and limbs with tenderness and intolerance of touch, even of the bed clothes, strongly suggest a peripheral neuropathy. If there is pain in a limb then it is important to establish if it is distributed within a dermatome. Such a distribution of the pain suggests a root lesion and back pain may also be present. Pain may cause the patient to limp and only the examination will establish whether there is true muscle weakness.

Another symptom which is often a subject of confusion is that of unsteadiness. Patients may use unsteadiness and dizziness interchangeably. Dizziness to the medical attendant may mean vertigo and it is not uncommon for patients with cerebellar disease, for example, to be referred to an ear, nose, and throat surgeon. Attention to detail is therefore necessary to ensure that the patient and doctor are on the same wavelength. Sudden movements, such as turning, are particularly impaired if there is a cerebellar disorder but it must be remembered that when proprioception is impaired the patient may also complain of unsteadiness.

Having obtained a description of the patient's difficulty in walking, the practitioner then must discover the distribution of the symptoms. Weakness affecting both legs, but not the arms, is unlikely to be the result of an intracranial lesion, whereas weakness of one leg, accompanied by weakness of the arm on the same side, is in favour of a cerebral lesion. Weakness of all four limbs can result from a cervical cord lesion, peripheral neuropathy, or muscle disease.

THE EXAMINATION

Although it may seem to be stating the obvious, any patient complaining of difficulty with walking or balance must have gait, stance, and posture observed. Occasionally, observation of these features will immediately suggest the diagnosis; the stooped posture, shuffling gait, expressionless face, and pill-rolling tremor of established Parkinson's disease is unmistakable. In other cases where the diagnosis is less obvious, the patient's ability to rise from a chair or the floor, to climb stairs, to stand on one leg, and to turn must all be assessed. Truncal ataxia may only be apparent when the patient walks heel to toe and occasionally it is so severe that the patient cannot maintain a sitting position unsupported. Romberg's test, hallowed by tradition and engraved in every undergraduate's memory, must be interpreted with great caution. It is certainly abnormal when proprioception in the lower limbs is impaired but it is positive, not infrequently, in anxious or neurotic patients, usually female, who have vague complaints of unsteadiness or dizziness. A positive Romberg's test as an isolated finding is unlikely to signify organic neurological disease.

The history and observation of the gait will probably indicate that the patient's complaint does have a neurological basis and may even localize the lesion but a formal systematic examination of the nervous system, including the cranial nerves and upper limbs, will be necessary in most cases before making a differential diagnosis. Although a large number of pathological conditions cause gait disorders, there is only a small number of patterns to be recognized and this can be done by simply observing the patient walk. These patterns and some of the commoner causes will therefore be described.

THE SPASTIC GAIT

The pattern of gait disorder in spasticity will depend on whether there is unilateral or bilateral involvement. If only one leg is affected the patient may drag the leg and the disability may become more troublesome the longer the patient walks. In spasticity other upper motor neuron signs, such as hyper-reflexia, extensor plantar responses, and weakness, should be sought on examination. The muscle weakness affects the flexor muscles rather than extensors so that initially the patient notices the toe catching, and eversion of the ankle is affected before inversion, causing the ankle to give way. If hip and knee movements are affected then the patient swings the extended leg from the hip, abducting rather than flexing the hip.

Upper motor neuron signs confined to one leg may result from a spinal cord or intracranial lesion. In the latter case, usually but not invariably, upper motor neuron signs, perhaps only hyper-flexia, will be found in the ipsilateral arm. An ipsilateral upper motor neuron facial palsy clearly indicates an intracranial lesion. As already indicated the sudden onset of weakness in a hemiplegic distribution usually presents little diagnostic difficulty and cerebrovascular disease is the likely explanation. A slowly progressive hemiparesis suggests the presence of an expanding mass lesion and a primary or secondary tumour will have to be excluded.

Sometimes a patient will complain of problems with one leg only and observation of the gait may appear to confirm this. However, on examination upper motor neuron signs may be found in the asymptomatic leg. It is important to remember that spinal cord disease may present with asymmetrical spasticity or weakness. More commonly both legs are equally affected producing the characteristic stiff-legged jerky gait with the feet scarcely clearing the ground. The toes may scrape the ground and inspection of the patient's shoes reveals wear at the toes, particularly along the outer aspect. Spasticity of the adductors of the hip produces the classical scissoring gait but this is unusual when spasticity develops in adults. Likewise walking on the toes is more characteristic of childhood spasticity. When there are bilateral upper motor neuron signs, spinal cord disease is highly probable, but then the level of the lesion must be determined. If there are neither symptoms nor signs in the upper limbs the lesion is below the cervical enlargement. In compression of the thoracic spinal cord a sensory level on the trunk may be found. A lesion at the lower end of the spinal cord produces a characteristic picture with increased knee jerks, absent ankle jerks, and extensor plantar responses. Careful examination may disclose sensory loss in the saddle area. The patient may complain of perineal numbness. A spastic paraplegia, accompanied by symptoms and signs in the upper limbs, indicates a cervical cord lesion between the fifth cervical and first thoracic segments. Anterior horn cells and pyramidal pathways may be involved producing a combination of upper and lower motor neuron signs in the arms. A classical example is syringomyelia. By analysis of the sensory loss and reflex changes it is often possible to localize the lesion to one segment of the spinal cord. A syrinx at C5/6 level will produce loss of biceps and supinator jerks but the triceps jerks will be preserved or even exaggerated from pyramidal pathway involvement.

From the history and examination the general practitioner may strongly suspect the cause of a spastic gait but only rarely will he be able to confirm it without hospital referral. Nevertheless he should be familiar with the likely investigations, their discomfort, risks, and limitations. If it is apparent that specialized neurological investigation is essential then there is little point in the general practitioner arranging the simpler investigations such as blood count, Wassermann reaction, and a plain chest X-ray which is essentail if metastatic disease is suspected. The importance of history taking and examination, when dealing with neurological problems, cannot be too strongly emphasized because the accuracy of the clinical diagnosis determines the plan of investigation. Failure to elicit an earlier history of retrobulbar neuritis, for example, in a patient who has a spastic paraplegia, will lead to unnecessary myelography. The importance of good record keeping in general practice should therefore be obvious. Investigations that may be undertaken after the patient is referred to hospital will be mentioned briefly as the commoner causes of spasticity are considered.

CEREBRAL SPASTICITY

It is usually possible on clinical grounds to determine whether a slowly progressive spastic weakness of one leg is the result of a cerebral or spinal lesion but occasionally the localization is not possible on clinical grounds. Fortunately this problem has been simplified by the introduction of CT scanning. A negative CT scan almost certainly excludes a cerebral cause of a spastic monoparesis affecting the lower limbs. Angiography is still necessary for the investigation of cerebrovascular lesions such as aneurysms, arteriovenous malformations and arterial stenosis or occlusion but it is rarely necessary now in the investigation of suspected intracranial tumour. Ventriculography and lumbar air encephalography have almost been eliminated with the availability of CT scanning.

SPINAL CORD COMPRESSION

A patient with suspected cord compression must be referred promptly to hospital and if there are bladder symptoms emergency admission should be sought. Once urinary retention has developed only a few hours are available if bladder function is to be restored. However, before urinary retention has developed, neurological symptoms have usually been present for some weeks at least. Cord compression may result from metastatic disease, the common primary sites being bronchus, breast, and prostate gland. Removal of a solitary spinal metastasis is still worthwhile providing the paraplegia is not complete and urinary retention has not been present for more than 24 hours, and hence the importance of early referral. Primary tumours such as meningioma and neurofibroma are less common and the latter can be suspected on clinical grounds if there are cutaneous lesions. Plain X-rays of the chest and relevant areas of the spine will be carried out but the definitive investigation is myelography. In suspected metastatic disease the ESR is still of great value. If there is spinal cord compression, laminectomy and removal of the tumour is the first step. In prostatic carcinoma and myeloma, radiotherapy and chemotherapy will usually be necessary. If there is inoperable metastatic disease, radiotherapy apart from providing pain relief is rarely of value. Spinal tuberculosis, which in Britain is more likely to occur in the immigrant population, responds well to chemotherapy and surgical treatment is rarely necessary.

MULTIPLE SCLEROSIS

This is the commonest chronic disorder of young people in the United Kingdom and not infrequently presents with spastic weakness of one or both legs with or without sensory symptoms. The peak age of onset is approximately 30 years of age and onset of the disease before the age of 15 and after the age of 50 years is unusual. In these age-groups the diagnosis should be accepted with reservation and only after other conditions have been excluded. Lesions disseminated in place and time are still the basis of the diagnosis which remains very much a clinical one. Careful enquiry to elicit previous episodes of neurological involvement, such as unilateral blurring of vision, diplopia, and paraesthesiae or weakness in a limb, is necessary. When there is a history of relapses and remissions of scattered lesions the diagnosis is almost certain on clinical grounds. Too early investigation has disadvantages. The diagnosis, if confirmed, may have to be disclosed leading to anxiety or sometimes hysterical elaboration which may be very difficult to differentiate from genuine relapses. Claims that the disease can be modified have not yet been substantiated and early diagnosis offers no great advantage except possibly in women of child-bearing age who may wish to limit or even forego pregnancy. Patients with minimal disability who are admitted to hospital, and in particular to a neurological unit, may well encounter patients with advanced forms of the disease which leads to unnecessary pessimism. Investigations which may help to confirm the diagnosis include examination of visual- and auditory-evoked responses, which, if abnormal, provide evidence of multiple lesions. Abnormal results are not pathognomonic of multiple sclerosis. Examination of the cerebrospinal fluid may be helpful in some cases by revealing an elevated cell count and IgG content. It is in cases of late-onset spastic paraparesis, with no historical or clinical evidence of lesions elsewhere, that problems arise and myelography will be necessary to exclude cord compression.

CERVICAL SPONDYLOSIS

After the age of 45 years cervical spondylosis is a significant cause of a spastic paraparesis. The mechanism is still controversial but direct cord compression from a spondylotic bar and interference with the blood supply are probably both involved. Pyramidal tracts and anterior horn cells are affected but root compression from foraminal encroachment complicates the clinical picture. The common presentation is a slowly progressive spastic paraparesis. There may be few sensory symptoms but a careful history may elicit paraesthesiae of the fingertips. Weakness of the arms and hands may develop later. Pain in the neck or arms in a root distribution may or may not be present. A painless presentation is not uncommon. The spondylotic process is most frequent at the C5/6 level and a typical patient shows wasting and weakness of the intrinsic hand muscles, possibly from interference with blood supply to the anterior horn cells, loss of biceps and supinator jerks as the result of root compression, and increased triceps jerks and a spastic paraplegia indicating pyramidal tract involvement. Sensory loss may be found in a root distribution but sometimes severe proprioceptive loss is found in the hands as a result of damage to the posterior columns. It cannot be too strongly emphasized that the diagnosis of cervical spondylosis producing a myelopathy can only be diagnosed by myelography. Plain X-ray degenerative changes are extremely common and a frequent error is to attribute any neurological symptom or sign to these changes which in fact are incidental. Some patients with a spastic paraplegia may be helped by surgical treatment which aims to produce more room in the spinal canal whichever surgical technique is used. Progression of the disease may be halted but a return to normal cannot be expected.

MOTOR NEURON DISEASE

This condition may be confused with cervical spondylosis because it may present with a combination of upper and lower motor neuron signs. However, wasting of the legs cannot be explained by a cervical cord lesion and often the lower motor neuron signs in the upper limbs are very asymmetrical, whereas in cervical spondylosis, symmetrical signs are the rule. Sensory signs are never found in motor neuron disease and careful examination to exclude sensory loss is essential. Fasciculation is an important lower motor neuron sign but it is not pathognomonic of motor neuron disease.

There is no investigation which will confirm motor neuron disease. The daignosis is made on clinical grounds but nerve conduction studies, electromyography, and occasionally myelography, will be necessary to exclude other conditions. Most neurologists admit such patients for investigation as no matter how certain the clinician may be of the diagnosis, patients, particularly in this technological age, find it hard to understand that such a diagnosis, which in effect carries a death sentence, can be made without resort to investigation. No treatment influences the relentless march of this disorder but much can be done by appropriate aids and alterations of the home to help both patient and relatives to cope with this condition, the most distressing of neurological diseases.

SUBACUTE COMBINED DEGENERATION OF THE CORD

The importance of this condition lies not in its frequency but in the fact that it is an example of treatable neurological disease. Automated methods in haematology have led to the earlier detection and treatment of pernicious anaemia so that the neurological complications are prevented. Nevertheless it is important to keep the condition in mind in any patient with a spastic paraparesis of gradual onset and slow progression. Paraesthesiae, first in the feet and later in the hands, precede the spastic weakness. Walking may also be affected because of proprioceptive loss producing a sensory ataxia. It is now accepted that some of the sensory features are attributable to peripheral nerve damage. This may explain the absence of ankle jerks in patients with established subacute combined degeneration of the cord, which is one condition producing the combination of absent ankle jerks and extensor plantar responses. Proprioception and vibration sense are usually impaired. The diagnosis requires a blood count although on rare occasions the blood count may be normal. Macrocytosis in the peripheral blood certainly demands that the serum B_{12} is estimated but the tendency to request this investigation in patients with any neurological disease should surely be resisted. Neurological complications of pernicious anaemia are invariably associated with a very low serum B_{12} level. Bone marrow will show megaloblastic change and there is gastric achlorhydria. Subacute combined degeneration of the cord should be preventable but if neurological involvement develops it can be arrested. The extent of recovery depends on the stage at which the diagnosis has been made. Spastic paraplegia and severe proprioceptive loss rarely remit, underlining the importance of early diagnosis. The general practitioner therefore has an important role to play and a blood count should be obtained in any patient with persistent paraesthesiae of the lower extremities.

THE SHUFFLING GAIT

A shuffling gait is likely to be encountered in patients over 50 years of age. The most important cause is parkinsonism. It is uncommon for parkinsonism to present with walking difficulty as an isolated symptom and other features, namely tremor and rigidity, are present in the majority of cases. However, in a significant proportion of patients these signs are absent and the diagnosis may not be considered. Patients may complain of dragging or heaviness of one leg before the characteristic shuffling, the result of bilateral involvement, becomes apparent. The shuffling may be particularly obvious when the patient attempts to turn. The manoeuvre is carried out by a series of shuffling steps and on occasions the patient may even stop temporarily. Loss of arm swing, even when tremor is absent, is an early and useful sign. At present idiopathic parkinsonism is the commonest variety but drug induced disease is often overlooked. Many patients with parkinsonism may initially complain of unsteadiness which is misinterpreted as vertigo so that prochlorperazine (Stemetil), a phenothiazine derivative, is prescribed. Such drugs should not be given to any patient with Parkinson's disease unless there is some strong psychiatric indication. There is no investigation which will confirm the clinical diagnosis of parkinsonism and if the general practitioner is confident of the diagnosis, referral to hospital is unnecessary. The long-term problems associated with the use of L-dopa preparations are becoming ever more apparent and unless there are contra-indications, such as glaucoma and prostatism, an anticholinergic drug should be prescribed. There is no good evidence that one anticholinergic is better than another and the general practitioner should familiarize himself with one or two drugs at the most. Particular care is required when prescribing these drugs for the older patient as an acute psychosis may be induced. If L-dopa has to be prescribed a combined L-dopa and decarboxylase inhibitor (Madopar or Sinemet) should be used but the dose should be kept as small as possible. It is important that these drugs are taken after food to reduce the incidence of gastrointestinal side-effects. Patients who cannot tolerate L-dopa in an adequate dose or who develop involuntary movements should be referred for specialist advice.

A condition which may superficially resemble parkinsonism is seen in an older age-group and often against a background of hypertension. It is thought these patients have diffuse cerebrovascular disease or even multiple infarctions. The gait is small-stepped and shuffling but the leg reflexes are often exaggerated and the plantar responses may be extensor. Tremor and rigidity are absent. Emotional lability and dementia may be present. This gait is often described as *marche à petits pas*. Sometimes the condition is called arteriosclerotic parkinsonism but most neurologists regard it as a separate condition and the distinction is important as patients with the condition are often very intolerant of anti-parkinsonian drugs.

ATAXIC GAITS

Ataxia has no satisfactory definition and may result from weakness, sensory loss, or cerebellar dysfunction. Classically two varieties are recognized, cerebellar and sensory. Before ataxia can be attributed to a cerebellar defect, muscle weakness and proprioceptive defect must be excluded. In practice, muscle weakness and cerebellar ataxia may co-exist and the examiner must decide whether the ataxia is commensurate with the muscle weakness or out or proportion to it. Cerebellar ataxia is unchanged with the eyes closed whereas sensory ataxia is aggravated.

CEREBELLAR ATAXIA

The patient with cerebellar ataxia complains of unsteadiness and as already indicated this symptom is often mistaken for vertigo. The latter is episodic and uaually brief with a definite sense of rotation, whereas cerebellar ataxia is almost invariably a persistent feature. With unilateral cerebellar lesions the patient tends to veer to the affected side. Ataxia can be demonstrated by asking the patient to move the heel up and down the shin bone and the ipsilateral upper limb is ataxic on the finger/nose test. Other features of cerebellar damage such as nystagmus and dysarthria may be present. The gait is wide-based and the ataxia becomes more pronounced on sudden turning or walking heel to toe. When both cerebellar hemispheres are involved then all four limbs are ataxic. In midline cerebellar lesions on the other hand, there may be no heel/shin or finger/nose ataxia on formal testing but the patient is unable to stand, let alone walk. Such a gait disorder may be attributed to hysteria because it appears the patient's complaints are out of proportion to the physical signs. Most patients with cerebellar ataxia will require referral for specialist opinion but the general practitioner if not able to make a specific diagnosis should be able to place the patient within one of several broad categories of cerebellar disorder.

Midline and unilateral cerebellar disorders are nearly always caused by tumours. Secondary tumours, most commonly from a bronchogenic primary, are characterized by a short history of unsteadiness and symptoms of raised intracranial pressure appear early. In contrast, with primary tumours, which in adults are usually benign, raised intracranial pressure is a late development. Long-standing deafness, often overlooked by the patient and not volunteered, accompanied by cerebellar ataxia suggests a cerebellopontine angle lesion. If there are associated ipsilateral sixth and seventh cranial nerve palsies there can be little doubt about the diagnosis although the development of these signs suggests a tumour of considerable size. The introduction of CT scanning has greatly simplified the investigation of posterior fossa tumours. The method is of value not only because it identifies the presence and site of the tumour, but in many cases the pathological nature of the tumour can be accurately predicted. Other ataxic states stem from diffuse cerebellar damage. Toxic substances, among which alcohol and drugs must be included, produce ataxia of all four limbs. A patient with epilepsy who develops ataxia is more likely to have developed phenytoin (Epanutin) intoxication than an intracranial tumour. Cerebellar ataxia accompanied by dementia may be a non-metastatic complication of bronchial carcinoma. The ataxia may antedate the appearance of the bronchial carcinoma by months or even years. However, the condition is rare and the mechanism is obscure. An equally rare but treatable condition which causes a diffuse cerebellar ataxia is myxoedema and in any obscure ataxia thyroid function should be investigated as the hypothyroidism may not be very striking clinically. Familial cerebellar ataxias were more easily recognized in former days but with smaller families and greater social mobility, this category may go unrecognized for several years. Multiple sclerosis in its advanced stages often causes ataxia of the upper limbs, but it is unusual for multiple sclerosis to present as an ataxic gait without other features. Sadly, in older age-groups, no specific cause can be identified and one is left with the unsatisfactory diagnosis of sporadic late-onset cerebellar degeneration of unknown cause.

SENSORY ATAXIA

Impaired proprioception in the lower limbs causes a wide-based stamping gait. The patient complains of loss of control over the legs and feet – 'they will not do what I want them to do'. Darkness and eye closure aggravate the gait disorder. Other sensory symptoms may be present. Proprioceptive loss affecting the ipsilateral leg and arm is likely to result from damage to the parietal lobe whereas bilateral proprioceptive loss in the legs results from damage either to the posterior columns of the spinal cord or the peripheral nerves. The distinction can be made simply from the examination because with posterior column damage other signs of spinal cord disease will usually be found, namely hyper-reflexia with extensor plantar responses, whereas in a peripheral neuropathy the reflexes are absent or at least diminished. Spinal cord diseases have already been discussed and a peripheral neuropathy is one cause of a high-stepping gait which will now be considered.

THE HIGH-STEPPING GAIT

A high-stepping gait results from weakness of the distal leg muscles. There is a flaccid paralysis, particularly of the dorsiflexors of the toes and ankles, nearly always accompanied by weakness of eversion. A useful test for detecting early foot drop, which can be carried out even by the busiest general practitioner, is to ask the patient to walk on the heels. The differential diagnosis of foot drop is wide as the lesion may be anywhere between the anterior horn cells and the muscles themselves. Observation of the gait is therefore insufficient for localizing the lesion. With the virtual elimination of poliomyelitis the only numerically significant disease of the anterior horn cells which may present with a foot drop is motor neuron disease. Not infrequently, the foot drop initially is unilateral and the onset is insidious. Wasting will not be prominent in early cases and the diagnosis may be extremely difficult until the passage of time reveals the true nature of the problem. The combination of muscle wasting and fasciculation but exaggerated reflexes should always alert one to consider motor neuron disease.

Compression or infiltration of the fifth lumbar root causes a foot drop. Back and leg pain in the appropriate dermatome accompained by a foot drop suggests a prolapsed lumbar disc and a unilateral presentation is the rule. Compression of the cauda equina by primary or metastatic tumour is often misdiagnosed as a disc protrusion. The presence of back and leg pain often results in these patients being referred to orthopaedic or rheumatological clinics. Careful examination usually shows that the wasting and weakness are too diffuse to be explained by a single root lesion. A foot drop resulting from an L4/5 disc protrusion should not be accompanied by an absent ankle jerk which indicates an S1 root lesion. Proper attention to the details of the examination should prevent overdiagnosis of prolapsed lumbar discs. Infiltration of the lumbosacral plexus in the pelvis by metastases from rectum or uterus may present a similar clinical picture to compression of the cauda equina and may also mimic a prolapsed lumbar disc. Once again there is usually diffuse muscle wasting and weakness which cannot be explained by a single root lesion.

Compression of the common peroneal nerve produces a unilateral foot drop and often a history of trauma or compression of the nerve can be obtained. Nerves damaged for some other reason are more susceptible to pressure and this explains the higher incidence of common peroneal palsies in alcoholic and diabetic patients, for example. Sensory loss may not be as prominent as some textbooks suggest.

Bilateral foot drop is a common presentation of a peripheral motor neuropathy. Pure motor neuropathies do occur but a mixed sensorimotor neuropathy is more common and the patient may complain of numbness and paraesthesiae especially in a sock or stocking distribution. By the time of presentation the ankle jerks, at least, are nearly always lost. In chronic polyneuropathies the weakness is invariably more severe distally but in acute polyneuritis of the Guillain–Barré type the weakness may be more marked proximally. However, the acuteness and severity of this neuropathy inevitably results in prompt referral to hospital. This is the only neuropathy in which cranial nerve palsies are common and bilateral facial weakness of lower motor neuron type is common in severe cases. Weakness of bulbar and respiratory muscles may develop at an alarming rate but with assisted ventilation the mortality, which previously was so high, is now very low and most patients make a complete recovery. The diagnosis is based on the clinical features and most cases have the typical CSF findings of an elevated protein with a normal cell count. Steroids are often prescribed for this condition but there is no good evidence that they are beneficial. Recently it has been reported that plasmapheresis can bring about dramatic improvement and curtails the duration of the illness.

A chronic peripheral neuropathy which has progressed to foot drop can be recognized on clinical grounds without resort to investigation. The cause of the peripheral neuropathy demands investigation unless the patient has some obvious condition which is a well-recognized cause of a neuropathy. Diabetes, for example, is probably the commonest cause of a chronic peripheral neuropathy in the United Kingdom and while the neuropathy will develop with established and long-standing diabetes, occasionally the neuropathy may be the presenting feature although inquiry may reveal symptoms such as weight loss, polyuria, and polydypsia. Urinalysis is therefore essential in patients presenting with a foot drop but the blood sugar, and if necessary a glucose tolerance test, will be necessary to exclude the diagnosis finally. With access to laboratory services this diagnosis comes within the province of the general practitioner. Rigorous control of diabetes is essential if neurological complications develop. Another cause of chronic peripheral neuropathy which the general practitioner may be in a better position to discover is excessive alcohol intake. The story, apocryphal though it may be, of finding the evidence in the patient's dustbin is worth remembering. Early recognition is important because despite abstinence and vitamin supplements the condition once developed rarely recovers completely and weakness and sensory deficit persist. The latter often takes the form of very unpleasant paraesthesiae and burning sensations which are resistant to all treatment. There are many other causes of peripheral neuropathy and evidence of systemic disease, for example renal failure, bronchial carcinoma and collagen-vascular diseases, to name but a few should be sought. Toxins such as drugs and chemicals should not be overlooked and a careful history of the patient's occupation and leisure activities is essential. Familial neuropathies occur but are rare. Unfortunately, once such obvious conditions as outlined above have been excluded, there remains a large number of patients with a peripheral neuropathy the cause of which cannot be established, despite the most careful and painstaking investigations. Nerve conduction studies and nerve biopsy, while revealing the pathological process, rarely, if at all, provide any clue to the aetiology.

THE WADDLING GAIT

A waddling gait is the result of proximal muscle weakness. Commonly the weakness results from primary muscle disease but the muscles may be affected by disease in the spinal cord or peripheral nerves. A useful clinical guideline is that a peripheral neuropathy produces a distal weakness whereas muscle disease results in proximal weakness so that the patient has difficulty in rising from a chair or the floor and the classical manoeuvre of 'climbing up the legs' may be seen, but it is not diagnostic of any particular muscle disease. Careful examination will distinguish in most cases whether the proximal weakness is the result of a primary muscle disease (myopathy) or whether it results from spinal cord disease or peripheral neuropathy. Sensory loss is never found with myopathies but areflexia is not uncommon when muscle wasting and weakness are severe. All forms of primary muscle disease are rare in general practice and if the diagnosis is suspected, referral for specialist opinion and investigation will be necessary. However, certain broad categories of muscle disease can be recognized from the history and examination. Proximal muscle weakness developing in a boy with pseudohypertrophy of the calves suggests Duchenne dystrophy and if there is a family history of disease confined to males then the diagnosis is certain. In any patient with suspected muscle disease a complete family tree is essential, not only for diagnosis but also for genetic counselling as there is no treatment which will modify or prevent muscular dystrophy. Survival beyond 20 years of age is unusual in Duchenne dystrophy but the other familial muscular dystrophies have a more benign course. They are classified on the basis of clinical features and genetic pattern. Dystrophia myotonica, although considered to be a primary muscle disease, is not usually characterized by a waddling gait because in this condition the muscle wasting and weakness disobey the general rule being more marked distally. The condition is worthy of special mention because it is so often of such insidious onset and slow progression that it may go unrecognized even in families where there is a strong history of the condition. Apart from the myotonia other notable features are the involvement of the facial and neck muscles and evidence of systemic involvement such as premature cataracts, and frontal baldness and testicular atrophy in the male.

Inflammatory disease of muscle, so called myositis, is probably commoner than is recognized. Apart from proximal muscle weakness, pain and muscle tenderness may be present in the early stages. Some cases are associated with skin disease or more widespread collagen-vascular disease. A myositis in older patients is sometimes associated with neoplasia. Some metabolic diseases cause proximal muscle weakness. In thyrotoxicosis minor degrees of proximal weakness can be found in a large percentage of patients on careful examination but occasionally the patient may present with a quite florid myopathy producing difficulty in walking. Tremor, agitation, tachycardia, and palmar erythema may suggest an alcoholic problem if the eye signs of thyrotoxicosis and a goitre are not marked. Osteomalacia, for example in immigrants or following partial gastrectomy or associated with renal failure, is another important metabolic condition to be considered.

Myasthenia gravis is a disorder of the neuromuscular junction which may present with difficulty in walking, rising from a chair, or climbing stairs. If there are no obvious cranial nerve signs such as ptosis or facial weakness this important and treatable cause of limb weakness may be overlooked. Unless one looks carefully for fatigue-ability of muscle strength, a young woman in particular may be regarded as hysterical because the weakness cannot be detected on a rapid routine examination, yet the patient may be insistent that she has severe problems.

When patients with suspected muscle disease are referred to hospital the diagnosis is made principally on clinical grounds but various investigations help to confirm the diagnosis. Estimation of the serum creatine kinase is helpful. In Duchenne dystrophy very high levels are found whereas in the other familial muscle diseases, the values are normal or only slightly elevated. When polymyositis is present the creatine kinase can be considerably elevated although never reaching the values seen in Duchenne dystrophy, but in advanced disease the result may be normal. Thyroid function and calcium metabolism will be investigated. Electromyography is useful in determining whether the muscle weakness is the result of primary muscle disease or anterior horn cell disease which causes spinal muscular atrophy. Muscle biopsy, particularly when newer histochemical methods and electronmicroscopy are used, is of increasing value in excluding rare metabolic disorders. Myasthenia gravis should always be excluded in any obscure myopathy by carrying out an edrophonium (Tensilon) test. Unless full resuscitative facilities are available, a general practitioner is well advised not to carry out this test in his surgery. In addition interpretation of the result is often not as simple as is sometimes claimed. False positives may occur unless a placebo is also used. Apart from polymyositis and myasthenia gravis none of the muscle disorders likely to be encountered by the general practitioner have any specific treatment. Steroids with or without immunosuppressive drugs control, if not improve, myositis in most instances. Treatment of myasthenia gravis is difficult and the general practitioner is unlikely to see sufficient numbers to gain the experience and confidence to manage these patients without close liaison with specialist services. Younger patients with disease of recent onset are increasingly being recommended for thymectomy. If despite thymectomy and anticholinesterase drugs, the myasthenia is not controlled, steroids and immunosuppressive drugs will be tried singly or in combination.

HYSTERICAL GAIT DISORDERS

Psychiatrists often maintain that the *grande hystérie* of Charcot's day is now rarely seen. While this may be true in psychiatric practice it does not apply to neurological practice where severe hysterical disorders of gait are not uncommon. It is doubtful if any practitioner whether primary physician or specialist ever feels comfortable with the diagnosis of hysteria and in any patient presenting for the first time in whom he suspects hysteria, the general practitioner would be unwise to make the diagnosis without specialist advice. Reference has already been made to truncal ataxia and myasthenia being dismissed as hysterical disorders. The general practitioner often does have the advantage of being familiar with the patient's previous behaviour and response to illness. He is also likely to be more aware of the patient's personal, family and work problems which may provide the explanation for hysterical illness. A diagnosis of hysterical gait disorder should rest on positive evidence and should not depend simply on exclusion of organic disease. The gait pattern may strongly suggest the diagnosis. Postures are often extremely abnormal so that the patient walks with a severe tilt, lordosis, or even kyphosis. Violent lurching, often conveniently near some support so that falling is rare, is often demonstrated. It is important, therefore, that adequate space should be available to allow the patient to walk without being able to hold on to walls or furniture. Another feature is the attempt to impress upon the examiner how much effort is being put into the attempt to walk. The face may have an anguished look and there may be sudden spasms of apparent pain when the patient clutches the affected area. Some patients attempt to lift the 'paralysed leg' with their arms. Despite these features noted in observing the gait, other positive signs should be sought and inconsistencies between patient's function and examination performance should be noted. Although the patient may still stand and walk even in an abnormal manner, formal muscle testing with the patient on the couch may show complete 'paralysis'. The patient who will not flex the hips against gravity can nevertheless be persuaded to carry out the heel/shin test. Sensory loss which does not conform to anatomical patterns can often be found. However, it should be remembered that a sock or stocking sensory loss can be seen in peripheral neuropathy as well as hysteria. In hysteria the mistakes made on testing proprioception are not random. The patient maintains that the toe is consistently opposite from the direction in which it has in fact been moved. Apart from the positive signs of non-organic disease, there will be no positive signs of organic disease except in that difficult situation when organic neurological disease and hysterical elaboration are combined in the same patient. This is not too uncommon in patients prematurely informed that they have multiple sclerosis, for example. When a hysterical gait disorder develops against a long background of personality disorder with well documented attendances for numerous and varied symptoms for which no organic cause has been found, the general practitioner may be justified in not referring the patient to hospital. There is no doubt that referring such patients, who may be over-investigated, may lead to strengthening of the patients' conviction of bodily disease and may even determine future conversion symptoms by exposing them to other disease states. The knowledge that patients with hysteria must die of some organic disease and the fear that 'wolf' will be cried once too often will persuade most general practitioners to refer such patients so that the guilt of organic disease being 'missed' can at least be shared. Once the patient is referred to hospital, investigation and its extent will vary on the attitude and philosophy of the neurologist. Some neurologists feel that organic disease must be excluded at all costs whereas others will feel more confident in their clinical judgement and limit investigations to non-invasive procedures so that the patient is convinced the problem has been taken seriously. Completely ignoring the symptoms will lead to more bizarre and dramatic attempts to persuade the physician of the genuine nature of the problem, while investigation of every symptom will only lead to a new set of symptoms being produced as soon as one disease has been excluded by investigation. A middle course has to be steered and as long as physical examination does not show positive signs of organic disease, enthusiasm for investigation should be tempered. If the history is short acceptance of the condition by the physician and physical treatment of some sort, such as physiotherapy with or without placebo drug treatment, can often bring dramatic recovery. When the history is long and multiple symptomatology has been produced over the years, the prognosis is poor and remissions are only temporary. Even if the patient should recover from an hysterical paraplegia, a new set of symptoms cannot be too long delayed. The neurologist will pray that some other system will be involved so that he can have the merciful release that is denied the general practitioner who is expected to stay the course.

CONCLUSION

Neurological disorders of gait cover a wide spectrum of disease of the nervous system. The elucidation of these disorders still largely depends on the traditional clinical skills of history taking and examination. For the general practitioner willing to exercise these skills and to acquire some knowledge of the commoner conditions, there is a satisfying role to play. To detect a neurological basis for a gait disorder in the many patients who present with walking difficulty and to select those patients with treatable or preventable disease for early specialist referral requires no little skill. Diagnosis is not always academic but even when drug and surgical treatment is not applicable, the general practitioner has an important contribution to make to the management of chronic neurological disease. Patients with such disease require a great deal of advice and counselling and families need support. Fortunately the problems and the needs of the disabled are receiving greater attention and the informed general practitioner has a responsibility to see that all the help and resources which are at present available do in fact reach the disabled person.

REFERENCES

Bickerstaff, E. R. (1976). *Neurological examination in clinical practice,* 3rd edn. Blackwell, Oxford.

Matthews, W. B. (1975). *Practical neurology,* 3rd edn. Blackwell, Oxford.

Patten, J. (1977). *Neurological differential diagnosis.* Springer, New York.

13 The limping child

C. S. B. Galasko

INTRODUCTION

A limp is an important clinical sign in childhood. Even if intermittent it may be the only indication of a major underlying disease and always requires thorough investigation.

The two main functions of the locomotor system are mobility and stability. Abnormalities of gait (i.e. mobility of the lower limb) produce a limp. The requirements for a normal gait are shown in Table 13.1 and the main causes of a limp in Table 13.2.

There are four phases to normal gait:

  (i) heel strike with the knee extended and the hip flexed;

 (ii) the stance phase during which weight is transferred from the heel to the forefoot and the hip is extended;

(iii) push-off with dorsiflexion of the toes, particularly the great toe, propulsion of the limb, and flexion at the hip and knee;

(iv) swing phase during which there is a pendulum swing through of the limb with flexion at the ipsilateral hip and abduction of the contralateral hip so that the pelvis is tilted proximally on the non-weight-bearing side.

MECHANICAL CAUSES OF LIMP

LIMB LENGTH INEQUALITY

The main causes are shown in Table 13.3. Limb lengthening inequality may either be due to overgrowth (Fig. 13.1) or shortening. The causes of congenital shortening are multifold. At the one extreme they include total absence of the limb or part thereof and at the other slowed growth of a single bone which

Table 13.1. *Requirements for normal gait*

Joints	Stable and congruous
	Full range of pain-free movement
Muscle	Normal power and tone
	Co-ordinated activity
Nervous control	Intact cerebral cortex
	Intact cerebellar cortex
	Intact ocular and vestibular balance mechanism
	Intact spinal cord and nerve roots

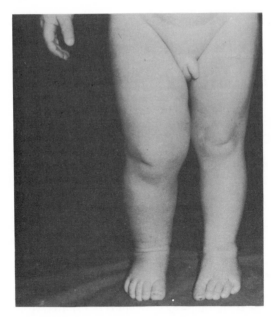

Fig. 13.1. Hemihypertrophy of the right lower limb.

otherwise is normal. Between these two extremes any variant can occur. Damage to the growth plate as a result of infection or trauma may produce shortening as can fractures associated with overlap or angulation. The latter will produce deformity as well as shortening (Fig. 13.2). Paralysis also results in slowed growth (Fig. 13.3).

True limb length is measured from the anterior superior iliac spine to the medial malleolus, each limb being placed in the identical position during measurement, e.g. if there is a 15° fixed flexion deformity of the left hip the right hip must be placed in 15° of flexion before the true limb length can be measured. If there is a 10° fixed adduction deformity of the right hip, the left hip must be placed in 10° adduction whilst the limb is measured. Differences of up to one centimetre may be normal, the longer limb being on the dominant side.

Depending on the cause the limb length inequality may change during growth. Differences due to a congenital abnormality, damage to a growth plate, or paralysis usually increase during the growth period. Shortening as a result of fracture that has not involved the growth plate often diminishes due to the compensatory overgrowth in the traumatized limb, but this does not always occur. In general terms the younger the child the better the prognosis. The maximum correction that can occur is 2–3 centimetres. Occasionally a fracture results in a longer limb when there has been no shortening but the trauma has resulted in a temporary increased growth spurt.

Table 13.2. *Causes of limp*

Mechanical	Limb length inequality
	Instability
	Contractures
	Ankylosis
	Deformity
Pain	Joints
	Bones
	Muscles
	Tendons
	Soft tissues
	Nerves
	Blood vessels
	Miscellaneous
Spinal and abdominal lesions	Flexor spasm
	Cord/nerve root compression
	Cord/nerve root traction
Neuromuscular	Compression
	Weakness
	Spasticity
	Inco-cordination
	Sensory neuropathy
	Ataxia
	Hysteria
	Malingering

Because the limb length inequality is often progressive, serial measurements over months or years are essential in order to determine the eventual limb length discrepancy and the optimum form of treatment. By placing the malleoli together, when the patient is lying supine with the pelvis squared and the hips and knees flexed it is possible to assess whether the shortening is mainly femoral or tibial (Fig. 13.4). Radiological assessment is more accurate than clinical examination.

Treatment

Differences of one centimetre or less require no treatment. For differences of one to three centimetres the best form of treatment is a heel raise leaving a difference of up to one centimetre. Surgical correction should be considered for discrepancies greater than three centimetres. It is possible to estimate the remaining growth in the upper tibial and lower femoral epiphyses from the Anderson–Green tables.

If the patient is likely to be tall a limb shortening procedure may be considered. Providing that sufficient growth exists the upper tibial epiphysis, lower femoral epiphysis or both epiphyses may be fused or stapled. The optimum timing of the procedure can be calculated from the growth tables. In older adolescents with fused epiphyses, up to four centimetres can be removed from the femoral diaphysis but it takes many months for the muscles, in particular the quadriceps, to adjust. Removal of excessive amounts of bone may result in permanent weakness. Non-union and delayed union occasionally occur.

In most patients it is preferable to leave the normal limb alone unless there is a contraindication to lengthening the shortened limb. With the newer limb-lengthening devices it is possible to gain up to seven or eight centimetres (Fig. 13.5) and if necessary the procedure can be repeated, both the femur and the tibia being lengthened. The complication rate is high although most of the complications are minor, e.g. pin tract sepsis. If union is delayed a secondary bone graft may be needed. The limiting factor is the neurovascular bundle and to a lesser extent the muscles. The device is fixed to the bone by percutaneous pins above and below the osteotomy. The majority of devices allow the limb to be lengthened at a daily rate of 1 to 1½ mm and the patient must be examined carefully each day. If there is any suggestion of neurovascular impairment or muscle contracture the lengthening may have to be stopped, either temporarily or permanently. With the modern devices successful procedures have been carried out in patients in their late teens and early twenties but it is preferable to operate before the age of 15.

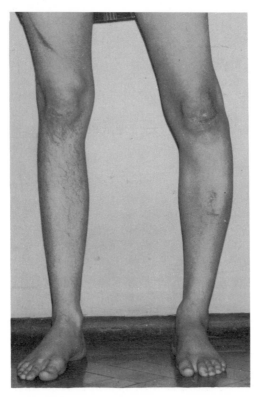

Fig. 13.2. A 17-year-old male with a shortened, deformed left lower leg due to malunion of a fractured tibia.

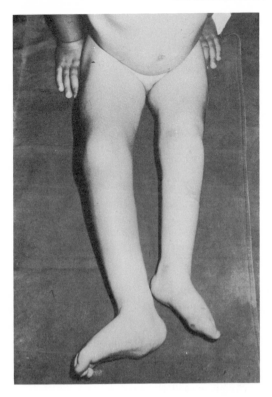

Fig. 13.3. Shortening due to neurological disturbance. Child with spina bifida, whose left lower limb has been affected to a greater extent than the right lower limb.

Apparent limb length inequality

Apparent limb length inequality is the result of a contracture. There is no loss of bone or cartilage. If the contracture is released the limbs will be of equal length. Total limb length is measured from the xiphisternum to the medial malleolus without attempting to compensate for any contractures. The apparent limb length inequality is the difference between the total limb length inequality and the real limb length inequality.

INSTABILITY

Instability of any joint in the lower limb will produce a limp. The main causes of instability are:
    (i) ligamentous damage due to trauma;
    (ii) ligamentous laxity associated with skeletal dysplasias, the resultant deformities being due to the combination of growth disturbances as well as ligamentous instability;
    (iii) congenital dislocation of the hip or knee.

Table 13.3. *Limb-length inequality*

Overgrowth	A–V fistula
	Hemihypertrophy
	Fracture
Real shortening	Congenital
	complete
	partial
	Fracture
	overlap of fragment
	angulation
	loss of bone
	Damage to growth plate
	fracture
	infection
	immobilization
	Paralysis
	Ischaemia
Apparent shortening	Contracture

Congenital dislocation of the hip

The most important cause of limp in early childhood as a result of instability is a missed dislocation of the hip. Despite careful neonatal screening approximately one-third of dislocated hips are not recognized at birth either because the congenital dislocation was missed or the hip dislocated during the first year of life. The examination of the hip is an essential part of the examination of the neonate as well as the infant.

At birth a congenital dislocation of the hip may sometimes be recognized by an extra thigh crease, shortening of the thigh on the affected side and prominence of the buttock (Fig. 13.6). In bilateral cases the perineum is often widened. The child is placed on a firm surface, the hips flexed to 90° and then abducted. Limitation of abduction indicates an irreducible dislocation. This occurs very rarely at birth but is a common sign after a few months. If the femoral head can be felt to slip into the acetabulum as the hip is abducted (a positive Ortolani sign) the hip is dislocated but reducible. The Barlow test diagnoses an unstable, dislocatable hip. The hip is flexed to 90° and abducted to 45°. The limb being tested is held with the thumb over the lesser trochanter and fingers over the greater trochanter whilst the pelvis is steadied with the opposite hand. Pressure is exerted over the greater trochanter in an attempt to dislocate the hip and over the lesser trochanter to reduce it. Instability of the hip will be detected during this manoeuvre.

The majority of 'clicking' hips are not due to instability of the hip, and the click is different to that felt during a positive Ortolani or Barlow test. It is a harmless condition which is of no pathological significance, provided that instability of the hip can be excluded.

Eighty per cent of late dislocated hips, i.e. dislocated hips diagnosed after the age of six weeks are detected during routine well-baby screening and, therefore, these tests should be carried out whenever an infant is examined. After the age of 3-4 months telescoping of the hip can often be elicited by longitudinal traction of the limb.

Once the child can stand a positive Trendelenburg test becomes apparent. Normally, the pelvis rises on the non-weight-bearing side due to contraction of the hip abductor muscles on the weight-bearing side. If the hip is dislocated the pelvis will dip on the non-weight-bearing side when the child is asked to stand on the affected limb (Fig. 13.7). Other causes of a positive Trendelenburg test include weak or paralysed hip abductor muscles, coxa vara, and un-united fractures of the femoral neck in the older child or adult.

The child walks with a waddling gait, i.e. the pelvis dips on the non-weight-bearing side during the gait cycle.

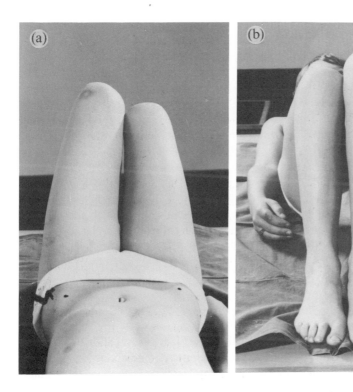

Fig. 13.4. Assessment of femoral and tibial length. The patient lies on a flat surface with the pelvis horizontal, the hips and knees flexed, and the medial malleoli together. (a) Viewed from the above. The right femur is shorter than the left. (b) Viewed from below the right tibia is shorter than the left. This patient's limb length inequality is due to a combination of femoral and tibial shortening.

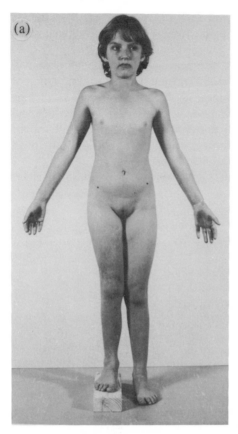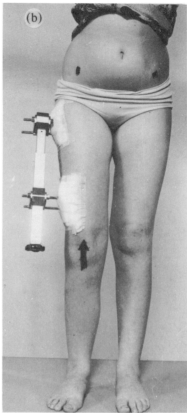

Fig. 13.5. A 12-year-old girl with congenital shortening of her right lower limb. (a) Prior to surgery. Black spots have been placed on her anterior superior iliac spines. Despite the large block under her right foot her pelvis is still not horizontal. (b) Following femoral limb lengthening the limbs are of equal length. The Wagner apparatus is fixed to her femur via percutaneous pins.

Treatment

The treatment depends on the age of the child. If a dislocated or dislocatable hip is diagnosed at birth the infant should be nursed in an abduction splint for 10 to 12 weeks (Fig. 13.8). Double nappies are not as effective since they allow the child to adduct the hips ever time they are removed. An abduction splint is contraindicated in the presence of an irreducible dislocation.

There is some controversy about the treatment of a dislocated hip diagnosed at the age of three to six months. Reduction under anaesthesia is associated with avascular necrosis of the femoral head in 40–45 per cent of cases and has been abandoned. In order to minimize the force required to reduce the hip an adductor tenotomy is carried out prior to reduction. Recent studies have shown that this is still associated with a high incidence of avascular necrosis and, therefore, many surgeons prefer either waiting or nursing the child in a Pavlik harness until the age of seven to nine months. If the hip is still dislocated, the child is admitted to hospital, placed on traction, and once the femoral head has been brought down to the acetabulum the limbs are gradually abducted. Frequently the hip is reduced and the child is then placed in a plaster of Paris hip spica until the acetabulum has developed. If gradual traction with abduction fails an open reduction is carried out.

Congenital dislocation of the hip is often associated with femoral anteversion (see below). If this is marked the hip will only be stable in excessive internal rotation and a secondary upper femoral derotation osteotomy may be required.

The greatest stimulus to acetabular development is the presence of the femoral head within the hip joint. The ability of the acetabulum to develop following reduction of the dislocated hip decreases with time and after the age of two and a half to three years reduction of the dislocated hip may not be sufficient to stimulate acetabular development and a pelvic osteotomy may be necessary to reshape the acetabulum.

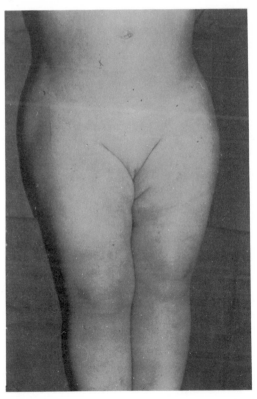

Fig. 13.6. Neonate with CDH. The left femur is shorter than the right, there is an additional thigh increase and the left buttock is prominent.

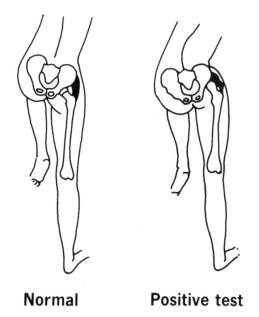

Normal   Positive test

Fig. 13.7. The Trendelenburg test. Normal abduction of the hip joint requires an intact hip joint, an intact femoral neck, and normal abductor muscles. In the presence of a dislocated hip, coxa vara, un-united fracture of the femoral neck or weak or paralysed abductor muscles the hip dips on the contralateral, non-weight bearing side when the patient is asked to bear weight on the affected side.

There is some debate about the treatment of dislocated hips after the age of six to seven years, particularly if they are bilateral, painless, and do not interfere with the child's ability to partake in normal activities. Surgery at this stage will require shortening of the femur as well as an open reduction and pelvic osteotomy. The results are not as good as in the younger child and may convert an unstable but pain-free hip to a stable but stiff one. Many surgeons will only operate on a unilateral dislocation at this age because the functional disability is greater.

After the age of 10 or 11 treatment is not advised unless secondary degenerative changes in the false joint lead to severe pain. Total replacement arthroplasty is often indicated even though the patients are only in their third or fourth decades (Fig. 13.9).

The earlier congenital dislocation of the hip is diagnosed and treated, the less the damage to the femoral head and the smaller the risk of subsequent osteoarthritis.

CONTRACTURES

Contractures of the hip, knee, and ankle will produce a limp. These are commonly due to muscle imbalance. Irrespective of the individual power of the muscles imbalance between the agonist and antagonist will produce a contracture of the joint in the direction of the more powerful muscle. The common contractures are adduction and/or flexion contractures of the hips, flexion contracture at the knee, and an equinus contracture at the ankle joint (Fig. 13.10) but any deformity may occur. In the initial stages correction of the muscle imbalance by tendon transfer, tenotomy, tendon elongation, or neurectomy will correct the deformity. However, if the imbalance has been present for any length of time secondary fibrosis of the ligaments and capsule occurs on the contracted side of the joint. At this stage correction of the muscle imbalance will not suffice; the joint contracture also must be released. In the growing child persistent muscle imbalance will produce a tertiary skeletal deformity if the muscles are attached below a growth plate. Osteotomy will be required to correct the deformity in addition to releasing the contracture.

The other main cause is fibrosis. This may result from localized compartmental ischaemia. In the lower limb Volkmann's ischaemic contracture most commonly affects the calf muscles producing an equinus

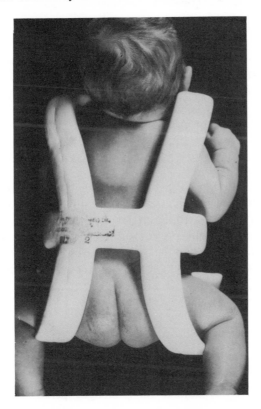

Fig. 13.8. Neonate with congenital dislocation of the hip being treated in an abduction splint. This holds the hips in abduction, prevents adduction, but allows some hip movement.

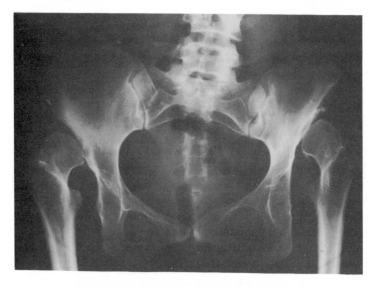

Fig. 13.9. Patient aged 28 with commencing pain due to congenital dislocation of her hips. The acetabulae are very poorly formed.

deformity of the ankle and flexion deformities of the toes. Fibrosis due to trauma, including intramuscular injections may also produce contractures. Contractures occur more readily in the growing child than in the adult, since growth of the limb tends to aggravate any difference in muscle power or muscle elasticity.

Treatment

Early contractures are best treated by physiotherapy in the form of stretching exercises supplemented by night splints where applicable. Reversed dynamic sling traction may help a fixed flexion deformity of the knee. Serial plaster of Paris casts may be of benefit, providing serial X-rays confirm that subluxation is *not* developing. Contractures that are more severe or resistant to conservative measures are treated by surgical release, the extent of the release depending on the contracture.

ANKYLOSIS

There are two main causes for ankylosis

1. Articular pathology, e.g. Still's disease, septic arthritis, resulting in damage to the articular cartilage and stiffening of the joint.

2. Immobilization of a joint for prolonged periods. This probably is associated with some degree of chondrolysis. It may occur in a normal joint which has been immobilized as part of treatment for another condition, e.g. the knee is often included in casts used to immobilize the hip or femur.

Treatment

Intensive physiotherapy is required but the results are often poor. Manipulation under anaesthesia may be helpful.

DEFORMITY

Deformity is a major cause of limp in childhood, and there are two main categories:

1. Deformities that usually are reversible.
2. Irreversible deformities.

Reversible deformities

Persistent femoral anteversion

This probably is the commonest cause of abnormality of gait under the age of five years. If a femur is laid on a flat surface, with the condyles and greater trochanter lying on the surface the head and neck of the bone will be seen to make an angle with the horizontal – the angle of anteversion (Fig. 13.11). The normal angle is about 20–25°, is greater in intrauterine life and often is increased in congenital dislocation of the hip.

Many children are born with an increased angle of anteversion which spontaneously corrects during the first few years of life. To accommodate the

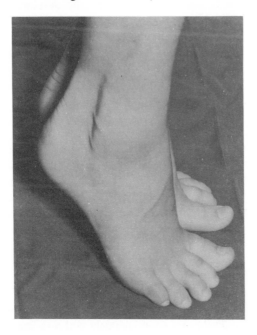

Fig. 13.10. Bilateral marked equinus deformity in a girl with cerebral palsy.

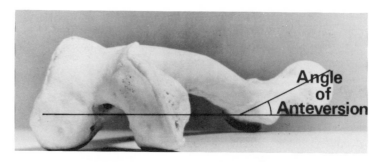

Fig. 13.11. Femoral anteversion. The angle of anteversion is the angle the head and neck of the femur make with the horizontal when the femur is placed on a flat surface, resting on the condyles and greater trochanter. In the normal adult it is approximately 20 to 25°.

increased angle of anteversion the femora are internally rotated – the greater the degree of anteversion, the greater the amount of internal rotation at the hip joint.

The condition is diagnosed by examining the hips in extension. Internal rotation is increased, often to 90° and external rotation is decreased often to 20° or less. Flexion, abduction and adduction are normal. The condition is associated with an intoeing gait and the children frequently fall.

The intoeing gait is usually improved by the development of secondary external torsion of the tibiae. If the child is asked to stand with the patellae facing forwards the degree of external torsion of the tibia can be observed from the position of the feet (Fig. 13.12). This external tibial torsion is often associated with a planovalgus deformity of the feet with flattening of the medial arches.

Treatment

In the young child no active treatment is required. However, it is important to explain the natural history to the parents. After the age of three the gait is helped by the provision of ³/₁₆- to ¼-inch medial heel wedges. They do not affect the underlying deformity but improve the gait and reduce the frequency of falls. Surgery is very rarely indicated except in children over the age of nine, with residual marked internal rotation of the femur, secondary external torsion of the tibia and painful planovalgus deformity. Less than 2 per cent of patients come to surgery.

External rotation deformity

Many children first stand and walk with an external rotation deformity of their lower limbs. This is physiological; the external rotation giving them greater stability. It usually disappears spontaneously within a year of the child first starting to walk.

Physiological genu varum and valgum

These deformities occur commonly in childhood. In the vast majority of instances the deformity is physiological, i.e. there is no underlying disease, and spontaneously is corrected. Genu varum (bow-legs) usually occurs in the younger child (aged one to three) whereas genu valgum (knock-knee) occurs in the slightly older child (aged three to five).

Treatment

In early childhood treatment is unnecessary and the vast majority of bow-legged deformities settle before treatment is considered. In the slightly older child with a valgus deformity, $^3/_{16}$- to ¼-inch medial heel wedges are sometimes used. If the deformity has not corrected itself by the age of four to five years mermaid night splints are often provided.

Calcaneovalgus deformity

Any deformity and any combination of deformities of the foot can occur in childhood. One of the commoner types is the calcaneovalgus deformity which

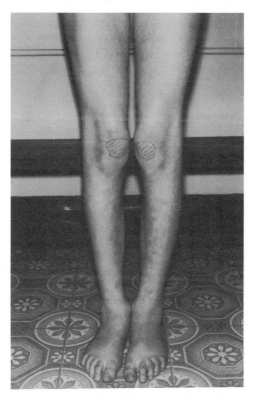

Fig. 13.12. Fourteen-year-old boy with persistent femoral anteversion. Standing with his feet held together. The patellae face inwards.

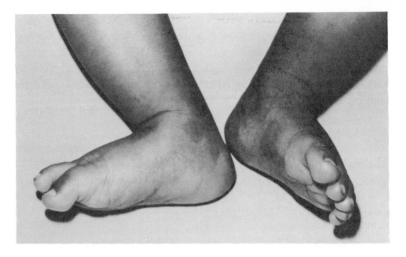

Fig. 13.13. Calcaneovalgus deformity. This usually is a benign deformity which responds rapidly to stretching.

is present at birth and is benign. It usually is a postural deformity, the foot is everted and dorsiflexed (Fig. 13.13) and responds readily to treatment.

Treatment

The parents are taught to gently stretch the foot into the overcorrected position by steady pressure over the dorsum. The stretching exercises are started immediately after birth and are carried out several times a day. Very rarely the deformity is still present after four weeks and more active treatment by holding the foot in the overcorrected position either with strapping or a plaster of Paris cast is required.

Equinus gait

The normal gait is a heel–toe gait, i.e. the individual puts the heel to the ground at the beginning of the stance phase. The loss of this pattern will produce a limp, the patient walking on his/her toes.

Some children never develop a heel–toe gait, although there are no contractures, movements in all their joints are full and there are no neurological abnormalities. Six weeks in a below knee plaster with the foot in 5–10° of dorsiflexion, using plaster boots or sandals for walking and followed by intensive physiotherapy usually is sufficient to correct their gait.

Irreversible deformity

These are deformities which will not correct spontaneously and which always require treatment. Any deformity can occur, most producing a limp but only few will be discussed in any detail.

Coxa vara

This term includes many conditions in which the neck-shaft angle of the femur is less than the normal of approximately 125°. The most important causes are:

   (i) congenital;

  (ii) slipper upper femoral epiphysis;

 (iii) fracture;

 (iv) softening of bone, e.g. rickets;

  (v) damage to femoral capital growth plate, e.g. infection.

Coxa vara leads to true shortening of the limb and in severe cases to a Trendelenburg 'dip' and a waddling gait. The treatment is the treatment of the underlying condition. In appropriate cases the neck-shaft angle can be corrected by an abduction osteotomy.

Genu varum and genu valgum

There are many causes of the more severe non-physiological type:

   (i) fracture of the femur or tibia with malunion;

  (ii) rickets of any type;

 (iii) uneven growth of epiphyseal plates, e.g. following injury, osteomyelitis, Blount's disease.

Unlike the benign genu varum of toddlers and the benign genu valgum of childhood these deformities are usually irreversible.

Treatment

The deformity can be corrected by lower femoral or upper tibial osteotomy, but, if it is due to unequal growth of the epiphysis, the deformity will recur

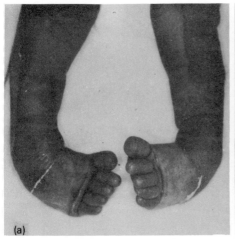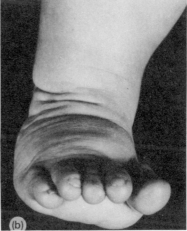

Fig. 13.14. Congenital talipes equinovarus. (a) Rigid congenital club foot which has not responded to conservative treatment with strapping and stretching. Child aged six weeks. (b) Child aged 18 months. At 10 weeks a posteromedial soft-tissue correction was carried out. A good result has been obtained.

during the growth period. Growth of the normal portion of the epiphyseal plate can be retarded by stapling or fusion, or the bony bridge can be resected and the space filled with a silastic spacer in an attempt to prevent further bridging.

Congenital club foot

The common form of congenital club foot is the congenital talipes equino varus. There are three aspects to the deformity. The hind foot is in equinus, the foot is inverted, and the forefoot is adducted (Fig. 13.14).

Table 13.4. *Painful causes of limp*

Joints	Injury Infection Haemarthrosis

Bone	Fracture
	Infection
	Tumour
	Infarct
	Osteochondritis/apophysitis
	  capital femoral epiphysis – Perthes' disease
	  os calcis – Sever's disease
	  navicular – Kohler's disease
	  metatarsal head – Freiberg's disease
	  tibial tubercle – Osgood–Schlatter's disease
	Slipped upper femoral epiphysis

Muscle	Inflammation, e.g. dermatomyositis
	Haematoma, e.g. haemophilia, trauma
	Rupture
	  partial
	  complete
	Compartment syndrome

Tendon	Rupture Paratendonitis Tenosynovitis
Soft tissue	Ligamentous tear Ligamentous strain Tender heel pad Plantar fasciitis Haematoma Fat necrosis
Nerves	Morton's metatarsalgia Compression syndrome, e.g. tarsal tunnel
Blood vessels	Arterial insufficiency Phlebitis Venous thrombosis
Miscellaneous	Plantar warts Foreign body in foot Callosities Abrasions Foot strain Ingrown toe nail

Treatment

Treatment should begin at birth and consists of weekly plaster changes or regular stretching and strapping correcting initially the varus and subsequently the equinus. If the deformity is not corrected by 6–8 weeks surgery is indicated. All the tight structures must be released posteriorly and medially, the talonavicular and subtalar joints being reduced. Post-operatively, the limb is held in plaster in the corrected position for approximately six weeks.

PAINFUL LIMP

Any painful lesion affecting the lower limbs may produce a limp. This is a particularly significant finding in the young child who may not be able to give a history of pain, nor be able to localize its site. Often the limp is intermittent even though it may be due to severe underlying pathology, e.g. Perthes' disease, slipped upper femoral epiphysis. Therefore, the history of a limp must be taken seriously and the child fully investigated even if the limp is no longer present.

Examination may be easier in the older child who can localize the site of pain but in all children a detailed clinical investigation is necessary. The examination must include gentle palpation of the limbs, back, and abdomen looking for any site of tenderness. It is essential that the palpation is gentle to avoid upsetting the child. The joints must be carefully examined as must the spine. A full neurological examination is important and X-rays should be obtained of any tender area.

The major causes of a painful limp are shown in Table 13.4. The majority of these conditions are discussed in other chapters but the more important ones that affect children will be presented in this chapter. The first two affect the hip joint, occur in childhood and may present with an intermittent limp and a normal examination. It is for this reason that the hip is often X-rayed in children with a history of a limp but a normal examination.

PERTHES' DISEASE

This is an osteochondritis of the femoral capital epiphysis. The aetiology is not fully understood but there is a temporary interference to the blood supply. The disease lasts two to three years, during which time the affected portion of the femoral head undergoes ischaemic necrosis, the dead bone is resorbed and new bone is laid down. The sequence of necrosis and subsequent bone formation is patchy and accounts for the radiographic appearance of fragmentation. During this period the femoral head is softened and may become deformed.

The disease usually affects children between the ages of five and 10 years although it can occur in younger and slightly older age groups. It is bilateral in 10 per cent.

Treatment

If the condition is associated with muscle spasm and limitation of movement, two to three weeks bed rest with bilateral skin traction is indicated. There is some debate about the subsequent management. One school advises weight relief. The rationale is that reduction of the force transmitted across the hip joint will minimize the risk of deformation but it is impossible to totally exclude all forces acting on the hip joint. Normal muscle tone, even if the child is confined to bed, will produce a considerable force on the joint. At its extreme the patients are confined to bed, in some cases with continuous skin traction for two or three years until the femoral head has consolidated. Alternatives include a variety of weight-relieving calipers, slings, or splints, which the child wears for up to two or three years.

The second type of treatment is based on the concept that deformation of the femoral head is not significant providing the femoral head deforms to the shape of the acetabulum, i.e. the development of a 'congruous incongruity'. The femoral head is best contained within the acetabulum when the limb is abducted and internally rotated. This position can be obtained by holding the limbs in plaster for two to three years or by a varus, external rotation osteotomy. The osteotomy heals within six weeks. Comparative studies indicate that the results of osteotomy are at least as good as those obtained by conservative measures but the effect on the general well-being of the child is much improved. After a short period in plaster he is able to return to all his activities whereas many of the conservative forms of treatment are likely to

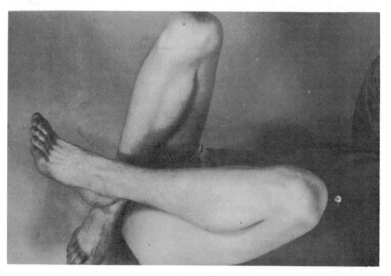

Fig. 13.15. Slipped upper femoral epiphysis. This 17-year-old boy has a major chronic slip of his right capital femoral epiphysis. When asked to flex his hips, the right hip goes into external rotation but flexion of the left hip is normal. The sign is characteristic of a slipped upper femoral epiphysis.

affect the general emotional, psychological, and educational development of the child. The author believes that today there is no valid reason for confining a child with Perthes' disease to bed with or without traction, for months or years. In addition prolonged immobilization in many of the slings and splints has a significant morbidity due to chondrolysis and stiffness of the knee or early epiphyseal closure.

It has also become apparent that not all patients with Perthes' disease require treatment. The severity of the condition is related to the degree of femoral head involvement, subluxation, lateral extrusion of the epiphyses and the age of the child. Where less than 50 per cent of the femoral head is involved the results are good, irrespective of treatment; whereas the results are universally poor when the entire femoral head is already grossly flattened and broadened. Osteotomy is not indicated in these two groups. In the intervening groups which include the majority of patients, the prognosis is related to age, children under the age of five having a better prognosis. In many of these patients osteotomy is the treatment of choice for the reasons given above. Some surgeons prefer a pelvic to a femoral osteotomy.

SLIPPED UPPER FEMORAL EPIPHYSIS (ADOLESCENT COXA VARA)

This condition usually occurs in the 10–15 age group although it can occur in younger and older patients. Boys are more commonly affected than girls. The femoral head is displaced downwards, medially, and posteriorly. The displacement usually occurs gradually but a major acute slip can occur. The greater the slip the greater the risk of osteoarthritis in adult life. Less than half the patients are overweight.

The patient usually presents with a history of intermittent pain which may be referred to the knee and/or a limp. The examination is characteristic. Flexion, abduction, and internal rotation are limited, particularly when the movements are carried out in the neutral plane. External rotation is often increased and the range of flexion and abduction is greater if the test is carried out with the limb externally rotated. If the patient is asked to flex or abduct his limb it often adopts an externally rotated position (Fig. 13.15).

The diagnosis is made on X-ray. The major direction of slip is posterior and may not be seen on an anteroposterior film. Therefore, if this condition is suspected a lateral X-ray of the hip must be obtained. The lesion should be suspected in every boy of 10 to 17 years, who complains of pain in the hip or knee.

Treatment

The treatment depends on the type of slip.

   (i) *Minor slip.* The position may be accepted and the femoral head and neck internally fixed to prevent further slipping until the epiphysis fuses.

   (ii) *Acute major slip.* This occurs very rarely. The hip is gently manipulated under anaesthesia to reduce the displacement and the femoral head and neck internally fixed. Manipulation must be gentle to avoid damage to the epiphyseal arteries and avascular necrosis.

(iii) *Chronic major slip.* Bone is formed where the femoral head has slipped off the neck making reduction impossible, yet the displacement it too gross to be acceptable. Resection of the new bone from the femoral neck, with reduction of the displaced head and internal fixation gives an accurate reduction but carries a high risk of avascular necrosis. Compensatory osteotomy is preferred by most surgeons. The epiphysis is internally fixed and after the growth plate has fused a wedge of bone is removed at the trochanteric level to compensate for the backward tilt and reduce the femoral head into the acetabulum.

The condition is bilateral in a quarter of patients, the contralateral slip often occurring months or years later. Therefore, it is particularly important to examine carefully the opposite hip whenever patients complain of pain or limp.

OSTEOSARCOMA

Primary osteosarcoma occurs in childhood and adolescence. In later life it is usually a complication of Paget's disease of bone. The commonest sites are the distal femoral and the proximal tibial metaphyses. Often the patient complains of pain for some weeks before a mass becomes palpable and occasionally the symptoms are precipitated by trauma.

Pain at the knee in a child or adolescent may be due to a malignant tumour and this must be borne in mind when examining the child. If there is the slightest suspicion an urgent orthopaedic referral is essential.

OSTEOMYELITIS AND SEPTIC ARTHRITIS

Today, the onset is subacute in the majority of patients. It is uncommon for the child to develop the classical features of the fulminating infection associated with marked pain; exquisite tenderness (sometimes to such an extent that the child cries when the bed is approached); and a marked constitutional illness with a gross pyrexia, toxicity, and even delirium. The child complains of a painful limp or will not bear weight on that limb. The patient is tender at the site of infection but in the region of the proximal femur it may be difficult to differentiate a septic arthritis from osteomyelitis affecting the proximal femoral metaphysis. The symptoms in the common subacute variety may be so mild compared with the classical acute fulminating osteomyelitis that the diagnosis was only considered by the referring physician in four of the last 14 patients the author has treated. If the condition is suspected, the child should be referred to an orthopaedic surgeon before instituting antibiotic therapy.

The most significant investigation is the erythrocyte sedimentation rate. This is virtually always raised in septic arthritis or osteomyelitis even though the white cell count may be within normal levels. In the early stages the X-rays are usually normal although an alteration in the soft tissue planes may be apparent. Radiographic changes are not seen in the joint or bones for 10 to 14 days. Skeletal scintigrams or scintigrams obtained after the intravenous injection of labelled leukocytes are more sensitive than the radiographs and indicate the site of infection within 48 to 72 hours.

Blood cultures and joint aspirates should be obtained before starting antibiotics. If the child is desperately ill the infected bone or joint must be drained immediately. However, the majority of patients can be treated with high-dosage antibiotics and immobilization once the relevant investigations have been obtained. If there is no improvement in the clinical features and the ESR does not fall within 12 to 24 hours surgical drainage is indicated. If daily examination indicates that the child is improving, and the ESR and white cell count (if elevated) are returning progressively to normal, conservative treatment may suffice. Once the erythrocyte sedimentation rate has returned to normal and all the clinical signs and symptoms have settled the child may be mobilized but antibiotics must be continued for at least six and usually 12 weeks.

NEONATAL SEPTIC ARTHRITIS

This is an important variant of septic arthritis. It occurs in the neonate and the hip is the commonest site. The epiphyseal vessels run in the capsule and increased pressure within the joint compresses these vessels sufficiently to produce avascular necrosis of the femoral head. The neonate is desperately ill, and often the infection is multicentric.

Treatment

It is essential to drain the hip joint before the capital femoral epiphyseal blood supply is damaged.

HAEMOPHILIA

There are several ways in which haemophilia may produce a limp in a child.

(i) A bleed into the hip joint if not diagnosed and drained may produce damage to the femoral capital epiphysis and capital growth plate resulting in shortening and pain.

(ii) Iliopsoas bleed is associated with pain in the right iliac fossa, spasm of the iliopsoas muscle producing flexion at the hip joint, and a femoral nerve palsy.

Treatment is expectant with immobilization of the hip in a comfortable position. As the spasm settles the hip is gradually allowed to extend.

(iii) Acute bleed into a knee or ankle joint. If the joint is tense it should be aspirated under factor VIII cover, providing that the patient does not have inhibitors. The joint should be immobilized until the acute bleed subsides.

(iv) Haemophilic arthropathy. Recurrent bleeds into a joint produce an arthropathy with intermittent effusions and repeated small haemorrhages. The patient requires an orthosis to protect the joint during weight bearing.

(v) Intramuscular bleeds will produce a painful limp and may result in fibrosis and contracture as a result of intracompartmental compression.

(vi) Superficial haematomata may produce a limp until the pain settles.

SPINAL CONDITIONS

Many spinal conditions may cause a limp in a child, usually due to pain although not necessarily so. Most of these conditions will be discussed in Chapter 9.

PROLAPSED INTERVERTEBRAL DISC

Less than 2 per cent of prolapsed discs occur in adolescence. They usually present with back pain and sciatica and the straight leg raising test is limited.

INTRASPINAL TUMOURS AND CYSTS

These lesions may present with back ache although the child or adolescent may develop sciatic symptoms, ascending paraparesis and bladder disturbance if there is a bleed with rapid enlargement of the lesion and compression of the cord or cauda equina.

SPONDYLOLYSIS AND SPONDYLOLISTHESIS

This usually presents with back pain but if there is neurological involvement due to compression of the cauda equina the patient may develop sciatica, neurological signs in the lower limb, and a limp.

DIASTEMATOMYELIA

This is an intraspinal bony or fibrous septum. It may be associated with increasing traction on the spinal cord, as the vertebral column grows in length, with progressive neurological impairment including the development of a limp.

SCOLIOSIS

This usually presents as a deformity. Pain and neurological involvement are extremely rare. However, neurological impairment may occur following correction, particularly of a congenital scoliosis due to traction on the cord or nerve root.

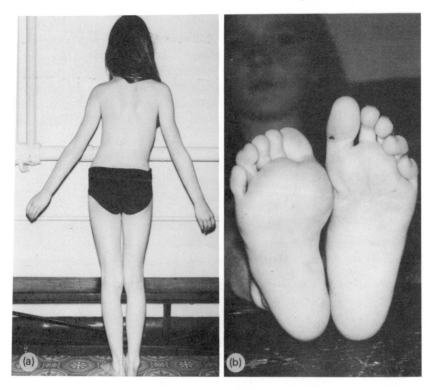

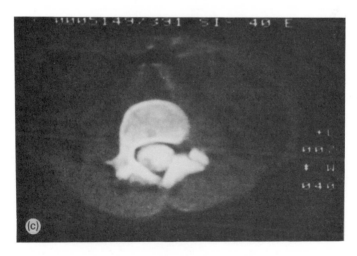

Fig. 13.16. Eight-year-old girl with an enlarged adherent filum terminale. (a) She has a scoliosis and a slightly shortened and thinner right lower limb. (b) Her right foot is much smaller than her left foot. (c) CT scan taken with metrizamide in the dural sheath. The large filum terminale can be seen. It is adherent to the right side of the spinal canal.

TRACTION ON A NERVE ROOT

This may occur in spina bifida occulta if there are adhesions to a nerve root or in association with an adherent thickened filum terminale. The vertebral column grows more rapidly than the spinal cord, the spinal cord 'ascending' within the spinal canal during growth. As a result adhesions to the nerve roots will produce traction. This usually presents with a growth disturbance, often a smaller foot or shortening of the entire limb, rather than with pain (Fig. 13.16).

NEUROMUSCULAR CAUSES

The main neuromuscular causes of limp are shown in Table 13.2.

NERVE COMPRESSION

Nerve compression syndromes can occur in the lower limb. The commonest is the tarsal tunnel syndrome, the posterior tibial nerve being compressed as it

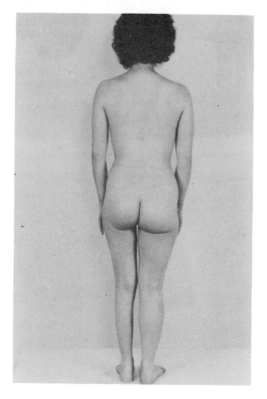

Fig. 13.17. Patient with poliomyelitis, which occurred after growth had ceased. She has been left with a wasted and weak left lower limb. This has resulted in a limp which is improved by an above knee caliper.

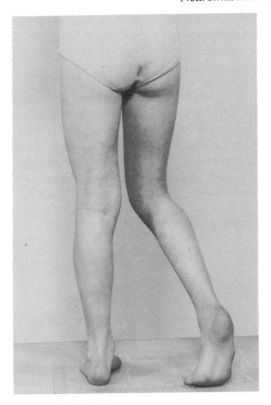

Fig. 13.18. Patient with spastic monoplegia showing the characteristic features of cerebral palsy. The hip is flexed, adducted, and internally rotated; the knee is flexed; and the ankle is in equinus.

runs behind the medial malleolus. The patient presents complaining of pain or paraesthesiae usually radiating along the medial border of the foot and involving the great toe.

Compression syndromes can occur at other sites, the clinical presentation depending on the site of compression.

WEAKNESS

Weakness of part or the entire lower limb will produce a limp. In the more severe forms the child will not be able to weight bear on the affected limb (Fig. 13.17).

Treatment

Orthoses can be used to support the weak or paralysed muscles, e.g. foot drop can be helped with a below-knee caliper and foot drop spring or a passive moulded splint which fits into the shoe and cosmetically is more acceptable.

Below-knee or above-knee calipers can be used to mobilize many children who are unable to walk as a result of a variety of neuromuscular diseases.

SPASTICITY

There are three main ways in which cerebral palsy can affect gait. The commonest is due to spasticity and the deformity is characteristic (Fig. 13.18). The hip is flexed, adducted, and internally rotated; the knee is flexed; and the ankle is in equinus. Initially the deformity is due to the muscle imbalance, the joint being contracted in the direction of the more spastic muscle. Later secondary soft tissue contractures develop. Physiotherapy and muscle relaxants are the main forms of treatment although surgery is sometimes required but is

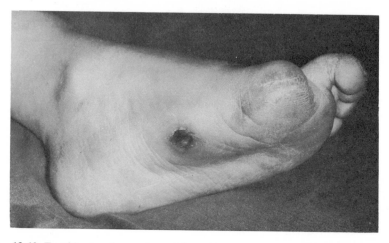

Fig. 13.19. Trophic ulcer in a patient with a mixed sensory and motor polyneuropathy. He has already had an amputation of his great toe for a trophic ulcer.

only part of the therapeutic regimen. There are a great variety of surgical procedures depending on the individual circumstances. Typical procedures include:

1. Tenotomy of the hip adductor tendons for severe adduction deformity. This may be necessary even in a grossly disabled and mentally retarded child for toilet purposes and may be combined with neurectomy of the anterior branch of the obturator nerve.

2. Elongation of the tendo Achilles for an equinus deformity.

3. Transfer of the tibialis posterior to convert it to an ankle extensor.

4. Soft-tissue release for a secondary contracture of the capsule and ligaments of a joint.

Inco-ordination or athetosis may also cause a limp. Treatment is by physiotherapy.

INCO-ORDINATION

Lack of co-ordination of movement occurs in a number of neurological conditions. The treatment is by highly specialized physiotherapy to improve the co-ordination of the different muscle groups.

ATAXIA

Ataxia produces an unsteady, unco-ordinated wide-based staggering gait. It may be due to cerebellar dysfunction or posterior column disturbance.

SENSORY NEUROPATHY

Loss of sensation may also affect gait and may be responsible for the development of trophic ulceration (Fig. 13.19).

HYSTERIA

An hysterical gait occurs more commonly in the adult than in the child but nevertheless can occur. The diagnosis should not just be one of exclusion. In addition to the absence of underlying organic pathology hysterical features should be present.

MALINGERING

A bizarre gait can be produced by frank malingering. This usually occurs in the adult, often allied to 'compensationitis', but can occur in childhood. The patient may have developed an hysterical or a deliberate malingering gait in a subconscious or overtly conscious attempt to obviate problems at home, at school, or elsewhere in their environment.

SUMMARY

The development of a limp is a significant sign in childhood. Although the vast majority are benign, and often the result of a minor traumatic episode it may be due to a major disorder of the locomotor system and always requires thorough investigation.

ACKNOWLEDGEMENT

I would like to thank the Department of Medical Illustration, Royal Manchester Children's Hospital and Hope Hospital for the illustrations.

Section IV
The general practitioner approach

14 Record keeping for locomotor disorders in general practice

Raymond Million

INTRODUCTION

The difficulties of keeping good records in general practice are so substantial that there is no standardized procedure at present. With this fundamental fact in mind any recommendation would be an advance. It is a sad reflection upon the standards of practice and upon the philosophy of our planners and policy makers that general practitioners can still expect to use documents which were designed under the National Health Act of 1919. The role fulfilled by to-day's general practitioner bears no relationship to his function in the first half of the twentieth century, and it is an anachronism that he should still be using the same files for record-keeping.

It is no easy matter to formulate a programme of recording systematically, having due regard to the limitations upon the general practitioner in his day-to-day work. Such factors include grossly restricted consultation time, profound limitation of available space both for writing and storing records, a certain lack of clinical knowledge due to lack of expertise and challenge, and a limitation in the terms of reference within this field because of the panoramic scope of work which this physician is expected to cover.

With all these dilemmas and restrictions in mind there is substantial encouragement for change which one would hope would lead to an improvement in systematic thinking (Wood 1974). Contemporary professional activities are conducive to change right now, having regard to the enormous strides made by the Royal College of General Practitioners in the practice, planning, and research of medicine, and the mushrooming of formal and informal teaching programmes for undergraduates and postgraduates, both learning a new and refreshing old knowledge. There has been research recently into record keeping in general practice. Witness the development of the use of A4 records in selected centres up and down the country, such programmes having been sponsored in part by the Department of Health and Social Security. Regrettably this has not become standard policy perhaps due to intellectual torpor or economic stringencies. Further encouragement is derived from the recommendations by the RCGP to keep records of morbidity and the epidemiology of important disorders. Many doctors keep the renowned 'E' book as well as Age/Sex registers, and this encourages them to have a deeper understanding of the frequency of disorders and the manner in which they impinge upon the population. In addition there has been some standardization of documents in the management of other groups of disorders, the most notable being the keeping of antenatal records both in hospital and in general practice which have now reached some degree of sophistication, though not uniformity. Interest has been shown in the prospective recording of the diagnosis and management of hypertension in general practice. Furthermore problem orientation was given greater credence in latter years by the development by Professor Weed of his problem-orientated records system, which, while it has become largely applied in hospitals, has been the special province of the general practitioner for a great deal longer. It seems strange therefore, that we have not paid proper attention to the systematic keeping of clinical records in general practice much sooner (Avery Jones 1981; Petrie and McIntyre 1979).

POLICY

We now have a need for programmed records because we have an increased knowledge of the mechanisms and pathology of locomotor disorders, and general practitioners have much greater opportunities than ever before for the investigation and management of their patients, either independently or in association with their specialist colleagues. Furthermore co-operative schemes have developed within primary care teams (Chapter 15) which generate the need for systematic recording of information such that events and their sequelae can be made known to every member of the primary care team, and there is mutual comprehension of each other's responsibilities and functions as well as achievements.

RECORDS

It may be said, without any cynicism, that current FP5 and 6 documents made available by the Department of Health and Social Security are useless for all but minor events. Their small size necessarily dictates an unreasonable brevity in recording and the omission of essential information. They may therefore be regarded as dangerous, being conducive to sloppy thinking and ultimately to lack of effective action. The A4 system is an advance. The document in itself is merely a bigger one than the FP5. However, as there are separate sheets in the A4 folder for X-ray reports, health visitors reports, special investigations and immunization records, the first glimmering of systematic recording is beginning to appear (Cormack 1981; Elliott 1981). The next step forward as witnessed by a paper published by the RCGP in 1980 on computers in general practice, seems so enormously large as to leave an important gap to fill. It cannot be denied that in the use of computers in medicine, with constantly changing information sources and improving techniques for building miniaturized computers which are equally constantly changing, there is a field of uncertainty into which few of us would yet be prepared to wander with confidence.

NEEDS

If we are to plan systematic records for the locomotor disorders, we need to pay attention to certain basic principles:

1. Ease of completion.
2. Ease of interpretation.
3. Adequacy of information.
4. Avoidance of irrelevance.
5. Ease of follow-up.

Locomotor disorders comprise musculoskeletal disorders, neurological problems, and vascular diseases. An effort to design a document which will fulfil all these functions must be doomed to failure as the scope encompassed by these groups of disorders is so vast that one cannot hope to produce a systematic chart to satisfy every need. There is far too much variety of information to be documented. The forms which we can hope to use must also cater for the general practitioner's relevant inexperience and lack of training in this field. Such forms will help to educate the practitioner towards the need for identifying specific clinical events and the management problems which they generate. We also have to bear in mind the need to follow-up our patients. It is in the very nature of locomotor disorders that a fair proportion of them are long lasting, follow-up is protracted, and there is the risk of creating an enormous amount of paper which may make the file heavy and unwieldy and make retrospective review confusing. Somehow we have to reach a compromise. The record forms suggested in this chapter are used by the author in his practice and represent an attempt to provide adequate information in a readily accessible form. Although they will not suit everybody, I hope that experience in using these forms will enable others to construct more precise record sheets that eventually might become a standardized format in general practice.

The bulk of the rheumatic problems which we meet includes osteo- and rheumatoid arthritis, the variants of seronegative polyarthropathy, ankylosing spondylitis, Paget's disease, gout, non-inflammatory spinal disorders, and non-articular rheumatism. The vascular lesions are arteriopathy and venous disorders and their consequences. The major neurological problems are stroke, multiple sclerosis, peripheral neuropathy, mononeuropathy for whatever reason, and Parkinson's disease.

DOCUMENTS

The first record form is a basic statement of findings at initial presentation. This encompasses the principal clinical material in an easily assimilable form, preferably both spatial and documentary (Fig. 14.1). The second is a problem summary. The third records investigations. The next records recommended procedures of management, including evidence of referral within and without

(a)

MUSCULOSKELETAL CHART

NAME_____

DATE_____

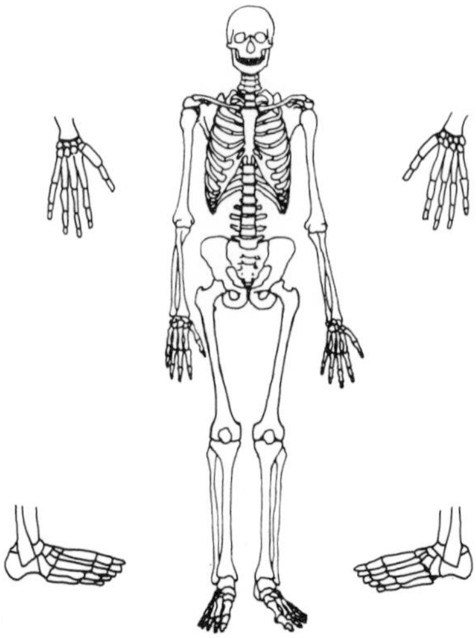

Indicate on diagram:

1. Joints or sites affected, by circle
   (red–synovial thickening
   black–bony enlargement)

2. Range of movement (major affected joints)
   (e.g. normal knee 5–0–150
   knee with 10° flexion and only flexing
   to 60° 0–10–60)

3. Sensory changes

Report clinical findings as relevant:

1. EMS (early morning stiffness) ....min or.........h

2. Ritchie index............

3. Grip strength (mm Hg) (L).......

   (R).......

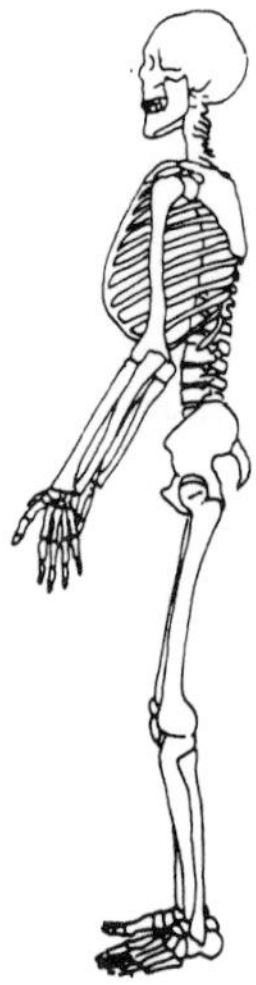

4. Spinal curves

   Normal ☐

   C ☐   Kyphosis ☐

   D ☐   Lordosis ☐

   L ☐   Scoliosis ☐

5. SLR   °R   °L

6. Stretch tests (tick if positive)

   Ely ☐ R   ☐ L

   Lasegue ☐   ☐

Diagnosis: ....

....

....

....

....

Fig. 14.1. (a) Musculoskeletal chart.

the primary care team. There is a progress chart for clinical findings and finally a follow-up chart for drug management in rheumatoid arthritis.

The group of disorders under discussion frequently require the services of the district nurse and health visitor and the A4 forms already designed for their use could well be incorporated into the scheme. The system encourages our colleagues within the primary care team to refer to our clinical notes, so improving the standards of management of all of us (Fig. 14.7).

CLINICAL REPORTS

A chart (Fig. 14.1a) is shown which aims to record, both graphically and pictorially, the bulk of the locomotor difficulties encountered. Many neurological problems such as strokes cannot be recorded on this form, and it is better to have a separate stroke document (Fig. 14.1b). The third form records vascular problems (Fig. 14.1c). These charts will not record every event, but I suggest them as a basis upon which further proposals can be built. They are simple to complete and summarize succinctly the major clinical findings.

MUSKULOSKELETAL CHART (FIG. 14.1a)

In using a skeleton with detailed drawings of the hands and feet, it is possible to record the majority of common events. Joint disorders are shown as a black circle for bony enlargement and a red circle for any inflamed joint. Thereby isolated joint problems or polyarthritis are easily depicted. The distribution of affected joints shown on the chart is an aid to diagnosis. The range of movement of affected joints is shown with reference to the neutral (0°) position.

Spinal deformities are difficult to record on a chart, but a simple programmed boxed system such as that shown will provide the bulk of information. Such a method has both brevity and reasonable precision. A lateral spinal view chart will reinforce the scripted information. Entrapment syndromes can be indicated by shading on the hand or foot the distribution of the nerve affected and the relevant sensory disturbance as well as by a written diagnosis.

The following are valuable parameters of activity of arthritis:

Morning stiffness

One of the critical symptoms of inflammatory disease in the arthritides is early morning stiffness. A simple historical statement from the patient of the duration of this stiffness can be used as a measure of change, and this can be recorded either in minutes or as fractions of an hour. The latter is better as the patient's assessment of the number of minutes of morning stiffness may be remarkably inaccurate, whereas to record change within intervals of a quarter of an hour may be both easier and more dependable.

(b)

NEUROLOGICAL CHART

NAME_____   Date_____

1. Indicate on chart

   (i)   Reflexes
   (ii) Wasting
   (iii) Loss of power
         or sensation
   (iv) Relevant features
         (other)

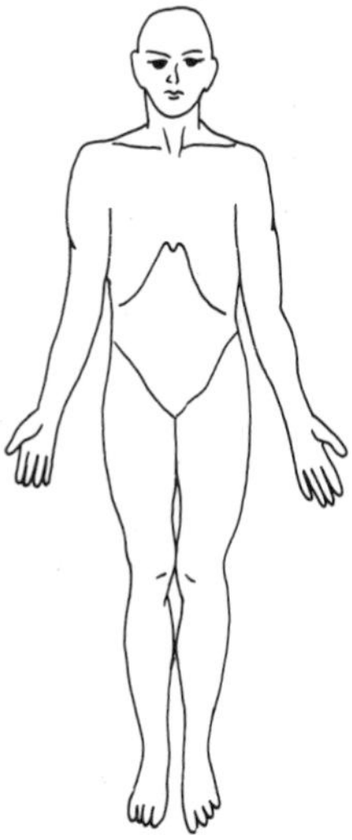

Paresis

	R	L
Face	☐	☐
Arm	☐	☐
Leg	☐	☐

Spastic ☐
Flaccid ☐

Grip

R....... mm Hg

L....... mm Hg

Urine   ☐ Glucose–

         ☐ Protein–

Fundi
(script)

....

....

....

Gait
(script)

....

....

....

Significant metabolic phenomena
(e.g. diabetes)

....

....

	Cont	Incont
Bladder	☐	☐
Bowel	☐	☐

Affected Functions

Vision   ☐
Speech   ☐
Deglutition   ☐
Pressure areas   ☐
Chest infection   ☐
Other (specify)   ☐

Diagnosis   ....

             ....

             ....

             ....

Fig. 14.1. (b) Neurological chart.

Ritchie index

A more detailed method for assessing the progress of inflammatory arthritis of a sytemic nature is to use the Ritchie index. Ritchie devised a system of assessing joint tenderness using a scale of 0–3, in which:

0 – represents no tenderness;
1 – tenderness on palpation;
2 – tenderness and the patient winces;
3 – tenderness and the patient winces and withdraws.

To do the test properly requires a standard clinical technique of examination and the same person to undertake each assessment. The technique should be applied to every joint being examined. To the inexperienced it appears to be a major, time-consuming technique. It is surprising how quickly the technique is learned and how easily it can be carried out. A trained and experienced observer will spend only a couple of minutes on this procedure and its comparative value is considerable. Thus for the patient suffering from inflammatory arthritis or localized lesions of the limbs, we now have a quantitative system for reporting disability, and for its follow-up.

Grip strength

In recording the progress of change in upper limbs, the measurement of grip strength is simple and repeatable. Measurements of grip can be used in all forms of locomotor disorders that affect the small bones or joints of the hands, wrists and musculature of forearm, whether this be inflammatory arthritis, osteoarthritis, nerve injury or stroke. The grip machine is a sealed bag (Figs. 14.8 and 14.9) inflated to 30 millimetres of mercury, using either a standard sphygmomanometer mercury column or an aneroid scale. This is known as the Davis bag. The patient takes the inflated bag firmly in the flat palm of the hand and squeezes firmly and consistently and the maximum pressure held for five seconds is recorded.

NEUROLOGICAL CHART (FIG. 14.b)

The second chart is constructed largely to record severity of stroke, as this is probably the commonest major neurological lesion in general practice. Again the picture of the patient is the central objective upon which abnormalities of reflexes, musculature, power, sensation and other fundamental neurological lesions will be documented. There is scope to record certain specific problems that may arise. These are selected as the findings or functions critical to reaching a diagnosis or used later as a basis for progress assessment.

VASCULAR CHART (FIG. 14.1c)

Similar objectives are laid down here as for the preceding charts, i.e. to document the fundamental relevant clinical findings in a simple form together with

(c)

VASCULAR CHART

1. Arterial

Palpable pulses (tick when present)   R   L

Femoral   ☐   ☐

Popliteal   ☐   ☐

Dorsalis pedis   ☐   ☐

Posterior tibial   ☐   ☐

Temp. of foot

Warm   ☐   ☐

Cold   ☐   ☐

Temp. gradient level
(script) ....   ☐   ☐

Trophic foot change
(script) ....   ☐   ☐

Limb skin defect
(script) ....   ☐   ☐

Walking time (50 metres) ....s (for claudicants)

Time to climb 10 steps   ....s (for claudicants)

2. Venous

2.1. Varices

Saphenous   Long   ☐   ☐

Above knee   ☐   ☐

Below knee   ☐   ☐

Short   ☐   ☐

Skin defect   ☐   ☐

(script) ....

2.2. Homan's sign positive   ☐   ☐

2.3. Superficial phlebitis

Above knee   ☐   ☐

Below knee   ☐   ☐

Diagnosis   ....

....

....

....

Fig. 14.1. (c) Vascular chart.

the initial diagnosis at the foot of the page. These form the basis of the essential parameters which will appear in further clinical follow-up forms and be documented on the progress chart. They are an important guide in coming to decisions about surgical intervention or conservative management.

Problem Summary

Diagnoses

Main ....

Secondary ....

Problems

Date	Principal (active)	Subsidiary (inactive)

Fig. 14.2. Problem summary chart.

PROBLEM SUMMARY CHART (FIG. 14.2)

Weed put a great deal of energy and intellect into designing problem-orientated record systems which would help the clinician to put a mass of historical, clinical, pathological, and social detail into perspective. He intended that the doctor should view the patient as a whole and that the record must show not only the physical ailment needing treatment, but how that illness interacted with the particular patient's environment, life-style, and circumstances.

General practice has anticipated this system for a long time and has recognized the need to put clinical conclusions in an appropriate perspective in all its thinking. In using clinically-orientated records for basic information, it helps to appreciate the significance of the findings by documenting the patient's problem as well as the diagnosis. For example, management of a patient with right-sided stroke who is left handed is quite different from one right-handed, and the one living in a bungalow different from one in a third-floor flat with no lift.

FURTHER INVESTIGATIONS

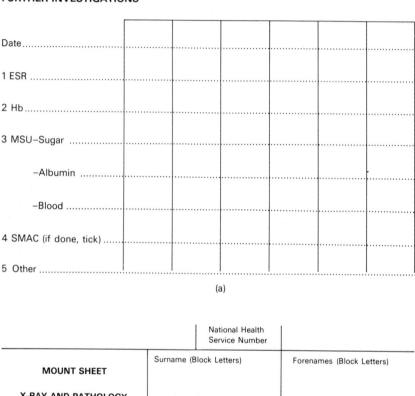

(a)

(b)

Fig. 14.3. (a) Further investigation chart. (b) Mount sheet, X-ray and pathology investigations. DHSS form FP111L, reproduced with permission.

An insulin-dependent diabetic, living alone, with a recent hemiparesis has a major problem. He can cope with the diabetes until the stroke occurs, but it then becomes a major difficulty.

These charts are used to record special investigations with the date of the record. Certain results are recorded separately from the laboratories and X-ray departments reports, which are affixed in chronological order (Fig. 14.3b). This makes all results easy to review. The ESR and presence or absence of sugar, albumen, or other abnormalities in a midstream specimen of urine is recorded directly onto this document.

Laboratory findings (Fig. 14.4)

In recording the treatment of inflammatory arthritis it is the follow-up of the use of toxic drugs which is so difficult, or is made to appear so. The worrying drugs are gold salts, penicillamine, azathioprine, and cyclophosphamide. The latter two drugs are uncommonly used. A follow-up chart showing the drugs used and the regular and cumulative (for gold) dosage with relevant clinical and biological effects is easily constructed so that progress can be assessed at a glance. This document should be used by the general practitioner if he is invited by the rheumatologist to follow-up treatment himself. It is easy for the practice community nurse to keep this chart up to date. However, each result must be inspected by the general practitioner in order to recognize the onset of toxicity. In such an instance charts need to be kept to record full blood and platelet counts, total white and differential counts, sedimentation rates and the presence of proteinuria or haematuria or adverse clinical events. In the follow-up of second-line management using gold salts and penicillamine the patient can carry a small tabulated document to be handed to the clinical observer at the time of reporting as well as the clinician keeping a record in his own premises. This is a pocket-sized version of the same chart. It can be shown to any concerned therapist, serving a similar purpose as a steroid card. General practitioner neurology and vascular work is far less demanding on the laboratory so that similar additional documents are unnecessary.

X-ray findings (Fig. 14.3)

X-ray reports are of necessity scripted and often opinionated and lengthy. The only reasonable way of filing them is to stick them serially on a chart, such as that described in Fig. 14.3b. The coding of X-ray changes is a highly sophisticated business largely for research workers and well beyond the scope of a general practitioner's talents or needs. If he is doubtful about the significance of apparent differences between two separate radiological reports, it is best to ask a single radiologist to review the two lots of films and rely upon his experience.

IMMEDIATE MANAGEMENT

Date

Acute phase

Drugs

1. ....
2. ....
3. ....
4. ....
5. ....
6. ....

Referral

A & E .... ☐
Orthopaedic surgeon .... ☐
Vascular surgeon .... ☐
Rheumatologist .... ☐
Neurologist .... ☐
Geriatrician .... ☐
Other (specify) .... ☐

Physiotherapy .... ☐
    Hospital .... ☐
    Domiciliary .... ☐
Occ. therapy .... ☐
Speech therapy .... ☐
Appliances officer .... ☐
Community nurse .... ☐
Health visitor .... ☐
Social services .... ☐

Maintenance assessment needs

Accommodation modification

Commode.................................. ☐
Bath seat.................................. ☐
Lavatory frame........................ ☐
Water bed................................. ☐
Monkey pole............................ ☐
Hoist.. ☐
Stairlift..................................... ☐
Ramp.. ☐

Aids to mobility

Tripod .... ☐
Walking stick .... ☐
Walking frame .... ☐
Elbow crutches .... ☐
Rollator .... ☐
Invacar .... ☐
Wrist splint .... ☐
Lumber corset .... ☐
Limb prosthesis .... ☐
Modified foot wear .... ☐
Wheelchair .... ☐

Domestic ADL

Tap handles .... ☐
Door handles .... ☐
Cutlery .... ☐
Others (specify) .... ☐

Cash benefits (DHSS)

Attendance allowance .... ☐
Mobility allowance .... ☐
Non-contributory pension .... ☐
Invalidity benefit .... ☐
Diet allowance .... ☐
Heating allowance .... ☐
Clothes allowance .... ☐
Invalid care allowance .... ☐
Others (specify), e.g. Supp. Ben ..

Alternative accommodation

Special school .... ☐
Part 111
    Resident home .... ☐
    Warden supervized
           bungalow .... ☐
    Day care .... ☐
    NHS geriatric day care.... ☐
    Private nursing home ...... ☐

Fig. 14.4. Immediate management chart.

TREATMENT FOR RHEUMATOID ARTHRITIS

NAME: ----------

Gold ☐   D-penicillamine ☐   Azathioprine ☐

Date	Dose (if gold cumulative)	Hb (g/dl)	Total (wcc $\times 10^9$)	Differential wcc		Platelets $\times 10^9$	Urine		Other e.g. rash, stomatitis, pruritus, diarrhoea	Signature
				N	E		Albumin	Blood		

Fig. 14.5. Treatment for rheumatoid arthritis chart.

IMMEDIATE MANAGEMENT (FIG. 14.4)

The summary chart will indicate the patient's individual needs. A brief block chart serves as an audit reminder to ensure that the appropriate service has been notified of need. It records the specific drugs recommended, whether referral has been made to a consultant colleague or the Social Services, and whether any aids to daily living have been provided.

Furthermore, any special care in respect of accommodation can be recorded. One often forgets that the handicapped patient may merit the special cash

PROGRESS CHART
(examples of parameters
but insert relevant features in first column)

DATE					
EMS (min)					
Ritchie index					
Grip strength (mm Hg) R					
L					
Walking time (s) to cover 50 m					
Stair-climbing (time to climb 10 steps)					
Finger/floor distance (cm) SLR R					
L					
Leg ulcer max. diameter (cm)					
Other (specify)					

Fig. 14.6. Progress chart.

benefits available through the DHSS and therefore a block to record this information is a useful *aide mémoir*. These benefits can help the patient cope with grave financial problems engendered by disability.

PROGRESS ASSESSMENT (FIG. 14.6)

Experience in clinical trials has led to the development of accurate techniques of measuring progress. These principles can be applied to clinical practice and used by the general practitioner. They can be used for clinical purposes on a less strict basis than those required for academic research, but will still have objective and comparative values. It is easy to keep records in terms of laboratory tests where the results are quantitative. It is less easy to use clinical methods of assessment which will stand comparison with each other at different periods of time. None the less, it is necessary to make these records if the clinician is to determine whether his patient is getting better, or worse, or staying the same.

NURSES' AND HEALTH VISITORS, RECORDS	National Health Service Number	
	Surname (Block Letters)	Forenames (Block Letters)
	Address	Date of Birth

Date	

Fig. 14.7. Nurses and health visitors records chart. DHSS form FP111M, reproduced with permission.

In order to minimize the volume of paper in the patient's records the progress chart can be adapted for several purposes (Fig. 14.6). The parameters suggested in the chart are probably the minimum required for follow-up of inflammatory arthritis, ankylosing spodylitis, sciatica syndromes, claudication, and vascular ulcers. Other parameters the practitioner might choose to follow can be included according to the syndrome treated. Routine measurement of movement in selected joints can be included though this requires the use of a goniometer to be accurate, the subjective guess work involved in assessing a range of movement being notoriously unreliable. The use of the goniometer requires a certain expertise and it seems unlikely that there would be sufficient time or opportunity for the general practitioner to indulge in these sophisticated measurements, though, with special interest, it is soon learnt.

Stroke

Determination of improvement in peripheral limb function after a stroke can be assessed by using the grip-strength method, walking distance, and stair-climbing ability. These are simple techniques of assessment which need not be carried out by the same observer every time.

SUMMARY

In general practice records the constant and urgent need always seems to be both to amplify and simplify information. In ensuring that one has probed a problem in detail, one has increased one's knowledge of the situation which can then better satisfy the needs of the individual. Proper documentation simplifies the analysis of the problem by laying down the findings systematically thereby crystallizing thought. It introduces an objective and scientific element to a mode of practice which until the last current quarter of a century has been criticized for appearing woolly, confused, and subjective.

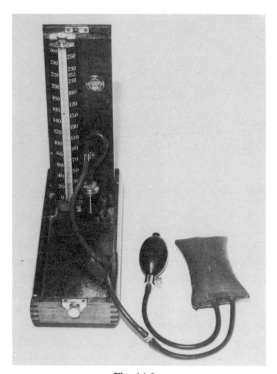

Fig. 14.8.

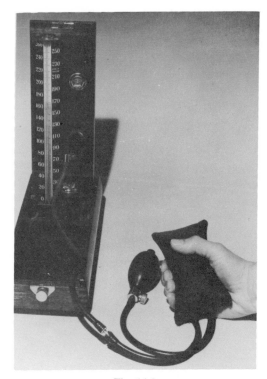

Fig. 14.9.

No claim is being made that the charts and documents outlined in this chapter are full, thorough, or comprehensive. However, they represent an effort to introduce system into documents which hitherto, in my view, are not even adequate for use to record elementary daily events.

I wish to give system to the recording of clinical material and to assist the physician in clarifying his own thoughts both from a diagnostic, therapeutic, and management point of view. With familiarity in the use of such charts, we hope not to consume a great deal more time in consultation than is reasonable in achieving our objectives, but at the same time provide useful retrospective and prospective detail in order to follow the progress of the individual patients. It may be that the next decade will become the age of the computer, in which case all present effort to record events will be looked upon as the flounderings of the unplanned, the unprogrammed, and the unscientific.

REFERENCES

Avery Jones, F. (1981). Innovations in medical records in the United Kingdom. *Br. med. J.* **282**, 1164–5.

Cormack, J. J. C. (1981). A4 folders – a step into the present. *Br. med. J.* **282**, 953–5.

Department of Health and Social Security (1980). Which benefit? Sixty ways to get cash help. DHSS leaflet F.B2/NOV.

Elliott, A. (1981). A4 record system and all that. *Br. med. J.* **181**, 1363–5.

Petrie, J. C. and McIntyre, N. (1979). *The problem orientated record.*

Royal College of General Practitioners (1980). Computers in primary care. Occasional Paper 13. RCGP, London.

Wood, J. P. (1974). Improved record keeping in general practice. *J. R. Coll. Gen. Practns* **24**, 865–74.

15 The responsibilities of the health care team

G. A. Griffin

The advances in medicine that have led to more specific treatment of disease have made possible changes in the whole mode of care of the locomotor diseases. The common rheumatic disorders previously treated in the out-patient department may well now receive the same treatment in the general practice surgery. Similarly the general practitioner, supported by the primary care team, can often undertake care in the family home of patients previously relegated to long-stay care in the hospital ward.

However, these decisions present new and complex problems to the general practitioner who must accept responsibility for initiating such changes. In this chapter we shall explore these problems by examining the role of the general practitioner as well as the resources available. We will then look at the other members of the primary care team, at their role and at the extra resources available to them. Equally important is the identification of the varying needs of the patients before the decision of the management and results of management of these individual needs are discussed. Lastly we shall explore the education and research needed to sustain and improve care.

ROLE OF THE GENERAL PRACTITIONER

Surveys show that a larger proportion of patients with rheumatic disease are referred to hospital than in almost any other branch of medicine. In 1969 Partridge and Knox reported that in the south east of Scotland nearly half of a sample of patients with rheumatic disorders had been referred to hospital for diagnosis or treatment. General practitioners attending courses at the RCGP from 1976 to 1978 had referred 20 per cent of their patients with muscle and joint pain for consultant opinion and 48 per cent for pathology tests or X-rays. In contrast to this, the Field of Work of the Family Doctor (1963) reported that 90 per cent of all medical episodes of disease are handled from start to finish by the family doctor *without reference to hospital.* In 1977 the British League against Rheumatism suggested that one-quarter of rheumatic patients referred to hospital could have been treated in primary care.

As more than 90 per cent of the rheumatic disease seen in general practice consist of local disorders lending themselves to simple local treatment, it would seem proper that such treatment should be offered in the surgery, particularly as such a change is to the advantage of the patient. The family doctor knows the intimate problems of the patient and of the patient's family so that the timing of treatment may be adjusted to meet patient need. The long waits for hospital appointments are avoided, as are the anxieties produced in the patient by such long waits.

Faced with such inconsistency it is proper to consider why such change of management has not evolved in the new general practice. First, it would seem to call for a significant change of behaviour by general practitioners which in turn would require a considerable improvement of knowledge and the recall of dormant skills. Apart from the presence or absence of such knowledge and skills, the attitude in both hospital and general practice is often that its treatments are not part of primary care. Finally there is much evidence that the failure to understand and diagnose these conditions results from a preoccupation of undergraduate education with the more serious types of rheumatic disease.

A simple example illustrates the problem; a rotator cuff lesion of the shoulder may lead to incapacity lasting many months. It may be cured by a single injection, given equally easily in general practice or in hospital. Part of the management, however, is an accurate diagnosis and explanation of the disease to the patient. This is important because if an incorrect diagnosis has been made, the pain is likely to be due to a capsulitis, which regardless of treatment given, may take up to a year or more for cure. Alternatively, it may be due to a cervical disc lesion which requires quite different treatment. The failure of diagnosis is seen in a survey of 100 patients referred to a district general hospital with pain in the neck or the shoulder. In only 50 per cent had any form of a diagnosis been offered and where offered 60 per cent were incorrect sufficiently to make treatment inadequate. Pain had been present for a median duration of five months; the longest 15 months. Half of these patients were cured of their pain at their first visit to the hospital or within a few days of their visit. Proper diagnosis and treatment given in general practice would have saved this group of patients an average of five months of pain.

Study of other records of patients referred to the same hospital confirms further the view that general practitioners tend not to attempt diagnosis or solution to problems. It is proposed that following appropriate postgraduate education solutions should normally be attempted and would be tested, either by confident investigation or by the equally confident initiation of treatment before referral to hospital is considered. Thus in a localized soft-tissue lesion, such as a tennis elbow, or a cuff lesion of the shoulder, testing of diagnosis would include a period of observation or of palliative treatment, followed where necessary, by injection of steroids. Management of an inflammatory disease would include the appropriate investigations such as pathology tests and X-rays. It might then include the results of treatment by first line drugs, such as aspirin or NSAI drugs in adequate doses.

By this time many minor problems will have been solved by cure and the direction of others will have been identified so that an informed diagnosis and discussion of treatment may take place. There will be illnesses where diagnosis is uncertain, where response to competent treatment is unsatisfactory or where clinical acumen shows the problem to be serious. In such cases a second opinion is properly sought. This second opinion may well be indicated in every case in which a second steroid injection is proposed for any condition. It may be asked in all cases of confirmed inflammatory diseases and in those cases that do not respond predictably to the first-line treatment.

Such confidence in the care of rheumatic disease will influence the care of all locomotor disease. With increased competence and confidence of primary care the flow of locomotor diseases to and through the out-patient department should be changed. The patient will remain more truly under the care of the family team with intervention of the hospital teams only on change of progress or circumstances. Such care will then involve a proper use of the resources available to the general practitioner initiating such treatment as well as the primary care team with the additional resources available to it.

GENERAL PRACTITIONER: RESOURCES

The relationship with the consultant and through him, with the hospital service, is twofold. A purely technical need is seen most clearly in requests for pathology tests and X-rays, but a special technical need underlies some referrals to consultants; for example to the rheumatologists for physiotherapy or for an appliance or even to the orthopaedic surgeon for arthroplasty or tendon transplant; to the urologist for management of incontinence by transurethral resection or urinary diversion; or ultimately to the hospital service for transfer of care and responsibility. In all such cases the general practitioner should endeavour to maintain contact with the overall direction of care. A truly consultative relationship exists when advice is sought on the main diagnosis, on other related disorders and on the direction of care. In such consultation the practice records of family and home will be among the factors determining management and in turn will identify lines of progress.

The day centre exists between the hospital and the home. On the one hand in offering hospital-based resources it can be seen as an extension of hospital and specialist care. On the other hand in offering support of a management strategy based on the community, it can be seen as an extension of care in the home. Referral of a patient to a day centre may reflect either of these roles.

THE HEALTH CARE TEAM

THE TEAM: ITS MEMBERS

The health care team is centred in the practice. However, the extension of this team wherever needed into the community gives it a wide range of power and resource. The entity of a central team is strengthened by the concept of attachment; the use of office space with individually kept records within the practice. The central team as the nucleus of care, comprises the general practitioner, the district nurse, and the health visitor supported by a reception staff. In a group practice accepting increased responsibility for the primary care of locomotor disease there is a good case for addition to the team of a physiotherapist employed part-time by the practice.

The district nurse is concerned with the general nursing needs of the patients as well as such special needs as nutrition, rest and mobility, hygiene, and control of bowels and bladder. Special knowledge of the wide range of treatments employed involves the recognition of complications resulting from them and the collection and management of pathology specimens for their control. As a regular visitor to the house, the nurse supports the patient and the family out of her intimate knowledge of them, of the effects of the disease, and of the disappointments and the joys of variable progress through it. With the physiotherapist or out of her own skill, she may tailor splints, slings, and other supports for the patient in the home where these will be used.

This intensely personal and physical contact with the patient is contrasted with and complemented by the entirely non-physical educational, psychological, and social approach of the health visitor. In a purely technical sense her function is concerned with mobilization of resources. However, these resources include the family of the patient who require both education and support from her in order to help. Thus not only does she mobilize support for the patient but also support for the patient's family with play groups for children, holiday play schemes, and other opportunities for children whose parents have lives disrupted by a crippling disease. It is often she who identifies the needs for grants and other needs supplied by Social Services and voluntary organizations.

Physiotherapy in the home may be offered by the practice physiotherapist or by one attending on invitation from hospital. Supports and splintage, positioning, and mobilization are best supervised in the presence of the family who will nurse the patient. Pain relief with ice or heat or special equipment, such as ultrasonic, is available where and when it is most required. All special problems of the invalid are met better within the home than in the distance of the hospital department. In this way also mobilization and early rehabilitation are suited to the patients' actual needs and resources as they are required. Aids may be installed in the home. Visits of the patient to the hospital for specialized treatment or rehabilitation may be arranged to the greatest economy and effect. An addition to the care in the home which may be offered by the practice- or district-based physiotherapist is the provision of treatment in the surgery itself. Patients may get to the surgery more easily, more quickly and less expensively than the ambulance journey to the hospital. Routine assessment of progress and function may be made at three-month intervals by a practice-based physiotherapist and treatment may be adjusted accordingly.

Each member of this team will have or will share a knowledge of the patient and the family, their strengths and weaknesses. Such knowledge may be shared at informal joint-assessment meetings. These factors added to their special skills place it in a unique position to manage chronic disease. Equally important however is the ability of each member to initiate and to use the services of the extended team as and when these are needed. This team includes all members of community and social services, voluntary or employed, but particularly:

social worker and occupations officer;
district (hospital) physiotherapist;
district or hospital occupational therapist;
members of voluntary associations and societies;
disablement rehabilitation officer.

Social work being a department of the local authority brings to care all the needed functions of that authority from housing to day centres and holidays, from home helps and meals on wheels, to transport and financial assistance. The occupations officer uses her training as an occupational therapist to identify and to effect for the patient alterations in the home and supply of wheel-chairs and carriages. The hospital occupational therapist and the occupations officer are the experts on the activities of daily living and on the aids required to support these. Between them lies the therapeutic recreation of the patient during illness and the vitally important rehabilitation of the patient after illness. The health visitor and the social worker combine in search of voluntary organizations with special interests in the problems which present. Preparation for work, special rehabilitation and assessment for work will develop because of their advice. They will involve as early as possible the advisory function of the disablement rehabilitation department and schools.

The day-to-day contacts in the small unit of a doctor's practice or informal joint assessment meetings may well be sufficient for adequate communication between members. Larger groups involving much extension of the team of care may be invited to case conferences to cover the complicated co-ordination of helping agencies.

THE TEAM: ITS RESOURCES

In an adequate group practice the team should have available to it a treatment room with equipment that will reflect its needs and its skills. There is also the hidden treatment reserve in the families, the homes, and the beds of the patients cared for by the practice. It has available to it the extension of resources available from the district nursing store, which can supply all equipment required in the nursing of the patient including such large items as hoists and special beds and mattresses.

Involvement of the social services, and the extended team makes available the supply of major aids and alterations of the house and its surrounds to suit specially identified needs.

Such organization is extended to national level with supply from specially interested organizations and in such international events as World Rheumatism Year and the International Year of the Disabled.

THE ROLE OF THE TEAM

The aim of the health care team is the achievement for the patient of maximum health with maintenance of the greatest independence. However, the care necessary to achieve this aim may not be supplied indiscriminately. The needs involved in the supply of this care require four levels of analysis:

ANALYSIS OF PATIENT NEED

(i) identification;
(ii) assessment;
(iii) management;
(iv) measurement of outcome.

Identification of need

The problem presenting immediately is to identify who defines the patient's needs. The patient will react initially to the changes due to illness and will present immediate biological needs. These may well conflict, for example, with the needs of the family: the patient may wish to stay at home and the wife may be unable to cope with the problems involved while caring for the children. As the patient and family come to accept illness so does their perception of need decrease. At this time must that of the health care team increase. However, different members of the team may themselves have differing views of what constitutes a proper need so that a consensus view is necessary. In this lies the value of developing a basic structure of the team care.

Assessment of the need

The assessment of need lies in the skill of the team in its relation of diagnosis and prognosis to the potential of the patient and to the services available. As the condition changes so does the need. Hence assessment must not be a static procedure but will require continuous evaluation and regrading. Similarly the presenting need must be related to the total of need to assess for instance whether the total support undermines the independence of the patient. Adequate practice records are a considerable part of this process.

Management

The team as a functioning unit is central to the management of need. It has a store of knowledge and experience applicable to the need itself, to the available resource, and to the mobilization of further resource. Much of the allocation and sharing of work by individual members will develop from this knowledge. However, the intimacy of the relationship of particular team members to a patient's need may influence counselling of the patient and family. Communication and records become still more important where such considerations influence care.

Outcome

An effective solution to a need is desirable. However, each solution attracts a cost either in use of a resource or in sacrifice of another need. It is the efficiency of the solution, in other words, the balance of effectiveness against cost, that must be measured. Waste of resource may easily be measured by non use or abuse of equipment in the house, but is not so easily seen in the waste of work of a member of the team. The measure of outcome is as essential to *each* need as are the other analyses.

Efficient team care requires a structure. Such a structure may be based on the meeting of specified needs. Maslow has suggested that people have a hierarchy of need.

THE HIERARCHY OF NEED

1. Biological: to maintain life
   A Mobility
   B Freedom from pain
   C Maintenance of body function
   D Control of movement
   E Speech
2. Psychological: to sustain intellect and personality
   A Independence
   B Self-respect
   C Autonomy
   D Communication – understanding
3. Social: to maintain social roles
   A Communication – mobility
   B Family support
   C Rehabilitation

1. Biological need

A. Mobility

Biological mobility or locomotion is threatened usually by neurological disorders or by disorders due to peripheral damage of soft tissues caused by rheumatic diseases, by trauma, or by degenerative joint disorders.

(i) Neurological disorders affecting mobility

Identification: Strokes are responsible for 60 per cent of all loss of mobility. The onset is sudden so that identification is made directly from observers in the home to the doctor, but the milestones of the progress of the patient must be the responsibility of the practice team. This will also be the case in neurological disorders of slow onset. Out of the records must come those criteria of prognosis on which the counselling of the patient will be based and on which further treatment will be planned.

Assessment: The immediate assessment of acute neurological disabilities has two functions; it provides a baseline for management and it identifies the immediate therapy to be planned while the progress of disease is assessed. Records should give an overall view of the site and size of the damage.

Management: On this is based the decision of place of treatment such as referral to hospital. If not for referral, immediate nursing, physiotherapy, and decisions about ancillary treatment such as speech therapy are made. The nurse will identify nursing needs, such as the height of the bed, the needs for a ripple mattress or a hoist. With the physiotherapist the nurse will decide and counsel the family on the positioning of the patient, the type of movement that is possible and if the patient is conscious, the exercises that should be taught. Such exercise includes moving in bed and care of life-support functions.

More definitive treatment is possible after a period of observation. At six weeks the needs of stroke patients are much more evident. The patient is probably aware of his state and how his life will be; there is more accurate knowledge of motor loss and of the sensory loss that interferes with understanding or acceptance of advice. There may be signs of recovery of function. The persona of the patient is becoming evident. At each stage the doctor has a responsibility to identify the disorder and its prognosis and to give to the patient some explanation of these findings and their significance.

(ii) loss of mobility due to trauma

Identification: Identified by ambulance attendant and by bystander. Identification by an unskilled onlooker is unsatisfactory.

Assessment: By ambulance attendant is confirmed usually by a casualty department but localized pain and loss of mobility should not be dismissed because of a negative assessment. In less severe trauma particularly we must accept personal responsibility for excluding fracture or disruption of ligament and tendons or nerve while assessing immediate tissue damage before swelling occurs. Within three days reduction of swelling should permit assessment for the earliest of progressive mobilization. This should be monitored.

Management: The injured area is managed by ICE: ice pack, then compression, and elevation for two days or until swelling is reduced. Mobilization is then urgent and decisions about splinting or bandages are related to this. Haematomata in joints or in soft tissues may be aspirated. Ultrasonics, short-wave diathermy, heat, or ice packs may relieve pain and allow mobilization. Drug therapy relates to pain and to primary or secondary inflammation. Significant local lesions may be treated at this stage by steroid injections, with or without local anaesthesia.

Long-term management

The supply of nursing aids in the home will need to be developed and possibly a ripple bed brought into use. With present effective district nursing, bedsores appear to be rare and if present, to be extremely well treated. Neurological disease rarely requires splinting. Splints impede function, though they may be necessary as a temporary measure to protect the use of weaker muscle groups. The nurse and physiotherapist can give the family instruction on positioning and mobility of the patient with the use of chairs, and of the basic functions of life.

The occupational therapist relates to the physiotherapist in re-educative exercises in the early stages of the disability. Associated with re-education is the early supply of aids for self-feeding and other needs of the patient. The occupational therapist also identify the links between balance and posture and daily functions such as washing and dressing. In diseases showing return of mobility, the occupational therapist's specific knowledge of the activities of daily living enables her to supply aids to daily living as an essential step in progress. This medical rehabilitation to fit the patient for activities in the home, must be followed as soon and as early as possible by decisions on occupational rehabilitation if this is to occur. At this stage the help of the disablement rehabilitation officer is best obtained. Where mobility is unlikely or is impossible, advanced aids such as the Possum should be supplied in order that the bedridden patient has maximum independence in control of the front door and of callers to the door.

Outcome

Of stroke patients 50 per cent become independent and 75 per cent walk unaided; there are better results in strokes affecting the dominant hemisphere. Better results occur when the patient is under 70.

B. *Freedom from pain*

Pain is such a basic biological phenomenon that its presence should be part of the knowledge of the general practitioner and the team supporting the patient. It has deep emotional significance for the patient and may suggest the further progress of the disease or of further damage. The occurrence of pain and its relationship to other factors will be reported to the doctor by the patient. Its effect on rehabilitation and on ordinary progress will be noted by the physiotherapist or by the nurse.

Timing of local pain is most helpful in diagnosis; pain continuing for more than an hour after rest and associated with stiffness suggests strongly inflammatory disease; pain after or during exercise or for a short time on resumption of activity suggests a degenerative or traumatic cause. Similarly change of the timing of pain may be significant; the change of pain occurring in a patient for the first two or three hours of the day to a pain occurring only on exercise suggests the ending of inflammatory disease with supervention of secondary degenerative disease. A claudication pain may not require treatment until it reduces the claudication distance to, for instance, the width of a main road that the patient cannot cross without pain stopping walking.

Referred pain – may indicate disease of any structure within the same dermatome as that in which the pain occurs. For instance pain in the knee may be referred from hip joint, pain in the arm referred from the shoulder joint. Isolated pain may occur in the pathway of the spinal nerve compressed by a disc or damaged by a traction injury as in the cervical plexes. It may be referred locally in an entrapment of a nerve.

Management: it may be an act of survival against the deep emotional relationships of pain, it may be a failure of medicine, or it may be an action merely of independence that makes the supply of pain drugs the largest component of 'over the counter' sales by chemists. The doctor must meet anxieties by achieving an understanding of the pain and its causes in order to explain it to the patient. This is particularly important in chronic disease. Failure to understand such pain demands a second opinion. Thus management of pain involves accurate assessment of cause and full treatment of the primary lesion as well as treatment of the pain itself.

It is difficult to overrate the importance of nursing care, splintage, and support in both short- and long-term prevention of pain of a disabled patient. The pysiotherapist's heat pads, ice packs, or electric treatment give immediate freedom from pain to permit increased joint movement and through this further control of pain.

A case has been made for full primary management of locally occurring pain by the general practitioner using drugs or steroid injections to permit early rehabilitation. The distinction between analgesic and anti-inflammatory drugs has become less clear with modern theories of pain. However, an anti-inflammatory drug is more expensive and carries more contra-indications to its use so that analgesic drugs should be preferred where inflammation is not a significant factor. The decision in any case is better made by the practice team with its knowledge of the patient.

The simplest interruption of the nerve pathways of pain is the local anaesthetic with or without steroid injection: for example, an intercostal block relieves the pain of malignant infiltration of the rib and is a skill possibly already acquired by the general practitioner in the management of more common rheumatic lesions. Pain clinics extend this interruption by intraspinal blocks of local anaesthetics, possibly with steroids, or by extrathecal or intrathecal injection of phenol or by cordotomy.

The pain clinic may also use vibration or percutaneous stimulation or acupuncture to stimulate pain inhibition. General practice is ideally situated to offer biofeedback control, relaxation, or hypnosis. Antidepressant drugs have a specific effect on pain mechanism as well as on the depression resulting from pain in long-term disability.

Outcome: Freedom from pain itself is a right and healthy desire of a patient. However, it may be more important that the patient should be independent than entirely free of pain; should be an autonomous being who feels capable of controlling in his own way significant levels of pain. There is a danger in removal of the pain that is an indicator of disease until the disease has been assessed. It may, for instance, be better to treat obesity than to relieve the pain of an osteoarthritic knee. In such cases a second opinion may be sought before further intervention.

C. Maintenance of natural body function

Failure of a biological function should be anticipated in the management of the patient at risk. Sometimes these complications are avoidable but if they do occur they may be severe and serious threats to progress and even to the life of the patient. Identification of these needs is a function shared probably between the doctor and the nurse.

(i) Control of the bladder

Assessment: Acute spinal cord lesions result in retention of urine for days or weeks. Cerebral lesions and, particularly, strokes, lead generally to urinary incontinence.

The retention due to spinal cord lesion may slowly resolve allowing the development of periodic micturition. In an upper motor neuron lesion this periodic micturition will occur at times spontaneously by reflex action. Complications may occur at any stage. Catheter drainage is liable to cause urinary tract infection. Repeated urinary tract infections are serious complications in that they result in renal damage. Constant soiling with urine on the other hand leads to skin maceration and bedsores which are a potentially even more serious complication.

Management: Retention of urine must be treated by catheterization within a few hours. This catheterization may be suprapubic or urethral though usually urethral drainage is performed. In specialized units intermittent urethral catheterization is frequently preferred so that bladder control may be assessed but in the home it is usually effected by a permanent indwelling catheter such as a Foley. When periodic micturition occurs in a lower motor neuron lesion the patient may be taught manually to express urine from the bladder.

Collection of the urine in the male is usually by permanent use of a condom with Paul's tubing draining into a bedside bottle or a bag worn by the patient. Appliances for a female rarely fit satisfactorily, so that there is potentially always a leakage of urine. In these cases absorbent pads with waterproof pants may be used. Despite the improvement of such pads, skin maceration may still occur and a permanent indwelling Foley catheter may be used.

Where return of natural function does not occur in a stroke patient, or where incontinence develops in other patients, a local lesion should be suspected. Urological investigation might identify a prostatic or other bladder neck obstruction. In such cases transurethral resection or division of the sphincter may be helpful. Where catheterization is required the risk of infection is increased and urine cultures must then be performed to assess the infection and to identify antibiotic need. In some cases long periods of antibiotic prescription may be required. Pure irritability of the bladder occurs and results in unsatisfactory urgency and frequency of micturition. This may be relieved by the prescription of flavoxate or emepronium.

If permanent catheter drainage is threatened there may be occasion for urinary diversion into an ileal loop or more rarely into the colon by ureterocolostomy.

(ii) Control of the bowels

Bowel control is lost in spinal cord lesions and it must be checked carefully in patients with stroke.

Assessment: With a spinal cord lesion the patient loses sensation and control of the bowel. Occasionally reflex defecation develops. This rarely occurs with low cord lesions and there is often much greater difficulty with such cases. Diarrhoea may occur; this must be distinguished from the spurious diarrhoea due to constipation.

Management: With care a combination of relative constipation with laxative produced diarrhoea on alternate days produces a sort of reflex action. Alternatively constipation is encouraged and manual evacuation of the rectum is performed by the district nurse twice a week.

If severe constipation develops slow and gentle treatment is required. Dioctyl sodium suphosuccinate is given by mouth as a softener. This is followed by repeated small enemeta. Split mattresses and mattresses designed for incontinence patients may be used.

(iii) Nutrition

It is important to remember that obesity is a threat to mobility and moreover is likely to lead to complications such as chest infection and bedsores. Apart from this the problems of nutrition are the feeding of the unconsious patient or of the patient who cannot swallow. In any patient where there has been unconsciousness or failure of nutrition for eight hours, fluids must be given by nasogastric tube. Automatic swallowing may permit the presentation of food or drink to the oropharynx. Feeding may be required by nasogastric tube in cerebral palsy. The foods to be given and the quantities of fluid required are matters of careful assessment.

(iv) The skin

The greater the loss of mobility the greater is the threat to the integrity of the skin. Immobilization of the stroke patient leads to pressure and the development of bedsores. Part of the essential nursing of all patients with strokes is the moving of the patient to avoid pressure while maintaining positioning to avoid contracture. The paralysed patient in a chair is at less risk than the patient immobilized in bed but both require movement and care of the pressure areas. Massage of areas under pressure is no longer considered good treatment but frequent cleaning, drying, and powdering of pressure areas is essential.

Bedsores appear to occur rarely in cases nursed by district nurses and if they do occur they appear to heal excellently under this skilled care. In all cases where mobility is impaired an early decision is required as to the type of mattress. A ripple bed is probably required. North pads may be helpful. A developed bedsore failing to respond to treatment may require hospital care with water beds or saline baths.

(v) The chest

Frequent movement in bed and early sitting of the patient in an armchair protects against complications in the chest, as well as on pressure areas. For this reason also smoking should be discouraged. Where infection occurs tipping for drainage of the affected lobe may be advised. Appropriate antibiotics should be given where indicated.

(vi) Cardiovascular disease/elevated blood pressure

Opinions appear to be divided on the treatment of hypertension in patients with strokes. In 2000 patients treated at the Passmore Edwards Rehabilitation Unit it was not considered that control of hypertension affected prognosis or survival. There is, however, a growing opinion that diastolic pressures in excess of 110 mm Hg should be controlled as part of the treatment of the stroke. Treatment of hypertension should be gradual and not sudden. Anticoagulation does not now have a place in the management of strokes. It may be considered advisable to exhibit peripheral vasodilators or drugs which will improve tissue nutrition.

The control of platelet aggregation by aspirin or dipyridamole is under trial.

D. Control of movement

Abnormalities of movement and their control are areas which demand much of specialist knowledge both at the level of diagnosis and of treatment beyond simple drug and therapeutic intervention. In many cases the initial identification is by observation in the surgery by the doctor. Such abnormality frequently develops very slowly, but it may be made out of observation and records of the practice team in their care of the patient.

Assessment: Impairment of movement control may be central or peripheral; those of peripheral neurological origin occur as: vertigo with its disorders of the vestibular apparatus; peripheral neuropathy with its many difficult to diagnosis palsies; muscular dystrophies; and disorders of the neuromuscular junction.

Any case may be due to iatrogenic disease so that overtreatment of the primary disease with drugs must be considered. Anticonvulsant or sedative drugs may act on the cerebellum to produce ataxia, or streptomycin may attack the vestibular apparatus to cause deafness or vertigo; peripheral neuropathy may result from diabetes or alcohol excess. The records may identify the development as well as showing the cause.

Management: Early referral for diagnosis, an initial plan of treatment and, later, referral for a consultant decision on advanced treatment as necessary in this area of confused diagnosis and complicated neurological pathways. However, there is specific drug treatment for the control of many of these disorders, for example, the use and choice of the dopa drugs in ataxia cinnarizine specifically for pure vertigo or prochlorperazine where there is also an anxiety component, baclofen in the spasms of multiple sclerosis, prednisolone in the stiffness of polymyalgia, prednisolone and anticholinergic drugs in myasthenia. In each case the choice is probably best determined by the knowledge of the individual doctor of the effects of a drug and its possible maximum dosage within a known range.

The physiotherapist and the occupational therapist liaise in this management. Each disorder responds to specially developed co-ordinative activities such as exercise designed to restore balance by increase of muscle and joint sense in vertigo or to increase muscle control in ataxia and some of the more peripheral diseases. Each has identifiable special abnormalities responding to equally specific management. The inco-ordination of ataxia may respond to the carrying of weights on the arms or the legs; the stiffness of walking of patients with vertigo responds to special relaxation exercises; training in the use of the eyes and sensitivity of muscle and joint sense increase the security of the patient with vertigo in light or in dark respectively. Further consultant advice is requested again where disease is advanced. In emergency a tracheostomy may be necessary in some acute episodes of diseases such as myasthenia. Timed intervention allows the surgery of middle-ear disease in vertigo, of thymectomy in myasthenia, of tendon transplants to balance muscle groups.

E. Speech

Its loss is sudden and is especially frightening to the patient suffering the horrors of paralysis due to stroke. In its major manifestations it is obvious to all though some of its minor occurrences require the most careful observation.

Assessment: Dysarthria may present as the slight slurring of the words that occurs in alcoholism or multiple sclerosis. Dysarticulation may make speech unintelligible in Parkinsonism. Dysphasia results in the inability to choose, to articulate, and to produce the right word in the sentence. Dysgraphia, the inability to write, though much less common may also be present and attracts greater problems of communication. It is, however, the sensory components which are more easily missed such as the inability to understand the spoken word which may be thought to be a disorder of understanding and of co-operation.

Management: Speech therapy should not be delayed. The patient with so great a disability requires reassurance and explanation. Early speech therapy leads to a better prognosis. Speech therapy can be developed with the physiotherapy and medical rehabilitation of the patient as a co-ordinated approach. Assessment must record the previous emotional history as well as past achievement of reading and of talking. Speech therapists use specialist methods of education by stimulation, by programmed instruction and by use of group learning.

2. Psychological needs

The aim of the care team is to achieve the maximum health for the patient. In its widest context this demands consideration of the patient's psychological needs, particularly independence, self-respect, and autonomy.

Assessment: A single observation by any contact of the patient may give the first indication of such needs. It is, however, in the sensitivity of the doctor and the team to such observations and to changes of mood or progress of the patient that the need is indicated and assessed. It is a direct function of the health visitor. Physiotherapists and nurses carrying out their physical treatments spend long periods talking with the patient and this creates the opportunity for dealing with the various psychological problems that often accompany locomotor disability. Appropriate training and orientation is increasingly provided so that it may be used in counselling. Even a home help may have a role to play in listening to the patients problems and concerns. In turn the doctor giving non-directive counselling offers the patients the opportunity to talk through their anxieties and worries.

Factors influencing need are:

(i) The premorbid personality, the present resilience and the determination of the patient, influence the response to treatment and to disability.

(ii) The illness itself presents quite obvious variables, as for example, the differing reactions of a football player to pain or to loss of mobility or of a woman in labour to pain or the safety of delivery.

(iii) The occurrence of the illness and its timing in the life of the patient must alter assessment and management.

Psychological reactions to disease: It is at this stage that the general practitioner who has accepted the responsibility for total care, must be available to offer help in understanding problems and co-ordinating different advices. It is at this stage also that there may be discussion of the place of self-care, of informal care, and, often, of potentially dangerous non-medical treatments. Here the value and the cost of this latter alternative medicine must be considered against known prognosis and set against the expectations of the patient and of the family.

(i) *Anxiety* – this may be the earliest reaction. The patient is unaware of the implications of the disease and is conscious only of the immediate results and the relation of these results to premorbid aspirations. This anxiety may result only in apprehension and fear. It may, however, influence the reporting of symptoms or the reactions to examinations. It will certainly influence the relationship of the patient with the supporting team, and this influence must be taken into account by the team both separately and together.

(ii) *Depression* – it is said that between 20 and 30 per cent of in-patients suffer depression. This is of particular importance in long-standing disease where the ability to cope, and the potential for positive drive and concentration are required. Depression will have also its social consequences on the relationship of the patient with his family and with the team. It may relate to an irritability in dealing with fellow sufferers.

Depression must be identified separately from the anger and frustration of slow progress. Unexplained and unexpected changes of behaviour may indicate unidentified needs such as deprivation of regular alchohol or even such physical need as sexual release.

(iii) *Denial* – this complex psychological process may well influence recovery in that the patient attempts to deny the presence of disability or even to deny the presence of the part that is disabled. It may lead the patient to attempting physical feats that are impossible or at least are ill-advised.

Failures of understanding and communication

Purely neurological disorders may give the appearance of psychological disability or even of dementia and they must be eliminated by careful assessment. Such disorders occur because of loss of sensory input and at their simplest leave the patient not understanding sufficiently to co-operate. The most simple is the loss of proprioceptive sensation and may be tested by asking the patient the position of a limb or digit in space with eyes closed. More difficult is loss of sensory hearing so that the patient does not understand the spoken instruction. It may be tested by asking the patient to close her eyes or to move a limb. More severe still is loss of awareness of the limbs and of the fact that this is impaired (anosognosia). Though the sensory pathways to and from the brain are intact, a lesion of the parietal lobe removes the ability of the patient to elaborate information. This is more commonly seen in strokes affecting the non-dominant hemisphere but may occur with other simpler lesions. It is probable that it is this fault which is responsible for the poor prognosis of lesions of this hemisphere.

Management

Management must be divided into the components of therapeutic and counselling skills. There is little doubt that the essential resource of the general practitioner is counselling skill. As the problem is one of identity, of self-respect and of self-perception, it is only the patient who can come to terms with his own disability. The doctor cannot tell him or her how to feel or how to react but having explained the problem and given what support he or she can give, the patient can then be guided through the various stages and crises which accompany chronic illness. As in the assessment of the state this counselling role is supported by the individual members of the team in their close relationship and understanding with the patient.

The therapeutic role relates primarily to the management of anxiety and depression. Simple sedation with drugs known to the physician may be given to support counselling in crises of anxiety. There is much to suggest that long-term use of sedative drugs is not of advantage where such co-operation is required. Conversely the use of antidepressant drugs over a much longer period may be indicated in a depressive patient. Tricyclic antidepressants also have a significant affect on the perception of pain and this may be valuable. Their anticholinergic actions, however, can be of great danger in upsetting bladder and bowel control. Rarely they may be of value in controlling the irritable bladder.

It is in this area of care that the differentiation of the two relationships of the general practitioner as 'doctor/friend' and 'doctor/physician' are most clearly defined.

3. Social needs

This is the area primarily of the health visitor and the social worker. The importance of the individual knowledge of a patient by any one member of the health care team and of the sharing of knowledge which leads to the total identification of social need has been noted. In this way the doctor acts as coordinator and facilitator. The needs may be identified in three separate groups; the need to communicate, the need for protection and the need for occupation.

The practice records will show the development of biological mobility of the patient. Out of this development will be the knowledge of need of the patient for aids to mobility. At its simplest will be the aids to walking; the supply of Zimmer frames, walking sticks, tripod, or quadrupod walking supports. The social worker will identify the type of wheeled carriage that is necessary. In this must be decided the permanence of the need and thus the excellence of the carriage: it may or may not be electrically propelled and indeed a modified motor car may be sufficient. The health visitor and the social worker identify the need for grants to enable transport. The health visitor may mobilize friends or pools of those helpful citizens who will drive disabled people. In the failure of social mobility lies the greatest need for support; the supply of an easily used telephone or a Possum aid which gives a bed-ridden or chair-ridden invalid control of the front door, communication with the caller at the front door, and control of other emergencies. In ultimate failure the Possum can be controlled by laryngeal impulse. Possum typewriters and other instruments are also available.

The support of the family and of the patient is achieved out of the attendance of the team and of the friends and helpers mobilized by the team. However, the financial catastrophe of dependants is lightened only by intimate knowledge of the grants available for such support; e.g. the constant attendance allowance, grants from special associations related to the disability, social security supporting grants.

We have identified the importance of medical rehabilitation running closely and cleanly into occupational rehabilitation. It is better that early in this process resettlement in occupation is considered. The department of the disablement rehabilitation office considers that its services are always requested too late to be of good value. Such resettlement must be supported by the availability of transport and of grants to support transport. Where such occupation is not possible due to physical or mental disability then occupation in the day centre or other social centre is necessary to support both the patient and family.

Outcome

Counselling is one area of management which might create problems of communication and liaison within the health care team. In counselling, the patient might require a high standard of confidentiality so that various intimacies might be relayed. By and large this confidentiality must be respected as the sharing of such items with the rest of the team may result in damage to the doctor–patient relationship with deleterious consequences for the patient's overall management.

It is in the areas of psychological and social need that the great threat to the patient's independence may occur. The mobilization of helpful supporters, the supply of an infinite variety of aids and supports, the supply of transport badges may add a further component to the initial disability of the patient. This disability is the primary deviance. The extra component which is so much to be avoided is secondary deviance. This secondary deviance is the removal of the independence of the patient by the labelling with the disease. Such labelling may well be an essential component of modern life. At its easiest it is seen as the sign for parking facilities on a disabled person's motor car. It exists to an extent in having a wheel carriage or a simple crutch. Such help and support may result paradoxically in the patient feeling truly incapacitated and dependent. Such loss of autonomy in its own way may be as damaging to the health of the patient as the original physical disability.

EDUCATION

The responsibility of the primary care team to increase its care of rheumatic disease carries an equal responsibility to educate itself to the level necessary to do so. The general practitioner as the member of the team responsible for initiating this direction of change has equal responsibility for initiation of the education.

There are other reasons for greater involvement of the general practitioner in the learning process. Firstly he is better able to respond to general practitioner identification of such learning needs as have been seen here. He works and therefore he learns in ways that differ from the specialist. His problem solving makes much greater use of probability. He attaches an earlier awareness to the psychological effects of illness and to social factors. He acquires a greater ability to tolerate uncertainty. His competence is acquired as much by practice as by learning, so that the more that the knowledge he gains relates to the problems he meets in daily work, the greater will be the learning from it, and the stimulus to further learning; the greater the relevance of knowledge to his work the more likely is such knowledge to be maintained and the necessary skills practised.

Other factors are of importance in planning this education. Further education for general practice has not been satisfactory. Courses, on the whole are poorly attended. There is much good evidence that the ordinary course produces little lasting change in patient management. The range of potential learning available to the general practitioner is so wide that choice will depend on the relevance of the education available to work need. In primary care the simple localized rheumatic disorders such as strains of tendon insertion, nerve entrapment, back pains, outnumber the more serious inflammatory diseases by 33:1. Thus it is proper that teaching and learning should concentrate on the recognition and management of these simple disorders which present so frequently while offering careful attention to the arthropathies as differential diagnosis. On such confidence improved management of the rarer, more severe and more crippling diseases may be built.

The results of such education were measured in the participants of the experimental courses for established general practitioners at the Royal College of General Practitioners. As confidence in diagnosis was increased, referrals for blood tests and for X-rays were reduced. There was reduction of referral for consultant opinion and treatment and this was balanced by increase of treatment by the general practitioner.

RESEARCH

Improved primary care of locomotor disease offers not only a new area of research but a need of audit of the work that general practice proposes for itself. General practice is the only place in which to answer much that is unknown or is obviously incorrect. It is wonderfully secluded for the study of advanced disease especially where such disease crosses the frontiers of speciality. It is the general practitioner who sees the life picture of disease striking in turn at different body structures and it is the general practitioner who feels the need to understand and to harness the differing cares of differing specialities on the same patient.

So much is self-treated and so much is simply suffered for want of better treatment that little is known of the epidemiology and natural history of much of simple muscle and joint pain. Little is known of the time for intervention or degrees of intervention that is required at each stage. Increased treatment in primary care will lead to greater increases of knowledge but possibly also to increase of unnecessary treatment. We must look carefully at the earlier and widespread use of such powerful intervention as steroid injection in general practice. It is time that general practice looked at the cost benefit of practice physiotherapy with the equipment required. It must explore greater access to hospital physiotherapy, the supply of appliances and use of diagnostic aids. The affects of such earlier interventions on disability and on national insurance must be significant and should be measured.

REFERENCES

Armstrong, D. (1980). *An outline of sociology as applied to medicine.* Wright, Bristol.

Barber, J. H. and Kratz, C. R. (1980). *Towards team care.* Churchill Livingstone, Edinburgh.

Griffin, G. A. and Barry, S. M. K. (1981). Muscle and joint pain. Design and evaluation of courses for general practitioners. *J. R. Coll gen. Practns* 31, 661–8.

Guttmann, L. (1976). *Spinal cord injuries. Comprehension, management, research.* Blackwell, Oxford.

Maslow, A. S. H. (1954). *Motivations and personality.* Harper, New York.

Marsden, C. D. (1975). Parkinson's disease and involuntary movements. *Medicine, Baltimore* 2146–55.

Marshall, J. (1976). *The management of cerebro-vascular disease.* Blackwell, Oxford.

Partridge, R. E. H. and Knox, J. D. E. (1969). Rheumatic complaints in general practice. *J. R. Coll. gen. Practns* 17, 144–54.

Wood, P. (1977). *The challenge of arthritis.* The British League Against Rheumatism, London.

16 The management of common handicaps

J. A. Muir Gray

Although the terms 'disability' and 'handicap' are often used as though they were synonyms it is conventional to distinguish between them (Wood 1981; WHO 1980).

A *disability* is any restriction or lack of ability to perform an activity in the manner or within the range considered normal for someone of that person's age; for example limited hip flexion resulting from osteoarthrosis.

A *handicap* is a disadvantage for an individual resulting from a disability that prevents him fulfiling a role which is normal for someone of his age and sex; for example unemployment due to physical disability.

Handicaps may be defined as the social complications of disabling diseases but are of importance to the doctor who is treating the disabling disease which is the cause of them for two reasons.

1. Because he may be the only professional in contact with the disabled person and therefore the only person who can suggest that she make contact with the services which can help her overcome her handicap.

2. Because the opinion of the general practitioner may be as important to her as the advice of the experts even if the disabled person is in touch with the appropriate professional or agency.

It is therefore important that the trainee general practitioner be encouraged to think about the causes of handicap and to learn about the services which can help the disabled person overcome her handicap.

CAUSES OF HANDICAP

Obviously the probability that a person will be handicapped is dependent on the degree of disability. The more disabled a person is the greater is the likelihood that she will be handicapped. However, two other factors are also important – psychological factors and environmental factors.

ENVIRONMENTAL CAUSES

Many of the handicaps which disabled people face are due to the design of our environment. To the majority of the population the whims and fancies of architects and designers pass unnoticed or cause only minor inconvenience. However, for many disabled people, and not just those confined to wheelchairs, it is the failure of architects and designers to consider that a significant minority of people have difficulty with steps or are so weak that smooth round door knobs, domed taps, hidden controls and sockets at ground level are impossible to use that is the principal cause of their handicap. Unfortunately the design of the British house creates many of the difficulties of disabled people.

Disabled people, of course, have housing problems, such as problems with landlords or repairs, which are unrelated to their disability except that disabled people have, in general, lower incomes than those who are not disabled and poverty causes and aggravates housing problems. However, certain housing problems are directly related to the disability. Inability to climb steps and stairs, inability to step into a bath safely, or, more commonly, inability to stand up from a sitting position on the bottom of a bath, inability to reach the toilet in time and difficulties in the kitchen are all common.

Sometimes it is necessary for the disabled person to move, but often when a general practitioner is asked if he will 'write a letter' to the housing department to help her housing application, rehousing is unnecessary. The possibility of solving the housing problem by adaptation should always be discussed with the domiciliary occupational therapist before a move is considered. Unless there are very good reasons for making a move to another area, such as the wish to move nearer a relative who is willing to help, the best plan is to stay put. The disabled person may underestimate the help she receives from the community in which she is living and the difficulty of making friends in a new community. The domiciliary occupational therapist should be consulted to see if she can solve the problem.

Other handicaps are also caused by the failure of architects and designers. Problems with employment, education, and leisure are often due to the design of the work places, colleges, pubs, and cinemas. Furthermore, the design of public transport vehicles places an additional obstacle in the way of many disabled people and it has been estimated that over two million adults in Britain find the bus step an impossible obstacle. However, environmental problems are not the only causal factors.

PSYCHOLOGICAL CAUSES

Some people in wheelchairs overcome all these environmental obstacles successfully, finding employment and becoming self-sufficient. Others who are no more disabled fail to overcome the obstacles and remain unemployed and dependent. The following factors influence the person's motivation to overcome obstacles and have to be taken into account when assessing a person who is handicapped.

1. Her personality before the onset of the disabling disease; what sort of person was she, someone who was easily discouraged or someone who was a 'fighter'? The person herself and her relatives should be asked to describe her

personality and the uses of terms such as 'a fighter' and phrases such as 'did she give up easily' will yield more valuable information than formal psychological testing unless the general practitioner has been trained to use personality tests.

2. Her reaction to her disability; this is usually predictable from a knowledge of her personality before the disability developed but enquiry should be made about the person's emotional reactions, whether she is anxious or depressed and whether she is optimistic or pessimistic about her future progress.

3. Her social circumstances; the possibility of financial gain can reduce the motivation to become independent, as can isolation (see p. 314). In addition the attitudes of other people also have to be considered for in some cases other people create or foster the dependence of disabled persons for their own reward (Blaxter 1976).

It may be possible for a general practitioner to make a full assessment of all these factors by himself but if other professionals are involved their views will be useful. If no other professional is involved and the general practitioner wishes further advice referral to a social worker with a special interest in disability, a rehabilitation assessment clinic, or the domiciliary occupational therapist will be helpful.

THE OCCUPATIONAL THERAPIST

The main objective of the general practitioner is to reduce the degree of a disabled person's disability by making an accurate diagnosis, prescribing appropriate treatment and by trying to ensure compliance. The physiotherapist also has an important part to play by helping the disabled person to avoid the effects of immobility – muscle weakness, joint stiffness, and loss of confidence – and helping her to develop the full potential of the parts of the body which are directly affected by the disabling disease. Once the person's disability has been reduced to as low a level as possible the occupational therapist tries to re-integrate that person with her environment.

In hospital the occupational therapist will be involved from the time of admission but it is only when the person's disability has stabilized by treatment and the prognosis can be assessed that she will be able to help the disabled person plan for her future. The occupational therapist will first try to teach the person new ways of performing the tasks she finds impossible. If this fails she may provide an aid, such as a bath seat or tap turner, and if the person is still unable to cope with the assistance of aids and special equipment she will consider the need for adaptation of the dwelling or for rehousing, seeking the support of the community physician who is adviser to the housing department if rehousing is necessary.

The domiciliary occupational therapist usually becomes closely involved with the disabled person and her family if the person is severely disabled and she is therefore often in the best position to learn about the reactions and motivation of all members of the household and to try to change them if they are a significant factor in causing the disabled person's handicap and dependence.

SELF-CARE PROBLEMS

Independence in the activities of daily living should be set as an objective for all disabled people. The multitude of activities of daily living can be considered under the following headings.

Dressing and undressing.
Washing all over.
Getting to the toilet in time.
Getting enough to eat and drink.
Doing light housework.
Doing light gardening.

The approach to all these problems is similar in principle although the practical aspects differ depending upon the particular handicap. This approach may be summarized in an algorithm.

THE HANDICAP ALGORITHM (FIG. 16.1)

This can be used as a check list when assessing someone who is unable to perform one or more of the activities of daily living. It is useful to do this before arranging for someone to perform the task which the disabled person finds impossible on a permanent basis, although it it often necessary to provide temporary help until the cause or causes of the handicap have been identified and overcome. It is important to emphasize this to the patient, for example by saying that 'we can arrange for someone to help with the housework until you are able to do it again by yourself' and to set a firm date for reviewing the person's problem and the need for help.

It is necessary to consider not only the handicapped person but also any other people on whom she has become dependent.

HANDICAP AND DEPENDENCE

A person who is handicapped in some aspect of self-care may be deprived of the benefits which that activity brings. Alternatively she may become dependent on other people to perform it for her.

Dependence creates problems for the disabled person and for those on whom she becomes dependent. The disabilities of the disabled person may increase as she becomes weaker and stiffer through inactivity. In addition her motivation to struggle for independence may decrease. This is a particular problem for isolated disabled people many of whom fear that they will become more isolated if they become so fit that they no longer need the help of the home help or district nurse.

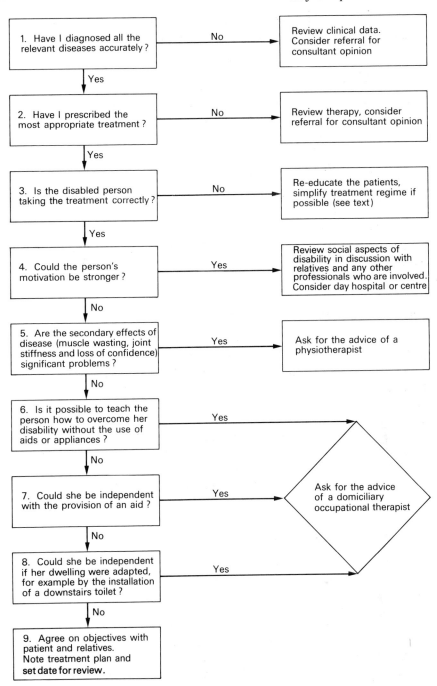

Fig. 16.1. The handicap algorithm.

Relatives and friends also create dependence, and not only do they create a greater degree of dependence than professionals but they are more likely to suffer by becoming over-involved. Furthermore the dependence may have developed to a significant degree before any professional becomes aware of its existence. When trying to help a disabled person to regain the ability to look after herself it is, therefore, important to consider the needs of both the disabled person and her helpers. Specific advice needs to be given to both about the objectives of rehabilitation and often it is helpful to discuss these with both parties simultaneously. It is also important to try to prevent dependence by telling both the disabled person and her supporters that if the former loses the ability to dress or undress, or to wash all over safely, or get to the toilet in time, or to get enough to eat or drink, or to do light housework or gardening this should be brought to the notice of the general practitioner, district nurse, health visitor, or occupational therapist (Gray and McKenzie 1979).

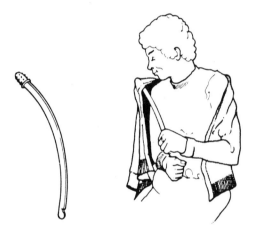

Fig. 16.2. Dressing stick (hanger with rubber page turner on end).

In addition, when the clinical condition of the disabled person is being reviewed or when relatives consult for some reason unrelated to the problems of the disabled person, the opportunity can be taken to review the degree of dependence of the disabled person. This can be done by relying on the memory of the disabled person and her relatives but it is possible to record the degree of dependence simply.

Assessment of dependence

Complicated methods of assessing and recording an individual's degree of dependence have been developed but these are of relevance primarily to research workers. For the busy general practitioner it is satisfactory to record the degree of dependence on a scale with three degrees.

0 Able to manage independently.
1 Requires some help.
2 Completely dependent on others.

If used to record the person's ability to dress, wash all over, reach the toilet in time, and get enough to eat a simple record of the person's degree of dependence can be kept.

COMMON HANDICAPS

There is no need for a trainee general practitioner to learn all the ways by which a disabled person can be helped to become independent in looking after herself. It is more important for him to learn how to manage disabling diseases effectively, to appreciate the psychological factors in disability, and how he can make best use of the skills of the physiotherapist and occupational therapist. To do this he has to be aware of the approaches which the occupational therapist might take.

Dressing and undressing

Most people who are having difficulty with dressing can be taught how to cope with their disability by the district nurse or occupational therapist. The first approach is to try to teach the disabled person to dress and undress using her ordinary clothes. If this fails her clothes may need to be adapted or she may need to buy clothes which are not specially designed for disabled people but which are easier to manage, for example a wrap-around skirt. Sometimes a dressing aid will be helpful (Fig. 16.2).

It may be necessary for the occupational therapist to help her buy her clothes from one of the manufacturers who make clothes which have been specially designed for disabled people. The Disabled Living Foundation (346 Kensington High Street, London) has a clothing adviser but the domiciliary occupational therapist is usually able to give appropriate advice. The Shirley Institute, Didsbury, Manchester M20 8RX produces a catalogue listing the types and suppliers of special clothing for disabled people.

If the person is dependent on other people to help her dress or undress she should be helped to become independent gradually, solving one problem after another. Reinforcement of her behaviour by praise from the person on whom whe has become dependent is important and the disabled person should be told that her efforts are a great relief to her helpers. The helper should also try to spend as much time with the disabled person while she is regaining independence in dressing as when she was completely dependent so that the 'reward' for independence is not isolation.

Washing all over

The occupational therapist will first try bath aids, such as hand rails or a pole to support the person when climbing in and out or a bath seat for the person who

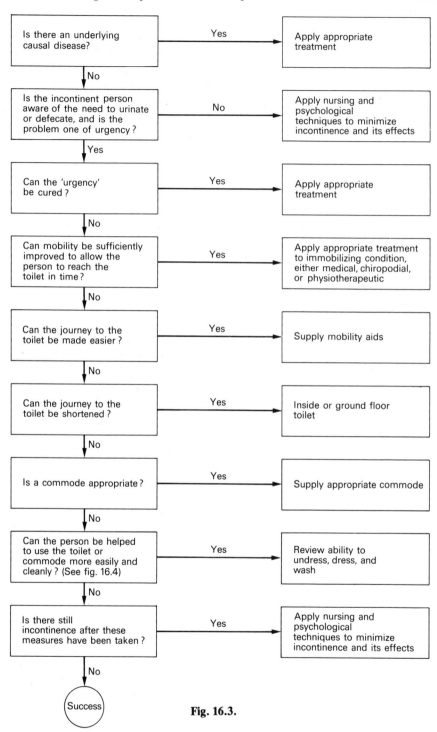

Fig. 16.3.

has difficulty in rising from a low sitting position. Sometimes it is necessary to install a shower if the person cannot get into a bath or if the bathroom is inaccessible. There are grants to help with this type of adaptation.

Because fear of falling is common it is often necessary for a relative or nursing auxiliary to be present when a disabled person is bathing even though the disabled person is physically able to step in and out. The provision of aids and structural adaptations are still relevant in such cases because they can make the task of the helper easier. Relatives and the disabled person may be embarrassed by bathing and it may be appropriate to ask the district nurse to assess the family's need for relief even though the relatives are physically capable of helping.

Reaching the toilet in time

This is a more complicated problem because there are three components in this activity each of which may be impaired by a disabling disease.

1. The period of time between the first sensation that evacuation is necessary and the moment at which the person can no longer contain herself, that is whether or not the person has urgency.

2. The time which the disabled person takes to reach the toilet.

3. The ability to undress and sit down, quickly and safely (Mandelstam 1977).

If the person does not have any urgency it is usually possible for her to avoid incontinence even though she has difficulty with walking or with undressing but when urgency is present the probability of incontinence is much higher. Obviously the problem created by urgency can be mitigated by the provision of a commode if the cause of the urgency cannot be treated, but a commode is a depressing and, to many people, offensive symbol of their disability which causes embarrassment and the objective should always be to help the person reach her toilet. The steps which can be taken can be summarized in an algorithm (Fig. 16.3).

Having reached the room in which the toilet is installed, the person's difficulties are not necessarily over. She may still be incontinent or make a mess of herself, which can be very degrading and depressing, especially if she has to rely on other people to wash her person or her clothes. Once more we use the algorithm approach analysing the process of using a toilet into its constituent parts and considering each in sequence (Fig. 16.4).

Getting enough to eat and drink

People who have difficulty in buying, preparing, cooking, or eating food are at risk of developing nutritional problems but it is usually only when the person becomes depressed or loses the motivation to eat that problems develop.

One useful measure of the person's nutritional status is the person's weight. The weight of a disabled person should be recorded on her notes, as with any other patient, and the person should be weighed from time to time. Loss of weight is an indication for further investigation but it is less common than an increase in weight. Obesity is the commonest form of malnutrition among

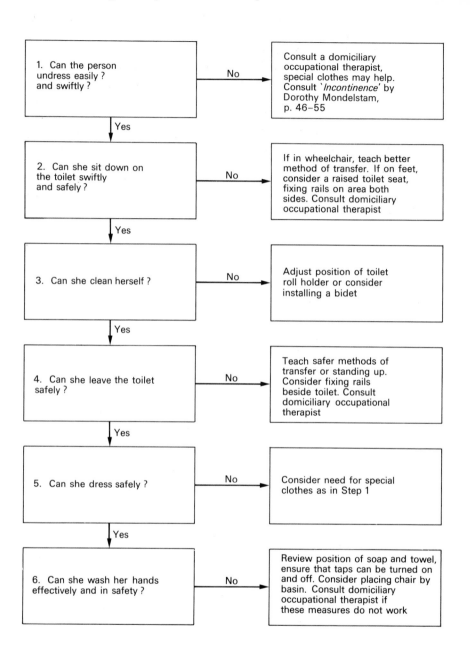

Fig. 16.4.

disabled people for whom obesity has a special significance because it may tip the balance between independence and dependence.

If a person cannot prepare sufficient food, meals-on-wheels or the services of a home help can be arranged but before this step is taken the precise nature of the problem – whether it is with shopping or food preparation or eating – should be identified. Then the disability which is the cause of that problem should be identified and treated. If the person is still unable to shop, or prepare food, or cook after medical treatment and physiotherapy the occupational therapist should be consulted. She may be able to provide aids for use in the kitchen or adapt the kitchen so that it may be used safely by the disabled person.

If the disabled person has become dependent on relatives it can be difficult to restore independence if the relatives believe, as many do, that she needs three cooked meals a day. Relatives have to be educated that nutritional needs can

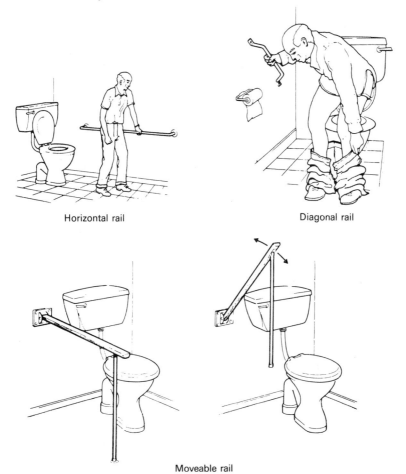

Horizontal rail   Diagonal rail

Moveable rail

Fig. 16.5. Rails can help an old person sit and rise. They are placed on the right or left side depending on the side which is stronger.

often be met much more simply, for example by fruit and sandwiches, and that the disabled person can be left to fend for herself providing that food and drink is within reach, even if she cannot cook. However, information alone may be insufficient to change the behaviour of relatives if their concern about providing food is not merely to meet the disabled person's nutritional needs but is a means of expressing their love, guilt or some other feeling. If the food is serving a symbolic, as well as a nutritional, function then it is necessary to try to help the relatives find other ways of expressing their feelings.

Doing housework and gardening

Inability to do housework depresses many disabled people because the disarray and decay of their environment provides a permanent reminder of their own disarray and decay. Furthermore it is a sign of disability that cannot be hidden from others except by discouraging friends and neighbours from visiting and this is one reason why some disabled people become isolated. The home-help services can help in many cases but, as always, should if possible only be brought in after every step has been taken to minimize the person's disability and the domiciliary occupational therapist has been consulted. If the disabled person depends on relatives to help with housework then they may welcome some help from the home-help service to relieve their burden.

If the disabled person can be helped to tackle her garden, or to take up gardening, morale is often improved and the fresh air and exercise may improve her physical wellbeing as well. Local authorities do not have the resources to build elevated garden beds but the occupational therapist may be able to help the disabled person by the provision of long-handled tools or by making suggestions about new ways in which tasks in the garden can be tackled.

IMMOBILITY AND ISOLATION

Mobility problems cause many of the difficulties with self care described in the previous section. However, many people are sufficiently mobile to manage inside their home but are still severely handicapped because they cannot get about outside. This leads to isolation, unemployment and to difficulties with shopping, going to the bank, seeing one's councillor or MP, visiting the environmental health department to ask about house renovation grants and all the other activities for which one must be mobile.

The main disabilities which lead to immobility outside the home are:
1. Blindness.
2. Joint stiffness and muscle weakness.
3. Disorders of balance.
4. Loss of confidence.
5. Major limb disorders.
6. Respiratory and cardiac disease.

In some disabled people a degree of agarophobia develops and even if the physical disability is overcome some people may find it very difficult to go out and meet people again. This is, however, relatively uncommon and most disabled people find no difficulty in going out if their mobility is improved.

WALKING AIDS AND WHEELCHAIRS

If a patient reports difficulty with walking or if the patient or a relative requests a wheelchair the first step is obviously to try to reduce the functional impairment to the minimum possible level. The advice of a physiotherapist is often very helpful to the patient and her relatives in these circumstances. When this has been achieved it may still be necessary for the person to use a walking aid, either a stick or a quadripod stick or a walking frame – a 'Zimmer' – but careful consideration has to be given before a walking aid is suggested or prescribed because the person who uses such an aid will alter her gait and this may create problems in other parts of her body. Before advising anyone who has difficulty in walking to purchase or use a walking aid it is, therefore, often helpful to have a second opinion from someone who is experienced in assessing the need for walking aids. A physiotherapist or remedial gymnast is trained for this purpose and can be consulted directly if the general practitioner is confident that his diagnosis and treatment are correct and if direct referral is possible. Otherwise referral to a consultant clinic will be necessary.

If a wheelchair is going to be used permanently referral to the nearest appliance centre is indicated.

SPECIAL CARS

The staff of the appliance centre will also be able to advise the disabled person about her suitability for a specially adapted car. In addition they will be able to advise her on the best way to purchase such a car, through the Motability scheme, and about the other concessions which are available to the disabled driver.

The domiciliary occupational therapist is also a useful source of advice on this subject.

THE MOBILITY ALLOWANCE

This weekly allowance may be paid to people aged between five and 65. The leaflet NI 211 – Mobility Allowance – says that the disabled person will qualify if three medical conditions are met. The leaflet tells the disabled person that he will be eligible if:

(i) you are *either* unable to walk *or* the quality of your walking outdoors is such that you are regarded as being virtually unable to walk;

or

(ii) the exertion required to walk would endanger your life, or would be likely to lead to a serious deterioration in your health.

You must be likely to remain unable to, or virtually unable to, walk for at least a year.

You must be able to make use of the allowance. A person who cannot be moved for medical reasons, or is in a coma, would not qualify.

There are other conditions which are explained in the leaflet which also contains the application form.

The person will be examined by a doctor who is employed by Social Security for this purpose and a second examination by a Medical Board may be necessary. If the person is refused the mobility allowance on medical, or other, grounds he can appeal against the decision.

HELPING THE HOUSEBOUND

Sometimes the barriers which prevent a disabled person from going out are not physical but psychological and even the provision of the full range of aids and services does not allow the person to go out to meet other people. Sometimes this difficulty will have been evident at the initial assessment but in other cases the disabled person initially appears to be keen to go out and about and it is only when the means of doing so have been provided that her difficulty becomes manifest. There are many reasons why a disabled person may be reluctant to go out. She may, for example, be embarrassed by her disability or she may feel that she has lost her social skills or she may simply be ashamed of her appearance, of clothes which are no longer fashionable and unkempt hair. This type of difficulty requires careful assessment and a programme in which the person is allowed to increase her social contacts gradually. The advice of a clinical psychologist will be helpful in difficult cases if it is possible to make a referral to a psychologist with a special interest in rehabilitation.

HELPING ISOLATED PEOPLE

Many disabled people will experience long periods of isolation even after every possible step has been taken to help them. The effects of isolation can be very serious and there are common consequences.

Depression and feelings of loneliness.

Confusion and disorientation, especially if there is blindness or deafness in addition to the physical disability.

Nutritional problems due to loss of appetite.

Dependence on those who come to help for their company as well as for the service they offer.

To mitigate the effects of isolation it is necessary to take three steps:

1. Try to arrange more trips out of the home, for example to the church or local pub.

2. Try to arrange more visitors in the home, for example student volunteer visitors.

3. Try to provide more stimulation for the disabled person in the home, for example by arranging for the delivery of library books or the loan of a television.

Isolation is one of the major problems of disabled people and its effects may be serious even if the disabled person does not appear to be distressed by feelings of loneliness.

EDUCATION AND EMPLOYMENT PROBLEMS

These problems are inter-related because it is often the poor level of educational attainment which is one of the principal causes of difficulty in finding employment (Darnborough and Kinrade 1977).

FURTHER EDUCATION

Educational opportunities for disabled people are improving, albeit slowly, and the objective is to try to provide the disabled person with the type of education which is suited to his educational needs and aspirations. In theory his disability should not influence the choice of course or institution but in practice there are still many universities, colleges, and polytechnics in which life is very difficult for a severely disabled person.

There are a number of organizations which can give advice to the disabled person about further education and the domiciliary occupational therapist will be able to advise the person where to seek expert advice and guidance.

FINDING EMPLOYMENT

In general disabled people experience greater difficulty in finding work than those whose abilities are not impaired by disease. There are a number of reasons for this.

Difficulty in travelling from home to work.

The architecture and design of many work places make them completely unsuitable for disabled people.

The employer's fear that the disabled person will be frequently off sick.

Embarrassment of the employer when faced by a disabled person.

Lack of appropriate qualifications.

The key person in the search for employment is the disablement resettlement officer (the DRO). He is trained to advise and help the disabled person who is experiencing difficulty in finding a job and is in touch with the full range of training and educational opportunities which exist to help newly disabled people retool with new skills for the job market.

POVERTY

Because unemployment is so common among disabled people and because the jobs they find are so often lowly paid poverty is a common problem. In addition the expenses of the household in which a disabled person lives are often greater than in the households in which there are no disabled members. Disabled people therefore need to be informed about the full range of financial benefits available to them. Three allowances are of particular importance.

The attendance allowance

The attendance allowance is for disabled people who have, in the words of the Department of Health and Social Security, 'needed a lot of looking after' for at least six months. There are no financial conditions attached to this allowance – people can apply whatever their income level and if the allowance is granted other social security benefits are not reduced.

There are two rates – a higher rate for those who need looking after both day and night and a lower rate for those who need day or night attendance. The financial benefits are considerable – the higher rate was worth about £1000 a year and the lower rate about £600 in 1981 – so disabled people should be encouraged to apply using the form in leaflet NI 205 – Attendance Allowance.

Injured at work

Everyone who has been in regular employment, paying National Insurance contributions, is entitles to receive help from Social Security if the disability becomes so severe that work becomes impossible. Sickness benefit is paid for the first 28 weeks, invalidity benefit thereafter (see leaflets NI 16 and NI 16A). If, however, the disability has developed as a result of an industrial accident or injury the disabled person is entitled to a higher level of Social Security payment. The main benefit for which they qualify is the Industrial Injury Benefit, which lasts for only 26 weeks, and the Industrial Disablement Benefit which starts after 26 weeks if the person is still disabled when the period of Industrial Injury Benefit expires.

Simple information about the benefits available is contained in the leaflet 'Which benefit – 60 ways to get cash help' (leaflet FB 2) but the system is complicated and the disabled person should seek advice from the Social Security office or from his trades union or the personnel office of his employer or former employer if he is in any doubt.

Injured in war

People who have become disabled while serving in the forces qualify for special benefits. Usually they are in receipt of all the benefits for which they are eligible because the War Pensions division of the Department of Health and Social Security performs much valuable welfare work in addition to its administrative functions. In addition the British Legion and a number of forces welfare organizations keep in touch with people who have been disabled while serving in the forces and with the dependents of those who have died while in service or as a result of their service.

If someone thinks he might be eligible for more help than he is receiving he should contact the nearest War Pensions Office, the address of which can be found from the local social security office.

The non-contributory invalidity pension

This is for people of working age who have not been able to work for at least six months and who have not paid enough National Insurance contributions to be eligible for invalidity benefit. Men and single women can apply, using leaflet

NI 210, and married women can also apply, using leaflet NI 214, if they are not only unable to work but also unable to do normal housework.

The invalid car allowance

Men and single women who stay at home to look after a severely disabled relative, that is one who is receiving an attendance allowance, qualify for this benefit which does not depend upon payment of National Insurance contributions. It is a valuable benefit not only for the financial benefit of the weekly payment but also because the person who receives it is credited with a National Insurance contribution as though he or she were in paid employment. Full details are given in leaflet NI 212.

Financial advice

Disabled people who have income problems are also eligible for help with rent and rates and, if their income is low enough, for Supplementary Benefit. The full range of benefits is extremely complicated and the disabled person may need more detailed advice than can be given by a health visitor, occupational therapist or social worker. If she does she can ask at the local social security office or at the Citizens' Advice Bureau or she can write to the Disablement Income Group, who are experienced in dealing with the most difficult types of problem.

THE DIRECTORY FOR THE DISABLED

This excellent book summarizes all the main services and organizations for disabled people and their families. It has sections on Social Security, mobility aids, housing problems, further education, holidays, sexual problems, and leisure activities. Every practice library should have a copy of this book which provides important information on all the common handicaps.

REFERENCES

Blaxter, M. (1976). *The meaning of disability: a sociological study of impairment.* Heinemann, London.

Darnborough, A. and Kinrade, D. (1977). *Directory for the disabled.* Woodhead-Faulkner, London.

Gray, J. A. M. and McKenzie, H. (1980). *Take care of your elderly relative.* Allen & Unwin, Beaconsfield.

Mandelstam, D. (1977). *Incontinence.* Heinemann, London.

Wood, P. H. N. (1981). The language of disablement. *Int. Rehab. Med.* **2**, 86–92.

World Health Organization (1980). *International classification of impairments, disabilities and handicaps.* WHO, Geneva.

17 Management of rheumatic diseases

E. C. Huskisson

Management is a good word to describe what doctors do for rheumatic diseases. Few are cured though some resolve naturally. Most are managed, the symptoms controlled, the progression limited, and the disability minimized. There are few patients who cannot nowadays receive the benefits of effective treatment. There is now a better understanding of the disease processes involved and their natural history. Patients benefit from understanding their diseases and there is no pain worse than one which is undiagnosed. There are more anti-inflammatory drugs, they are safer and their action lasts longer. Patients benefit from the choice, from the absence of side-effects and from the convenience of infrequent administration. There are some specific treatments like allopurinol (zyloric) for gout and penicillamine (distamine) for rheumatoid arthritis. There are surgical replacements for most joints. There are aids and appliances, shoes, specialists, and special clinics – all quite new.

There are lots of patients and lots of diseases to manage. I shall concentrate on those which are particularly common or important in general practice. These include soft-tissue syndromes like tennis elbow, frozen shoulder, and carpal tunnel syndrome which are common and easy to treat. They include acute joint problems which require urgent management such as gout, pseudogout, septic arthritis, and traumatic problems. They include serious conditions like polymyalgia rheumatica and benign syndromes like fibrositis, ankylosing spondylitis and psychogenic rheumatism. Osteoarthritis is so common that all physicians see it. Rheumatoid arthritis is much less common but the general practitioner is involved in its management even if the patient attends a specialist clinic. This list is far from complete.

WHO MANAGES?

There are no rules about what a physician can or cannot, may or may not, should or should not do. The general practitioner who feels competent to take on the management of a case of rheumatoid arthritis, treatment with penicillamine or corticosteroid injection can, may and should do so. If he is not competent and confident, he should make use of the facilities of his specialist colleague. But he is always involved. It is the general practitioner who is called in the night to a patient with rheumatoid arthritis who has pericarditis and who receives the first complaint of loss of taste from the patient on penicillamine, who wasn't warned of this complication at the hospital. Co-operation and collaboration are the ideals.

STAGES OF MANAGEMENT

Management is seldom as simple as the writing of a prescription, the skilful administration of an injection or any other single measure. There are six stages in the management of a patient with a rheumatic problem involving diagnosis, assessment, advice, symptomatic therapy, specific therapy, and subsequent supervision. Diagnosis is the first essential. Assessment is the next. At the assessment stage, it is useful to list the patients problems and to try to meet them with solutions. One patient's concern may simply be the possibility of progressive and crippling arthritis. He needs reassurance. For others, it may be morning stiffness, painful feet, disturbed sleep, or difficulty in carrying out some everyday task that troubles them. Management must be directed towards these problems. Every patient needs advice and no management plan is complete without it. The next stage in the management of most rheumatic disorders is the use of symptomatic therapy. Non-steroidal anti-inflammatory drugs are prominent at this stage but local injections and physical methods of treatment are also important. For some diseases there is also specific therapy, usually for the difficult case. Once the treatment is planned and instituted, appropriate supervision will be needed to assess response and to make appropriate adjustments. Complications arise in most conditions, either related to the disease or its treatment and there are a variety of problems to be solved. Management is often a complicated and ever-changing package, different for each patient according to his needs.

MANAGEMENT OF GOUT

Gout is easy. Once the diagnosis has been made, treatment should proceed with confidence. The diagnosis is not always easy. Gout doesn't always affect the big toe, isn't always monarticular, and doesn't always afflict the typical drinking man. The best diagnostic test, examination of synovial fluid under polarized light, is not usually available in general practice. The serum uric acid is too slow and too unreliable. The diagnosis must usually therefore be clinical and if in doubt, it's a good idea to save a little sample of synovial fluid. An exact diagnosis is comforting even in retrospect.

It is essential to separate the two phases in the treatment of gout, symptomatic therapy in the acute stage and specific therapy for long-term treatment in patients with appropriate indications. Allopurinol and uricosurics make acute gout worse. Indomethacin remains a good choice for the acute stage. An initial dose of 50 or 100 mg by mouth, depending upon the severity, is followed by a dose of 50 mg three times daily and then 25 mg three times daily, reducing as the condition improves. Some relief should be obtained within a few hours of the first dose and within days, the patient should be very much better. Colchicine for acute gout is inconvenient as well as toxic. Phenylbutazone must be avoided in the elderly but can be given to younger patients for short periods of time with minimal risk. Even this minimal risk I prefer to avoid. In the difficult patient, with a peptic ulcer for example, the newer non-steroidal anti-inflammatory drugs are a suitable alternative to indomethacin. Naproxen, piroxicam, and azapropazone are amongst those which can be used. If there is the slightest problem with an attack of acute gout, I use ACTH. An initial dose of 80 units by intramuscular injection is dramatically effective and sometimes sufficient. Another dose of 40 units may be required a day or two later. Intramuscular or intravenous use of a non-steroidal drug is unnecessary and worth avoiding. A little thought is required once the acute attack of gout is better – an assessment. Occasionally gout is secondary, to a myeloproliferative disorder for example. Serum uric acid is required at this stage and a full blood count should be requested at the same time. Patients with very high serum uric acid levels (greater than 0.55 mmol/l or 9 mg per 100 ml) along with those with frequent attacks and those with evidence of urate deposition around the joints or anywhere else need long-term treatment. Allopurinol (zyloric) is so straight-forward that for most physicians it has become the first choice of long-term treatment. It is given in a single daily dose of 300 mg. It must never be started within three weeks of an acute attack or before the acute attack has completely settled. It must always be combined with a prophylactic for the first three months. I use colchicine (0.5 mg twice daily) but an anti-inflammatory drug can also be used for this purpose. Once started, it must be continued forever. The patient must understand the nature and purpose of the treatment or he probably won't go on taking it. Uricosuric drugs are seldom required today. Allergic reactions to allorpurinol are very rare but these are occasional patients whose disease is not adequately controlled on allopurinol in whom combination with a uricosuric is useful. The serum uric acid should be checked after six months of treatment – a larger dose (600 mg) of allopurinol is occasionally required. At this stage, the disease is usually controlled and no more should be heard of it.

MANAGEMENT OF PSEUDOGOUT

Pseudogout is not quite so easy. It should be considered in a patient, usually elderly, with an attack which resembles gout but affecting the knee joint. Identification of crystals under polarized light is again the best diagnostic test but an X-ray may also be useful – linear calcification of cartilage is often seen. Many patients respond well to indomethacin or an alternative non-steroidal anti-inflammatory drug such as flurbiprofen or naproxen. Some do not. In such patients, it is useful to aspirate the affected joint, removing as much synovial fluid as possible and injecting a corticosteroid preparation such as triamcinolone hexacetonide (lederspan). A systemic steroid should not be used. Most attacks settle within a few weeks and alas, there is no allopurinol for calcium pyrophosphate deposition.

MANAGEMENT OF TRAUMATIC SYNOVITIS, SPRAINS, AND STRAINS

Trauma is a very common cause of joint pain, its mechanism ill-understood despite the obvious cause but outcome favourable. Joints and soft tissues have remarkable healing powers as evidenced by the high proportion of football players who reappear for their team within seven days despite their injuries. Treatment is empirical. Rest seems a good idea though not complete immobilization since this may induce stiffness. Anti-inflammatory drugs seem to be effective and there is often abundant clinical evidence of inflammation. Propionic acid derivatives such as naproxen or ibuprofen seem to be particularly effective in this situation. Local corticosteroid injection is often helpful but with one big caution. Soft-tissue injections should not be used in athletes because of the risk of tendon rupture. Various forms of physiotherapy are used to relieve symptoms. With these measures, the optimistic drive of the victims and a little rehabilitation, most patients are soon better.

SEPTIC ARTHRITIS

Septic arthritis is fortunately rare but its importance and urgency demand that it be mentioned. It should be considered in any patient with unexplained acute arthritis and is a particular worry in those with pre-existing joint disease like rheumatoid arthritis. A single joint which is suddenly painful may be a flare but may be an infection. Joint aspiration is required to make or exclude the diagnosis. It is urgent since on-going sepsis in a joint will lead to cartilage destruction within days. Some joint infections are undramatic, tuberculosis for example and gonococcal arthritis which bears particular consideration in young women. Joint aspiration is an important and very useful diagnostic manoeuvre in patients with acute arthritis. It must often be necessary to send patients to hospital for this procedure.

MANAGEMENT OF TENNIS ELBOW

Tennis elbow is a benign condition, an inflammatory lesion which arises in the common extensor origin at the lateral humeral epicondyle. It is very common, sometimes due to tennis or repetitive manual work but often mysterious in origin. Left alone, it gets better after a period which varies from a few weeks to a few years. Patients worry about it because it feels like arthritis. There is a sickening pain which affects the whole elbow. In mild cases, explanation and reassurance are all that is necessary. It is not arthritis and it will get better. Local corticosteroid injection is very successful and should be considered if the condition is worse than mild. General principles of corticosteroid injection are considered below. The technique for tennis elbow is easy. I use a long-acting preparation such as triamcinolone hexacetonide. I inject 0.5 ml (10 mg) in and around the point of maximum tenderness over the lateral epicondyle. This procedure is often very painful at the time and for up to two days afterwards.

Patients must be warned. Most are subsequently relieved. If not, the injection may be repeated but a third injection is most unlikely to be successful if the first two are not. Time heals most of the failures and the very few chronics require specialist help.

Corticosteroid injection is the treatment of choice for tennis elbow, frozen shoulder, and carpal tunnel syndrome. Intra-articular injection is a useful technique for a wide range of inflammatory arthropathies. Given competence and confidence, there is no reason why this procedure cannot be carried out in general practice. The biggest hazard of intra-articular injection is the introduction of infection. Sterile syringes and needles and a no-touch technique are therefore mandatory. As to the place for the injection, the patients home is probably ideal since there he is surrounded by his own friendly bacteria. In hospital, the bacteria are less friendly but infections are nevertheless very rare. The following equipment must be assembled for the procedure; sterile syringe, sterile needle for drawing up, sterile needle for injecting, material to be injected, sterile swab to clean the skin, dry swab to staunch the flow of blood after the injection and plaster to cover the hole. It is better not to use multidose vials which may become contaminated with frequent use and hang about for long periods of time in doctors surgeries. It is essential to use a corticosteroid preparation which is suitable for intra-articular injection. The technique is simple. Palpate the injection site and if necessary mark it. Wash your hands. Draw up the appropriate dose of corticosteroid and change the needle on the syringe. Clean the patient's skin and do not touch it again. No further palpation is allowed. Inject and remember to advise the patient appropriately. It is important that patients should know whether the injection is likely to be very painful afterwards, whether they should rest or exercise and when improvement is likely to occur. Patients should be warned to report at once after intra-articular injection if the joint becomes painful and swollen because of the risk of infection. Intra-articular injection probably carries a small risk of joint deterioration and patients should be warned to rest for 48 hours and not to increase their activities for a few weeks. In appropriate indications, joint or soft-tissue injection of corticosteroid is a straightforward and very successful procedure.

Frozen shoulder is a general term for a group of soft-tissue conditions which affect the joint. It may sometimes be a tendinitis, sometimes a bursitis, sometimes a capsulitis, and sometimes calcific periarthritis. In some cases it is possible to be more exact than just calling it frozen shoulder and in some cases it is not. The term frozen shoulder, though widely used, is not ideal for those cases in which there is little or no limitation of shoulder movement. It is sometimes reserved for cases in which the shoulder is completely immobile. It may be better to call this group of diseases 'painful, stiff shoulder'. The condition, like tennis elbow, is self-limiting but may last as long as two years and may leave a permanently stiff shoulder. In mild cases, without limitation of shoulder movement, no treatment may be required. A non-steroidal anti-inflammatory drug may be used in such cases to relieve symptoms. Since pain in the shoulder is often worst at night, a night time dose – perhaps Indocid-R 75 mg or flurbiprofen 100 mg – may be most appropriate. When there is restriction of shoulder movement, I recommend local injection of cortico-steroid and physiotherapy to mobilize the joint. Non-steroidal anti-inflammatory drugs may again be useful. This regime is empirical but seems to work well. General principles of injection technique are discussed above. The appropriate dose of triamcinolone hexacetonide (lederspan) is 20 mg. The coracoid process is the important anatomical landmark. The needle slips underneath it into the shoulder joint, headed in a posterior direction. Repeated injections are unwise. One should be enough. The exercises required are simple and in the absence of a physiotherapist, can be explained to the patient in a few minutes. It is important to restore full abduction and rotation of the shoulder joint and to maintain this movement with a regular exercise programme.

MANAGEMENT OF CARPAL TUNNEL SYNDROME

Carpal tunnel syndrome is another common soft-tissue problem which presents with pain in the arm and hand, associated with pins and needles in the fingers. The nocturnal exacerbation of symptoms is often a characteristic clue to the diagnosis. Physical signs are often absent, especially in mild cases. Sometimes there is subjective impairment of sensation over the lateral three-and-a-half fingers. Only in severe cases is there thenar wasting. In such cases surgical decompression of the carpal tunnel is required. For most cases, local injection of corticosteroid is ideal treatment. The dose of triamcinolone hexacetonide is 20 mg. The site of injection is at the distal skin crease, just lateral to the long flexor tendons. The needle enters at 90° to the forearm and to a depth of about two centimetres. A wrist splint is an alternative mode of treatment but is cumbersome and often uncomfortable. I hesitate to use diuretics for this indication. Anti-inflammatory drugs are not usually successful. Most patients respond to local injection of corticosteroid – those cases which fail to respond or recur require surgical decompression.

MANAGEMENT OF POLYMYALGIA RHEUMATICA

Polymyalgia rheumatica is another soft-tissue condition of enormous importance. It is important because it is very disabling and unpleasant, a little bit dangerous because of the underlying giant-cell arteritis, and very easily treated. It should be considered in any patient over 50 who presents with pains in or around the shoulder or pelvic girdles. Best clue to the diagnosis is often morning stiffness, characteristically so severe that the patient has difficulty getting out of bed. The physical signs are most obvious in the mornings and may have disappeared by night time. There is restriction of movement of the shoulders and/or the hips with local tenderness around the shoulder and pelvic girdles. The pain is muscular but the symptoms arise from joints and adjacent connective tissues. The pathological basis of the condition is giant-cell arteritis and some patients have associated clinical evidence of temporal arteritis. A very few patients with polymyalgia rheumatica will become blind so that treatment is urgent. Once a clinical diagnosis of polymyalgia rheumatica has been made, treatment with corticosteroids must be started at once. Non-steroidal anti-inflammatory drugs are much less effective and do not prevent blindness. Prednisolone should be started in a dose of 15 mg daily and the patient should report dramatic improvement within a few days. Smaller doses are occasionally sufficient. The full dose of prednisolone should be continued for 2–3 months when a cautious reduction process can be started. The first reduction should be by 2.5 mg daily to 12.5 mg daily and a month or two later a further reduction of 2.5 mg daily can be made to 10 mg daily, Thereafter, reductions should not exceed 1 mg daily and should not be made more often than once in two months. Patients should be advised that if after making a reduction in dosage, symptoms recur, they should immediately return to their previous dose. Wild swings of doses should be avoided. Most patients have at least one flare of symptoms during their reduction programme and most take 2–4 years to complete their treatment and stop corticosteroid therapy. Recurrences are very rare.

MANAGEMENT OF FIBROSITIS

Many doctors will have been taught that fibrous tissue does not become inflamed and that fibrositis does not therefore exist. The syndrome was debunked some years ago but is staging a revival. Of course, patients continued to have it when it did not exist and it is actually very common. Its revival is attributable to some interesting experimental findings. Patients with fibrositis have a characteristic sleep disturbance on EEG. One can even induce a fibrositis-like syndrome by repeatedly disturbing normal people during the night. Some people think that fibrositis is entirely psychogenic but I am not sure. I know that it exists. Patients present most characteristically with muscular pain, often between the scapulae or in the lower back, sometimes elsewhere. Disturbed sleep is a very characteristic part of the syndrome. Examination of such patients reveals localized, tender points, often paravertebral and occasionally the well-known nodules. Treatment is not easy. The syndrome can seldom be cured but it can certainly be made more manageable with a regime including heat, massage, non-steroidal anti-inflammatory drugs, hypnotics, and adequate explanation and reassurance. Local heat and massage are undoubtedly helpful and most patients can organise this for themselves at home. One of the milder non-steroidal anti-inflammatory drugs such as ibuprofen often helps and a short course is appropriate. It is important to break the vicious circle of sleep disturbance and exacerbation of symptoms and a small night time dose of a benzodiazepine such as diazepam should be given. Patients with fibrositis are often tense, anxious individuals. They have often seen a lot of doctors and collected a lot of diagnoses. An explanation of the nature of the syndrome combined with a simple guide for how to manage it often helps.

MANAGEMENT OF PSYCHOGENIC RHEUMATISM

Whether or not fibrositis is psychogenic, there is certainly a substantial group of patients whose joint or back symptoms appear to be psychogenic in origin. This diagnosis should not be made by exclusion. There are lots of patients with back pain for example in whom an accurate diagnosis cannot be made. This does not mean that their pain is psychogenic. There are many clues which should help in the diagnosis of psychogenic pain. The pain tends to be continuous and unvarying and is described in dramatic terms. It is often of long duration with lots of unsuccessful treatments. Analgesics do not usually help. Such patients are often frequent attenders for different problems and may have had unnecessary investigation and treatment of other disorders. They may have many other symptoms including headaches, perhaps the commonest of all psychogenic symptoms. Most patients have no definite psychological abnormality though it is important to exclude depression or anxiety states. The course of the illness is of course benign. Management is very difficult and despite the size of the problem, it receives little attention in medical education programmes. Amongst the positive aspects of the treatment are reassurance that there is no serious arthritis (sometimes sufficient treatment), explanation of the fact that pain does not always mean disease and discussion of possible factors in the development of the disorder. Apart from the occasional patient with depression or anxiety states, drugs like antidepressants or tranquillizers are of little value and should be avoided. There are lots of other pitfalls to avoid. It is important not to diagnose a non-existent disease or to encourage the patient to behave if he or she had some disease. It is usually best to avoid medication which is likely to be ineffective. It is certainly wise to avoid predicting extravagant success from any particular treatment – this is very unlikely. Many patients with psychogenic pain embark upon a succession of referrals in the vain hope of reaching an organic diagnosis and this does not help either. Though it is important to tell such patients that their symptoms are not due to arthritis, it is not helpful to tell them that their pain is imagined. They may go away, but only to begin again with someone else. One must take on the problem and hope that with a patient, optimistic, and reassuring attitude, the problem will at least be minimized and stabilized. It sometimes goes away.

MANAGEMENT OF ANKYLOSING SPONDYLITIS

Back pain is one of life's difficult and common problems. Most backaches seen in general practice are benign and disappear within weeks. The physician must of course be alert to the very occasional patient with something more serious such as a tumour. He must be alert to the patient whose backache is due to some surgically amenable cause like spinal stenosis or spondylolisthesis. He must think of disc disease and bone disease (Paget's disease now eminently treatable) and he must try to spot the psychogenic patient. He must also think of inflammatory back pain, especially in the younger patient. Ankylosing spondylitis is probably much commoner than was previously thought particularly in women. Women usually have ankylosing spondylitis mildly, often without raised ESRs or X-ray changes. The clinical findings are the only guide to the diagnosis. One should therefore be especially suspicious of back pain which is worse in the mornings, accompanied by morning stiffness, aggravated by rest and relieved by exercise. There are two aspects to the treatment, exercise and anti-inflammatory drugs. Since morning symptoms are the worst, it is the night time dose of anti-inflammatory compound which is often the most important. Useful possibilities include Indocid-R 75 mg, flurbiprofen (froben) 100 mg, or naproxen 500 mg. Sometimes the night dose is sufficient – sometimes day-time treatment is also necessary. There are lots of alternative anti-inflammatory drugs which are sometimes very effective for a particular patient. Upon diagnosis, the patient should spend a few minutes with a physiotherapist learning an appropriate exercise regime. He or she should also be encouraged to take part in as much activity as possible, swimming and playing any other favoured sport. Exercise and exercises relieve the symptoms of ankylosing spondylitis and also prevent deformity. While some limitation of movement may be inevitable, kyphosis is definitely preventable – patients should be examined from time to time to ensure that their posture remains good.

MANAGEMENT OF OSTEOARTHRITIS

Osteoarthritis has been neglected. It is surprising since the disease is about five times commoner than rheumatoid arthritis, affects at least 10 per cent of the whole population and about 50 per cent of an elderly population. Its neglect is probably related to the aging or degenerative concept which seemed to fit with the facts about the disease and was therefore widely believed. It did not seem unreasonable that after walking about on our joints for 50 years, they would begin to wear. The facts are really quite different. While osteoarthritis can sometimes be attributed to mechanical insults including menisectomy and obesity, most cases begin inexplicably at about the age of 50. The knees are the commonest site of the disease with the hands a close second. The disease is often a low-grade polyarthritis with plenty of evidence of inflammation. It is not therefore surprising that anti-inflammatory drugs are much more effective than analgesics in many cases of osteoarthritis. They make a good starting point for treatment. As always it is important to make a diagnosis first, to assess the disease and to advise the patient. Assessment is important because osteoarthritis is very variable. There may well be sub-sets of the disease and one of these may be osteoarthritis of the hip in men. For a patient who presents with severe disease of one hip, accompanied by marked restriction of movement and severe X-ray changes, surgery is the obvious choice of treatment. Patients with mild or intermittent symptoms may require only the diagnosis and some appropriate advice. We no longer have to tell patients that osteoarthritis is part of the aging process, certainly not what they wanted to hear. Nor is it necessary to tell them that it is an inevitable, progressive, and untreatable disease. I tell patients about the chemical changes that have recently been identified in cartilage and about the inflammation which is probably important in causing the symptoms. I then proceed to try to establish them on an optimal anti-inflammatory regime. There is a wide choice of non-steroidal anti-inflammatory drugs and these are discussed further below. It is important to be prepared to use a wide range of drugs in order to please a wide range of patients. Alas there is no specific treatment for osteoarthritis. Many patients will find an anti-inflammatory drug which suits them. Some will not and for them, simple analgesics taken on demand probably represent the best approach to drug treatment.

Physical therapy is another alternative and patients with osteoarthritis of the knee are still sent to physiotherapy departments for heat and exercises. This is a very expensive way of providing a little symptomatic relief and a recent study showed that acupuncture was rather better. Both are probably placebos. It is certainly worthwhile to keep the osteoarthritic joint mobile and to preserve muscles and prevent deformity. Suitable exercises can be done by the patient at home where heat is also usually readily available. A visit to a physiotherapy department may be useful when there are specific problems such as quadriceps weakness and wasting requiring a particular exercise regime. A walking stick may be useful for patients with disease of weight-bearing joints. Fat patients must get thin, which sounds easy but it is not. Even when they lose weight, they may not have a corresponding symptomatic improvement but an attempt in this direction must certainly be made.

Clinical trials do not support the view that intra-articular steroids are effective in osteoarthritis though many physicians use them and claim benefit. An injection into the first carpometacarpal joint often seems to help with disease which is localized there. Inflammatory episodes in the knees often seem to respond and one must remember that some patients with osteoarthritis have episodes of pyrophosphate deposition which could cause inflammation. Corticosteroid injection is not a routine procedure but one which should occasionally be considered.

Finally there is surgery. Usually this is for the wrecked joint which cannot be relieved by medical means and which shows severe changes on X-ray. One can consider double or tibial osteotomy for the knee at an earlier stage and this relatively simple procedure often relieves pain for long periods of time. Knees and hips can now be replaced with good expectations of success. Even the subsequent loosening or infection is less formidable since replacements themselves can now be replaced and infected prostheses reinserted with antibiotic cement. But each replacement makes potential future work for the orthopaedic surgeon and therefore makes him cautious. For the difficult first carpometacarpal joint, the trapezium can be removed or replaced.

MANAGEMENT OF RHEUMATOID ARTHRITIS

The management of rheumatoid arthritis is a long and complicated business. A patient with established disease will require supervision, even if the disease is mild, for the rest of his days. It is therefore important at an early stage to establish a satisfactory relationship between patient and physician, whether that physician is a general practitioner or a specialist. Whether the general practitioner decides to take on this role will depend upon his skill, knowledge, and interest – given all these, there is no reason why he should not. Diagnosis, assessment, and advice are again the first three stages in the management and nowhere more important than in rheumatoid arthritis. The disease is very variable and the patient usually presents with many different problems which require a package of remedies. Not only must the disease be assessed initially but it must be reassessed repeatedly to determine the degree of progression and to recognize complications which require treatment. Nowhere is it more important to advise the patient. The diagnosis of rheumatoid arthritis is to many patients a sentence of automatic disability. They need reassurance and advice about many different matters. The trend in the modern management of rheumatoid arthritis is away from rest and towards maintenance of all possible aspects of the patients' way of life. Rest may relieve symptoms but if it means that he loses his job or she loses her ability to manage at home, or they lose their social contacts, the price is too high. With a chronic disease, it is essential to preserve the patients' way of life as normal as possible. The disease will be much worse if it has to be endured in poverty and social isolation. There is no evidence that exercise is harmful to the rheumatoid joint or that patients who rest do any better. Patients should therefore be advised to remain as mobile and fit as possible within the limits of their symptoms. They should be advised about joint protection and maintenance. It is most important to maintain joint movement and certain joints like the shoulders particularly easily become stiff. It is important to maintain the position of joints and some easily become bent like the knees. Muscles are the best protection and the best means of maintaining the position and function of joints and these should be maintained with some kind of exercise programme. Patients should be aware of the aids and appliances which are available to make life easier in the event of difficulty. It is important not to use joints beyond their normal range of movement and to avoid excessive stress. Excesses of exertion should also be avoided along with the carrying of unduly heavy loads. None of this advice should be too restrictive. There is nothing that a patient with rheumatoid arthritis should be forbidden to do but advice may alter the manner in which he does it. Patients require advice about lots of other matters including diet, climate, and cures from the health food store suggested by friendly neighbours. At this stage, the patient needs time.

The next stage in the management is an attempt to control the patients symptoms and non-steroidal anti-inflammatory drugs are the major weapon in this task. One may also use intra-articular steroid injections, splints, and appropriate measures for other problems. For a very active case, a period of rest in hospital may be desirable and induces a remission which may last a long time. This is not always available and not always possible for the patient. It is however worth considering. The use of non-steroidal anti-inflammatory drugs is further discussed below.

Once the patient is established on an optimal anti-inflammatory regime with all possible measures taken to relieve his symptoms, the disease must be re-assessed. The object of this re-assessment is to identify patients with uncontrolled or progressive disease who require treatment with penicillamine or similar drugs. Treatment with these drugs is largely conducted by hospital clinics but often impinges on the general practitioner and is further discussed below.

The next phase in the treatment of rheumatoid arthritis is the regular supervision of the patient and the treatment of problems which arise. Many problems arise. They are articular, extra-articular, social, and domestic. The solutions are many and various. Joint problems arise either from active inflammation in joints or from anatomical changes as a result of the progression of the disease or from complications such as infection. The appropriate solution to active inflammatory disease may be an injection of corticosteroid. If there are severe anatomical changes, surgery may be required. The solutions are likely to be different in different anatomical sites. Painful feet can often be managed with suitable shoes such as space shoes and painful wrists may be helped by splints. The extra-articular problems of rheumatoid arthritis are many and various with solutions likewise. Dry eyes require replacement tears, nodules require reassurance (not cancer!), and ruptured tendons require repair.

Alas, some patients with rheumatoid arthritis progress inexorably and become disabled. The last stage in the management is therefore the management of the disabled patient and there are usually lots of things which can be done. Ways must be found to relieve pain, to help the patient to remain mobile and get out and about, to keep him in employment if possible, to help at home, to help financial problems and with the use of aids and appliances to remain as self-sufficient as possible. Always the role of the physician in rheumatoid arthritis is to care, comfort and sustain the patient in his illness. This role begins at the beginning and ends at the end.

HOW TO CHOOSE A NON-STEROIDAL ANTI-INFLAMMATORY DRUG

What a lot of choice there is. The available drugs are summarized in Table 17.1. Phenylbutazone and oxyphenbutazone have been omitted because it is seldom if ever necessary to use these toxic drugs. Many alternative preparations of aspirin have been omitted because with them all, the table would be too big.

There are really four ways in which non-steroidal anti-inflammatory drugs can be used. They can be given in regular dosage either for a course of a few weeks or months or indefinitely in chronic diseases. This is described in the table as the regular anti-inflammatory dose. They can also be given a large dose at night for the relief of morning stiffness, perhaps combined with another anti-inflammatory drug during the day. This is described as the night-time

Table 17.1. *A selection of non-steroidal anti-inflammatory drugs with doses for various indications*

Compound		Unit dose	Regular anti-inflammatory dose	Night-time dose	Initial dose for acute gout, etc.	Analgesic dose (p.r.n.)
		(mg)	(mg)	(mg)	(mg)	(mg)
Aspirin		300	900 q.d.s.	–	–	600
Azapropazone	Rheumox	600	600 b.d.	–	600	600
Benoxaprofen	Opren	300	600 o.d.	–	–	–
Choline Magnesium tri-salicylate	Trilisate	500	1000–1500 b.d.	–	–	–
Diclofenac	Voltarol	50	50 mane 100 nocte	100	–	–
Dilfunisal	Dolobid	250	250–500 b.d.	–	–	500
Fenoprofen	Fenopron	300	600 q.d.s.	–	900	300
Fenbufen	Lederfen	300	300 mane 600 nocte	600	–	600
Fenclofenac	Flenac	300	600 b.d.	900	–	–
Feprazone	Methrazone	200	200 b.d.	–	–	–
Flurbiprofen	Froben	50	50 mane 100 nocte	100	–	–
Ibuprofen	Brufen	400	400 t.d.s.	–	–	400
Indomethacin	Indocid	25	25 t.d.s.	75	100	–
Indomethacin	Indocid-R	75	75 daily	75	–	–
Ketoprofen	Orudis	100	100 b.d.	100	–	–
Mefenamic acid	Ponstam Forte	500	500 t.d.s.	–	–	500
Naproxen	Naprosyn	500	500 b.d.	500	500	500
Piroxicam	Feldene	10	20 o.d.	–	40	20
Sulindac	Clinoril	200	200 b.d.	–	200	–
Tolmetin	Tolectin	400	400 t.d.s.	–	–	–
Zomepirac	Zomax	100	–	–	–	100

dose. In acute self-limiting conditions like gout, a higher initial dose is sometimes used and this is given separately. Finally these compounds are all analgesics as well as anti-inflammatory drugs and they can be used on demand for the relief of pain. The appropriate dose is described as the analgesic dose.

Five factors need to be considered in the choice of an anti-inflammatory; efficacy, tolerance, safety, convenience, and cost. There is very little difference in the efficacy of these different compounds though some of them are more suitable for some diseases and this has already been discussed. The major factor which determines the response of a particular patient is individual variation. Some people respond to one drug and some to another. It is therefore necessary, especially in chronic diseases, to be prepared to use a number of different drugs to find the best for a particular patient. The same applies to tolerance. Though there are undoubtedly differences in the relative incidence of gastrointestinal side-effects and rashes with different compounds, individual variation is again the major factor which determines whether or not a patient can tolerate a particular drug. It is a matter of trial and error to find the drug which works well and which is well tolerated. There are very few patients who will not respond to and cannot tolerate at least one of these compounds. Most of these drugs are safe. Aspirin and indomethacin may cause gastric bleeding and aspirin is dangerous when combined with anticoagulants. Neither of these drugs should be given to patients with peptic ulcers. Convenience is another factor and many of the newer drugs can be given twice daily or even once daily, pleasing the patient. Aspirin is cheap – most of the newer drugs are much more expensive. One must consider whether the convenience of infrequent dosage and the absence of side effects is worth the extra cost. I think it is. There are now preparations of aspirin which can be taken twice daily and which cause an incidence of side-effects comparable to that of the other newer drugs. Choline magnesium trisalicylate is an example. Unfortunately with increasing sophistication, the cost advantage is lost.

Like most doctors, I have a few favourite compounds which I tend to use first in chronic conditions like rheumatoid arthritis and osteoarthritis. They include naproxen, piroxicam, diclofenac, fenbufen, and diflunisal. I favour indocid-R or flurbiprofen for night-time use. I like fenoprofen for its flexible dosage, as little as 200 mg as required for an analgesic effect and as much as 2.4 g daily for the full anti-inflammatory dosage. The new compound zomepirac is a useful analgesic and many other compounds including mefenamic acid and of course aspirin can be used in this way. If one doesn't work, keep trying.

CORTICOSTEROIDS FOR RHEUMATIC DISEASES

The use of intra-articular and soft-tissue injections of corticosteroids has been discussed in relation to particular conditions for which they are used. Systemic corticosteroids are seldom required for rheumatic diseases. They are always required for polymyalgia rheumatica and temporal arteritis. They are sometimes required in systemic lupus erythematosus, usually in polymyositis and polyarteritis nodosa. They are seldom required in rheumatoid arthritis. The use of corticosteroids is particularly disastrous in younger patients. In the elderly, surprisingly perhaps, the side-effects seem less terrible. There is a particular sub-set of elderly patients with rheumatoid arthritis who present with an explosive type of disease, a very acute onset but a much better prognosis than insidious cases, and who respond very well to small doses of prednisolone. It is my practice in such patients to start with a dose of 7.5 mg daily of prednisolone. If the effect is dramatic after one week, the treatment can be continued. If there is little or no benefit, the treatment can be stopped without risk. This type of rheumatoid arthritis behaves more like polymyalgia rheumatica and often settles down over the course of a few years. It is of course very important not to allow the dose of creep up – rather it should creep down. While daily doses of 5 or 7.5 mg daily are probably not very harmful in younger patients, they are not usually very effective and better avoided.

THE ROLE OF ANALGESICS IN THE RHEUMATIC DISEASES

Conventional simple analgesics have several important limitations in rheumatic diseases. The first is the absence of anti-inflammatory activity. Most rheumatic diseases have inflammation as part of their pathogenesis and anti-inflammatory drugs are therefore more effective. The classical experiment in rheumatoid arthritis compared aspirin and narcotic analgesics, finding aspirin much more effective. The same has now been demonstrated in osteoarthritis. In general therefore, non-steroidal anti-inflammatory drugs are preferred even though they are used in small doses for their analgesic effect. There are some patients, with osteoarthritis of the hip for example, who do not respond to anti-inflammatory drugs. Such patients will often achieve at least reasonable relief of pain from a preparation such as Distalgesic and this mixture of compounds cannot be ignored if only because of its popularity. There are now several clinical trials showing the combination of dextropropoxyphene and paracetamol to be more effective than either dextropropoxyphene or paracetamol alone. Paracetamol, combinations of codeine and dihydrocodeine with aspirin or paracetamol are alternatives. Another limitation of these compounds is their danger in overdose. Overdoses of paracetamol cause hepatic necrosis and the dangers of dextropropoxyphene and other centrally-acting drugs, especially when taken with alcohol, have recently been emphasized. The non-steroidal anti-inflammatory drugs are probably safer in this respect.

Analgesics are also used as a supplement to regular therapy. Even patients who are well controlled on a non-steroidal anti-inflammatory drug have bad times and need something to turn to. An analgesic is the obvious choice. For this purpose Distalgesic is again very popular amongst rheumatic patients.

SUPERVISION OF PATIENTS RECEIVING GOLD, PENICILLAMINE, AND
AZATHIOPRINE

Treatment with gold, penicillamine, and azathioprine is often conducted in
hospital clinics and forms a significant part of the rheumatologist's work load.
This is often shared with the general practitioner who by taking on part of this
work load can save the patient from long trips to hospital clinics. Even if the
patient is attending hospital exclusively, the general practitioner may still have
to make decisions because he will be the first to know when something goes
wrong. All these three drugs cause blood problems and full blood counts are
therefore required at least every four weeks. With azathioprine and gold, it is
the white cells which are most likely to fall. With penicillamine it is the platelets
and patients receiving this drug need a platelet count. Falls may be sudden or
gradual and because of the latter possibility, it is essential to compare successive
counts. A succession of small falls is easy to miss but may be just as important
as a large fall. As a rough guide, treatment with any of these drugs should be
stopped if the white count or platelet count fall by more than 50 per cent of
their initial value or if the white count falls below 3000/mm^3 or the platelet
count below 100 000/mm^3. The smaller falls or falling counts require more
frequent checks. Gold and penicillamine are both capable of causing proteinuria.
Patients having gold injections should have a urine test before each injection
and those receiving penicillamine should have their urine tested at the time of
each blood sample, therefore monthly. It is usual to stop gold treatment if
patients develop more than slight proteinuria. The same is not true for penicillamine. Mild proteinuria is usually benign and disappears spontaneously
though it may take a year to do so. Patients with significant and continued
proteinuria on penicillamine should therefore have a 24-hour urinary collection
to quantitate the abnormality. If the protein excretion is less then 3 g per day
and renal function tests remain normal, it is reasonable to continue treatment.
The aim of the treatment with drugs of this type is to achieve a long-term sup-
pression of the disease. If treatment is to be continued long-term, skilful
management of side-effects is essential. It is a tragedy when a well-controlled
patient stops treatment unnecessarily. Some side-effects are dangerous and
clearly require that treatment be stopped. Others are not dangerous and may be
self-limiting so that treatment can continue. A decision must be made in each
case in the light of the patient's response, the adverse reaction and its natural
history. It is particularly important to continue gold therapy whenever possible
since patients who are allowed to relapse do not respond again.

It is no longer appropriate to think of gold in terms of a 1 g course. The drug
should be given by weekly intramuscular injections until the patient responds
well, when the interval between the injections can be increased to two weeks. If
the response remains good, the intervals between the injections can be further
increases to three and then to six weeks. Since relapse may take a few months,
the intervals between the dosage reduction should never be less than three
months. Treatment should be continued indefinitely. Penicillamine is nowadays

usually given in an initial dose of 250 mg daily, increased to 250 mg twice daily after three months unless the patient is already responding. The usual dose of azathioprine is 50 mg twice daily.

WHO MANAGES? A MATTER OF CO-OPERATION

There are no rules about who does what in the management of the rheumatic diseases. Though hospital clinics are likely to be good places for the management of complicated cases of diseases like rheumatoid arthritis, the interested and informed general practitioner can do just as well. He may call on his specialist colleague for occasional advice. The hospital patient may also call on the general practitioner for advice and successful management will always involve co-operation. It is however essential that someone assumes responsibility for the long-term management of a disease like rheumatoid arthritis. Someone must have the responsibility for reveiwing the patient at regular intervals to assess the degree of progression and the need for alternations in drug therapy or the introduction of drugs like penicillamine. Someone must have the responsibility of supervising drug therapy especially with the long-acting drugs. Someone must be prepared to deal with the many problems which arise during the course of rheumatoid arthritis, the articular, extra-articular, social, and domestic crises. It does not matter who that someone is. Management is often a team affair and communication between members of the team is essential to success.

REFERENCES

Boyle, A. C. (1978). Injection techniques. In *Copeman's textbook of the rheumatic diseases,* 5th edn (ed. J. T. Scott). Churchill Livingstone, Edinburgh.

Hart, F. D. (ed.) (1982). *Drug treatment of the rheumatic diseases,* 2nd edn. Adis Press, New York.

Huskisson, E. C. (1979). Routine drug treatment of rheumatoid arthritis and other rheumatic diseases. *Clin. rheum. Dis.* **5**, 697–706.

Appendix

Useful addresses

1. For doctors—medical associations

1. BLAR   British League for Arthritis and Rheumatism
                 41 Eagle St., London WC1R 4AR
2. ARC   Arthritis and Rheumatism Council
3. BARR   11 St. Andrew's Place, Regent's Park, London NW1 4LE
                 British Association for Rheumatology and Rehabilitation
4. BAMM   62 Wimpole St., London W1M 7DE
                 British Association of Manipulative Medicine
5. Biological Engineering Society, c/o Biophysics Dept., University College, Gower St., London WC1 6BT
6. Back Pain Association, 31–33 Park Road, Teddington, Middlesex TW11 0BA

2. For patients—specific disabilities

1. British Rheumatism and Arthritis Association (BRAA), 6 Grosvenor Crescent, London SW1X 7ER
2. Multiple Sclerosis Society, 4 Tachbrook St., London SW1V 1SJ
3. Society of Parkinson's Disease (UK) Ltd., 36 Queens Rd., London SW19 8LR
4. Queen Elizabeth Foundation for the Disabled, Leatherhead, Surrey
5. Royal Association for Disability and Rehabilitation, 25 Mortimer St., London W1N 8AB
6. The Spastics Society, 12 Park Crescent, London W1N 4EQ
7. Multiple Sclerosis Action Group, 71 Gray's Inn Road, London WC1X 8TR
8. British Limbless Ex-Servicemen's Association, Frankland Moore House, 185–187 High Road, Chadwell Heath, Essex RM6 6NA
9. The Chest, Heart and Stroke Association, Tavistock House North, Tavistock Square, London WC1H 9JE
10. Spastics Society, 12 Park Crescent, London W1N 4LQ
11. British Diabetic Association, 3–6 Alfred Place, London WC11 7EE
12. Friedrich's Ataxia Group, Bolsover House, 5–6 Clipstone Street, London W1
13. Haemophilia Society, PO Box 9, 16 Trinity St., London SE1 1DE
14. Muscular Dystrophy Group of Great Britain, Nattrass House, 35 Macaulay Rd., London SW4 0QP
15. Myasthenia Gravis Council of the Muscular Dystrophy Group of Great Britain, Nattrass House, 35 Macaulay Road, London SW4 0QP
16. The Association for Spina Bifida and Hydrocephalus (ASBAH), 30 Devonshire Street, London W1 2EB
17. National Spinal Injuries Centre, Stoke Mandeville Hospital, Aylesbury, Bucks. HP21 8PP
18. Spinal Injuries Association, 120–126 Alberty Street, Camden, London NW1 7NE

3. General organizations

1. Age Concern, Bernard Sunly House, 60 Pitcairn Road, Mitcham, Surrey CR4 3LL
2. British Red Cross Society, 9 Grosvenor Crescent, London SW1X 7EJ
3. The Centre on Environment for the Handicapped, 120–126 Albert Street, London NW1 7NE
4. Disabled Living Foundation, 346 Kensington High Street, London W14 8NS
5. National Fund for Research into Crippling Diseases (Action Research for the Crippled Child), Vincent House, Springfield Road, Horsham, West Sussex RH12 2PN
6. The Royal Association for Disability and Rehabilitation, 25 Mortimer Street, London W1N 8AB

4. Aids

1. Rehabilitation Engineering Movement Advisory Panels (REMAP), Thames House North, Millbank, London SW1P 4QG
2. Possum Controls Ltd., 63 Mandeville Road, Aylesbury, Bucks. HP21 8AE
3. Possum Users Association, Copper Beech, Parry's Close, Stoke Bishop, Bristol BS9 1AW

5. Education

1. Association of Disabled Professionals, The Stables, Banstead, Surrey
2. National Bureau for Handicapped Students, City of London Polytechnic, Calcutta House Precinct, Old Castle Street, London E1 7NT

6. Employment

1. The British Computer Society, 29 Portland Place, London W1N 4HU
2. Home Opportunities for Professional Employment (HOPE), Oakwood Further Centre, High Street, Kelvedon, Essex
3. Department of Employment – Local Office
4. Department of Employment Rehabilitation Centre – Local

7. Getting around

1. The British School of Motoring, Disabled Drivers Training Centre, 102 Sydney Street, London SW3
2. Disabled Drivers Association, Ashwellthorpe Hall, Ashwellthorpe, Norwich NR16 1EX
3. Mobility International, 2 Colombo Street, London Sw1 8DP
4. Motability, The Adelphi, John Adam Street, London WC2N 6AZ

8. Entitlements

1. Disability Alliance, 96 Portland Place, London W1N 4EX
2. Department of Health and Social Security – Local Office
3. Citizens Advice Bureau – Local Office

9. Leisure

1. Sports Council, 70 Brompton Road, London SW3
2. British Sports Association for the Disabled, Stoke Mandeville Stadium for the Disabled, Harvey Road, Aylesbury, Bucks. HP21 8PP

3. Riding for the Disabled, National Headquarters, National Agriculture Centre, Stoneleigh, Kenilworth, Warwickshire CV8 2LY

10. Disabled child

1. Voluntary Council for Handicapped Children, National Children's Bureau, 8 Wakely Street, London EC1V 7QE
2. Handicapped Adventure Playground Association, Central Office, Fulham Palace, Bishops Avenue, London SW6 6EA
3. Physically Handicapped Ablebodied (PHAB) Association, 42 Devonshire Street, London W1N 1LN

11. Self-help equipment

1. Aids for the Disabled Homecraft Supplies (Fleet Street) Ltd., 27 Trinity Road, London SW17
2. Llewellyn Living Aids, Carlton Works, Carlton Street, Liverpool L37 ED
3. British Red Cross Society, Aids for the Disabled, 9 Grosvenor Crescent, London SW1X 7EJ

12. Agencies abroad

Australia

1. Civilian Maimed and Limbless Association, 159 Princes Highway, St. Peters, NSW, 2044.
2. Australian Cerebral Palsy Association, 5 Blake Street, North Perth, Western Australia 6006.
3. Multiple Sclerosis Society of Australia, 239 Mowbray Road, Chatswood, NSW 2067.
4. Australian Council for Rehabilitation of Disabled, Bedford and Buckingham Streets, Surry Hills, NSW 2010.
5. Arthritis and Rheumatism Council (New South Wales), 12th Floor, Wynyard House, 291 George Street, Sydney, 2000.
6. Canberra Arthritis and Rheumatism Association, P.O. Box 352, Woden ACT. 2606.
7. Queensland Arthritis and Rheumatism Foundation, 1st Floor, Holman House, Cnr. Main and Holman Streets, Kangaroo Point, QLD. 4169.
8. Rheumatism and Arthritis Association of Victoria, P.O. Box 195, Kew, VIC. 3101.
9. Rheumatism and Arthritis Foundation of Tasmania, 84 Hampden Road, Battery Point, Hobart, Tasmania, 7000.
10. South Australian Arthritis and Rheumatism Association, 24 King William Road, North Adelaide, S.A. 5006.
11. Western Australian Arthritis and Rheumatism Foundation, P.O. Box 7157, Cloisters Square, Perth, 6000.
12. Arthritis and Rheumatism Association of Northern Territory, 7 Worgan Street, Fanny Bay, N.T.
13. The Australian Council for Rehabilitation of Disabled, Acrod House, 33 Thesiger Court, Deaking, ACT. 2600.

Canada

1. Canadian Cerebral Association, 1 Yonge Street, Toronto, Ontario MSE 1E8.
2. Canadian Haemophilia Society, Chedoke Center, Patterson Building, Box 2085, Hamilton, Ontario, L8M 3RS.
3. Canadian Paraplegic Association, 520 Sutherland Drive, Toronto, Ontario M4G 3VG.
4. Amyotrophic Lateral Sclerosis Society of Canada, Mrs Doreen Konradis, Executive Director, 234 Eglinton Avenue East, Suite 305, Toronto, Ontario, M4P 1K5.
5. The Arthritis Society, Mrs S. L. McConnell, Secretary-Treasurer, Suite 420, 920 Yonge Street, Toronto, Ontario, M4W 3J7.
6. Canadian Cardiovascular Society, Dr James F. Symes, Secretary, 1455 Peel Street, Room M-31, Montreal, P.Q. H3A 1T5.
7. The Canadian Rheumatism Association, Andre Lussier, M.D. (FRCP(C)), Secretary, Rheumatic Diseases Unit, Centre Hospitalier Universitaire, Sherbrooke, P.Q. J1H 5N4.
8. Multiple Sclerosis Society of Canada, Shelia Kieran, Executive Director, 130 Bloor Street West, Suite 700, Toronto, Ontario, M5S 1NS.
9. The Muscular Dystrophy Association of Canada, Miss Mary Ann Wickham, Secretary, Suite 1014, 74 Victoria Street, Toronto, Ontario, M5C 2A5.
10. Canadian Neurosurgical Society, Dr F. E. LeBlanc, Secretary, 1403–29th Street, N.W. Calgary, Alberta, T2N 2T9.
11. Canadian Association of Physical Medicine & Rehabilitation, Dr M. G. P. Cameron, Secretary, Department of Rehabilitation Medicine, University Hospital, 339 Windermere Road, London, Ontario, N5A 5A5.
12. Canadian Physiotherapy Association, Nancy Christie, Executive Director, 25 Imperial Street, Toronto, Ontario M5P 189.
13. Canadian Association for the Mentally Retarded, Dr Hugh G. Lafave, Executive Vice-President, Kinsmen NIMR Building, York University Campus, 4700 Keele Street, Downsview, Ontario, M3J 1P3.

New Zealand

1. The Arthritis and Rheumatism Foundation of New Zealand, P.O. Box 10-020, Southern Cross Building, Brandon Street, Wellington, New Zealand.

South Africa

1. South Africa Rheumatism and Arthritis Association, Namaqua House, 36 Burg Street, Capetown, 8001.

United States of America

1. National Amputation Foundation Inc., 12–45 150th Street, Whitestone, New York, N.Y. 11357.
2. The Arthritis Foundation, 1212 Avenue of the Americas, New York, New York 10036.
3. United Cerebral Palsy Associations Inc., 66 East 34th Street, New York, New York 10016.
4. Friedreich's Ataxia Group of America, Inc., Box 1116, Oakland, California 94611.

5. National Haemophilia Foundation, 25 West 39th Street, New York, New York 10018.

6. National Multiple Sclerosis Society, 205 East 42nd Street, New York, New York 10010.

7. Muscular Dystrophy Associations of America, Inc., 810 Seventh Avenue, New York, New York 10019.

8. American Parkinson's Disease Association, 147 East 50th Street, New York, New York 10022.

9. Spina Bifida Association of America, 104 Festone Avenue, New Castle, Delaware 19720.

10. American Coalition of Citizens with Disabilities Inc., 1346 Connecticut Avenue, NW Washington, DC 20036.

11. National Association of the Physically Handicapped, 2810 Terrace Road, SE Washington, DC 20020.

12. President's Committee on Employment of the Handicapped, 1111 20th Street, NW Washington, DC 20036.

13. Rehabilitation International, 431 Park Avenue South, New York, New York 10016.

Index